Essential Molecular Biology
Volume Two

The Practical Approach Series

Related **Practical Approach** Series Titles

Basic Cell Culture
Proteolytic Enzymes 2/e
Human Cytogenetics: Malignancy and
 Acquired Abnormalities 2
Human Cytogenetics: Constitutional
 Analysis
Animal Cell Culture 3/e
RNA Viruses
Differential Display
Mouse Genetics and Transgenics
DNA Viruses
Gene Targeting 2/e
Protein Phosphorlyation 2/e
Crystallization of Nucleic
 Acids and Proteins
DNA Microarray Technology
Post-Translational Modification
Protein Expression
Chromosom Structural Analysis
Gel Electrophoresis of Proteins 3/e
In Situ Hybridization 2/e
Chromatin
Mutation Detection
Molecular Genetic Analysis of
 Populations 2/e
PCR 3: PCR In Situ
 Hybridization

Antisense Technology
Genome Mapping
Protein Function 2/e
Protein Structure 2/e
DNA and Protein Sequence
 Analysis
Protein Structure Prediction
Antibody Engineering
DNA Cloning 4: Mammalian
 Systems
DNA Cloning 3: Complex
 Genomes
Gene Probes 2
Gene Probes 1
Pulsed Field Gel
 Electrophoresis
Non-isotopic Methods in
 Molecular Biology
PCR 2
DNA Cloning 2: Expression
 Systems
DNA Cloning 1: Core Techniques
Molecular Genetics of Yeast
RNA Processing Volume II
RNA Processing Volum I
PCR 1
Plant Molecular Biology

Please see the **Practical Approach** series website at

http://www.oup.com/pas

for full contents lists of all Practical Approach titles.

No. 255

Essential Molecular Biology

Volume Two
Second Edition

A Practical Approach

Edited by

T. A. Brown

Department of Biomolecular Sciences
UMIST, Manchester, U.K.

OXFORD
UNIVERSITY PRESS

1002734822

OXFORD
UNIVERSITY PRESS

Great Clarendon Street, Oxford OX2 6DP

Oxford University Press is a department of the University of Oxford.
It furthers the University's objective of excellence in research, scholarship,
and education by publishing worldwide in

Oxford New York

Athens Auckland Bangkok Bogotá Buenos Aires Calcutta Cape Town
Chennai Dar es Salaam Delhi Florence Hong Kong Istanbul Karachi
Kuala Lumpur Madrid Melbourne Mexico City Mumbai Nairobi Paris
São Paulo Singapore Taipei Tokyo Toronto Warsaw

with associated companies in Berlin Ibadan

Oxford is a registered trade mark of Oxford University Press in the UK and
in certain other countries

Published in the United States by Oxford University Press Inc., New York

British Library Cataloguing in Publication Data
Data available

Library of Congress Cataloguing in Publication Data

1 3 5 7 9 10 8 6 4 2

ISBN 0-19-963645-1 (Hbk.)
ISBN 0-19-963644-3 (Pbk.)

Typeset in Swift by Footnote Graphics, Warminster, Wilts
Printed in Great Britain on acid-free paper
by The Bath Press Ltd, Avon

Preface to the First Edition

There are now a number of molecular biology manuals on the market and the editor of an entirely new one has a duty to explain why his contribution should be needed. My answer is that although there are some excellent handbooks for researchers who already know the basic principles of gene cloning, there are very few that cater for the absolute beginner. Unfortunately, everyone is a beginner at some stage in their careers and even in an established molecular biology laboratory the new research student can spend a substantial amount of time not really understanding what is going on. For the experienced biologist expert in a discipline other than molecular biology, and perhaps without direct access to a tame gene cloner, guidance on how to introduce recombinant DNA techniques into his or her own research programme can be very difficult to obtain. For several years I have run a basic gene cloning course at UMIST and I have continually been impressed by the number of biochemists, botanists, geneticists, cell biologists, medics and others who want to learn how to clone and study genes.

The contributors to *Essential Molecular Biology: A Practical Approach* were asked to write accounts that combine solid practical information with sufficient background material to ensure that the novice can understand how a technique works, what it achieves, and how to make modifications to suit personal requirements. Where appropriate, the reader is also given advice on more advanced or specialized techniques. In all cases the authors have responded to the challenge and produced chapters that make concessions to the beginner without jeopardizing scientific content or practical value. I hope that the result is a handbook that will guide newcomers into molecular biology research.

The book is split into two parts. Volume I deals with the fundamental techniques needed to carry out DNA cloning experiments. The emphasis is on coming to grips with the necessary practical skills and understanding the background in sufficient detail to be able to adjust to circumstances as the project progresses. In Volume II, procedures for preparing gene libraries and identifying genes are described, together with methods for studying the structure of a cloned gene and the way it is expressed in the cell. It is assumed that the basics from Volume I are now in place, but the procedures are still described in the same down to earth fashion with protocols complemented by background information and troubleshooting hints.

I must thank a number of people for their help with this book. First, I am grateful to the authors who provided the manuscripts more or less on time and were prepared in many cases to make revisions according to my requests. I would especially like to thank Paul Towner for stepping in at the last minute after I had been let down with one chapter. The participants and assistants on the recent UMIST gene cloning courses helped me formulate the contents of the book and my colleague Paul Sims provided valuable advice. The Series Editors and publishers made encouraging noises when the going got tough and sorted out a number of problems for me. I am also grateful to my research students for giving their opinion on the practical contents of the book. Finally, I would like to thank my wife Keri for proving that even an archaeologist can learn how to clone genes.

1990   T.A.B.

Preface to the Second Edition

The primary objective of *Essential Molecular Biology: A Practical Approach*, that the book should be accessible and informative to researchers encountering molecular biology techniques for the first time, is as valuable today as it was when the first edition was written 10 years ago. If anything, the need for a basic DNA manual has increased as molecular biology techniques have become more and more mainstream in disparate areas of research. Gene cloning, sequencing and PCR are now central to those various research fields grouped together under the heading 'molecular life sciences' and an ever-increasing number of new research students enter these fields every year. Equally impressive has been the spread of DNA techniques into areas of research such as molecular ecology and biomolecular archaeology, where the power of DNA analysis is contributing to scientific knowledge in disciplines that just a few years ago either did not exist at all or had little use for clones and PCR products. When planning the Second Edition of *Essential Molecular Biology: A Practical Approach* I therefore determined from the very start that the objective of the book should remain the same: 'to combine solid practical information with sufficient background material to ensure that the novice can understand how a technique works, what it achieves, and how to make modifications to suit personal requirements'.

Since the first edition was published there has been a dramatic expansion in the range and sophistication of molecular biology techniques, even in those basic techniques that are of greatest importance to the newcomer. To address these changes, every chapter has been revised and updated as appropriate, with some chapters completely rewritten. This is particularly true for the more advanced techniques described in Volume II but also applies to the very basic procedures covered in Volume I, the advances over the last 10 years even involving such fundamental techniques as DNA purification and agarose gel electrophoresis. Despite these changes, the two volumes are still organized according to the rationale of the first edition, with the procedures for DNA and RNA manipulation (purification, electrophoresis and the construction and cloning of recombinant molecules) in Volume I and those for isolating and studying individual genes (preparation and screening of libraries, polymerase chain reactions, DNA sequencing and studying gene expression) in Volume II.

An edited volume is only as good as the chapters it contains and I therefore wish to thank all the contributors for taking such great care to address the objective of

Essential Molecular Biology by making their chapters lucid and informative. I also thank Liz Owen and Lisa Blake of Oxford University Press for ensuring that the chapters became a book rather than languishing in my computer, and my wife Keri Brown for helping me find the time to organize and assemble these two volumes.

2001   T.A.B.

Contents

Protocol list

Abbreviations

A_{260}	absorption at 260 nm
ABM	*m*-aminobenzyloxymethyl
APT	*o*-aminophenylthioether
ATP	adenosine 5′-triphosphate
BACs	bacterial artificial chromosomes
BAP	bacterial alkaline phosphatase
BCIP	5-bromo-4-chloro-3-indolyl phosphate
Bis	bis(hydroxymethyl)aminomethane
bp	base pairs
CCD	charge-coupled device
cDNA	copy DNA
CIAP	calf intestinal alkaline phosphatase
CIP	calf intestinal alkaline phosphatase
c.p.m.	counts per minute
dATP	2′-deoxyadenosine 5′-triphosphate
DBM	diazobenzyloxymethyl
dCTP	2′-deoxycytidine 5′-triphosphate
ddA	2′,3′-deoxyadenosine 5′-triphosphate
ddC	2′,3′-deoxycytidine 5′-triphosphate
ddG	2′,3′-deoxyguanosine 5′-triphosphate
ddNTP	dideoxynucleotide
ddT	2′, 3′-deoxythymidine 5′-triphosphate
DEPC	diethyl pyrocarbonate
dGTP	2′-deoxyguanosine 5′-triphosphate
dITP	2′-deoxyinosine 5′-triphosphate
DMSO	dimethyl sulphoxide
DNA	deoxyribonucleic acid
DNase	deoxyribonuclease
dNTP	deoxyribonucleotide
DOC	sodium deoxycholate
d.p.m.	disintegrations per minute
DPT	diazophenylthioether

dsDNA	double-stranded DNA
DTT	dithiothreitol
dTTP	2'-deoxythymidine 5'-triphosphate
EDTA	ethylenediaminetetra-acetic acid
EGTA	ethyleneglycolbis(aminoethyl)tetra-acetic acid
FISH	fluorescent *in situ* hybridization
GST	glutathione *S*-transferase
HPLC	high-performance liquid chromatography
HPRI	human placental ribonuclease inhibitor
IAA	indole acrylic acid
IgG	immunoglobulin G
IPTG	isopropyl-β-D-thiogalactopyranoside
kb	kilobase pairs
kDa	kilodaltons
MCS	multi-cloning sequence
MES	2-morpholinoethanesulphonic acid monohydrate
mol. wt	molecular weight
MOPS	3-morpholinopropanesulphonic acid
mRNA	messenger RNA
NBT	nitroblue tetrazolium
NMR	nuclear magnetic resonance
nt	nucleotides
NTA	nitrilotriacetic acid
NTP	nucleotide triphosphate
OD	optical density
PACs	P1 artificial chromosomes
PEG	polyethylene glycol
PCR	polymerase chain reaction
p.f.u.	plaque-forming units
PIPES	piperazine-1,4-bis(2-ethanesulphonic acid)
poly(A$^+$)	polyadenylated
PPO	2, 5-diphenyloxazole
RACE	rapid amplification of cDNA ends
RAPD	random amplified polymorphic DNA
RFLP	restriction fragment length polymorphism
RNA	ribonucleic acid
RNase	ribonuclease
rRNA	ribosomal RNA
r.p.m.	revolutions per minute
RT-PCR	reverse transcriptase-polymerase chain reaction
SDS	sodium dodecyl sulphate
ssDNA	single-stranded DNA
TCA	trichloroacetic acid
TdT	terminal deoxynucleotidyl transferase
TLC	thin-layer chromatography

TEMED	N,N,N',N'-tetramethylethylenediamine
TMAC	tetramethylammonium chloride
Tris	tris(hydroxymethyl)aminomethane
tRNA	transfer RNA
UV	ultraviolet
YACs	yeast artificial chromosomes
X-gal	5-bromo-4-chloro-3-indolyl-β-D-galactopyranoside

Chapter 1

Strategies for research in molecular biology

T. A. BROWN

Department of Biomolecular Sciences, University of Manchester Institute of
Science and Technology, Manchester M60 1QD, UK

1 Introduction

The techniques described in Volume I of this book form the foundation to research work in molecular biology and, once they have been mastered, any of the
more specialized procedures can be approached with confidence. However,
successful use of these specialized procedures requires more than just the ability
to string together basic manipulations. As in any scientific endeavour the experimental programme has to be planned in accordance with the questions being
asked and the goals being sought. There are two or three different way of doing
just about anything in molecular biology, and choosing the best experimental
approach requires an understanding of the requirements, scope and limitations
of the different procedures, as well as a clear conception of what the overall
research programme is intended to achieve. As with the procedures described in
Volume I, the newcomer to molecular biology should approach the chapters in
Volume II with the dual aim of understanding both the technical and conceptual
aspects of the methodology.

I begin this introductory chapter by outlining the alternative strategies that
can be followed for isolation of an individual, predetermined gene from an
organism, expanding on the points made in Volume I, Chapter 1, Section 4 in
which the basic features of gene cloning and the PCR were described. I then
summarize the choice of procedures available for studying the structure and
activity of the isolated gene. Finally, I make some general comments on the
interpretation of data obtained from molecular biology experiments.

2 Strategies for gene isolation

The two different approaches to gene isolation were described in Volume I,
Chapter 1, Section 4:

- by PCR, which involves enzymatic amplification of the desired DNA fragment

- by DNA cloning, which involves insertion of the desired DNA molecule into a cloning vector, followed by replication of the cloning vector in a culture of *Escherichia coli* bacteria

The success or failure of a PCR is determined by the design of the primers, the pair of oligonucleotides that anneal either side of the DNA segment to be amplified (see Volume I, Chapter 1, *Figure 2*). If the primers are correctly designed, so that they are specific for the gene of interest and work effectively in the PCR, then gene isolation becomes a straightforward process. Primer design is described in detail, along with other aspects of PCR, in Chapter 7. The success or failure of a cloning experiment is less easy to ensure and requires much more strategic planning. This is because the procedures used to obtain the gene of interest depend on the nature of that gene and the size of the genome from which it is being isolated. The following section describes the problems and explains how they can be overcome.

2.1 Different approaches to gene isolation by cloning

In general terms, there are three ways by which a desired gene may be obtained via DNA cloning:

- by isolation of a DNA fragment that carries the gene of interest, followed by cloning of just that single fragment
- by direct selection of the desired gene when transformants are plated out
- By isolation of the gene from a clone library

Each of these strategies will be examined in turn.

2.1.1 Cloning a single restriction fragment

If a restriction fragment carrying the gene of interest can be identified in an agarose or polyacrylamide gel, then that fragment can be purified from the gel (using one of the procedures described in Volume I, Chapter 6) and ligated into a cloning vector, with the result that virtually all of the clones obtained after transformation and recombinant selection will contain the gene of interest. It is 'virtually all' rather than 'all' because, in practice, a band in an electrophoresis gel always contains a small number of contaminating DNA molecules (mostly smaller restriction fragments that have become entrapped in the larger band during the initial sorting stage of electrophoresis). Normally, however, the desired recombinants form the vast bulk of the clones obtained and their identities can be confirmed by checking the size of the fragment carried by each one.

Unfortunately, this simple strategy for obtaining a desired clone has few applications, at least in the initial stages of gene isolation. To be of use the genome under study must be small enough for all the fragments produced by restriction to be distinguishable after gel electrophoresis. This may be possible with some unusual genomes (e.g. the smaller mitochondrial and chloroplast genomes, some viral genomes and also plasmid molecules) but it is not achievable with the central genomes of prokaryotes or eukaryotes. These central genomes give rise to so many different restriction fragments that a digest appears as a continuous smear after electrophoresis, and individual fragments cannot be distinguished.

Despite these limitations the strategy is important in the later stages of a cloning project. Often, a gene is obtained initially as a part of a relatively large piece of cloned DNA, for instance in a 20-kb fragment carried in a bacteriophage λ vector. This fragment may well contain other genes in addition to the one being sought, in which case a second round of restriction and cloning ('subcloning') will be required. One way of doing this would be to restrict the recombinant λ molecule, fractionate the individual fragments in an agarose gel, and then use Southern hybridization (see Chapter 5, Section 2.6) to identify the one that contains the gene of interest. This fragment would then be excised from the gel and cloned individually to provide the isolated gene.

2.1.2 Direct selection

The strategy which can loosely be called 'direct selection' requires that the cloned gene confers a selectable phenotype on the host cells used for transformation. A simple example would be the cloning of a gene that confers antibiotic resistance on *E. coli*, which could be performed as follows:

1. Prepare a sample of DNA from an *E. coli* strain resistant to the antibiotic.

2. Restrict the DNA and ligate the mixture of fragments into a suitable cloning vector.

3. Transform a strain of *E. coli* that is sensitive to the antibiotic.

4. Plate out the transformants at high density on an agar medium containing inhibitory amounts of the antibiotic.

Only cells containing the cloned antibiotic resistance gene will be able to grow on the selective medium, so the desired clone will be directly selected during the plating-out step.

This approach is relevant to many genes in addition to those coding for resistances to antibiotics and inhibitors. It has been of greatest value in cloning genes involved in biosynthetic pathways for which auxotrophic mutants can be obtained. For example, the *E. coli* gene for tryptophan synthase can be cloned by transforming a tryptophan-auxotrophic *E. coli trpA* strain and selecting recombinant prototrophs on a minimal medium (*Figure 1*). In theory, direct selection by marker rescue could be used in this way to clone genes from any transformable species for which auxotrophic mutants are available, and success has been achieved with bacteria other than *E. coli* and with microbial eukaryotes such as *Saccharomyces cerevisiae*. Heterologous experiments are also possible, with some genes from higher eukaryotes functioning sufficiently well in *E. coli* or *S. cerevisiae* to rescue auxotrophs and allow the desired recombinants to be selected.

2.1.3 Preparation of a clone library

The two approaches described so far—cloning an individual restriction fragment and direct selection—are applicable to some genes but not all. If these strategies

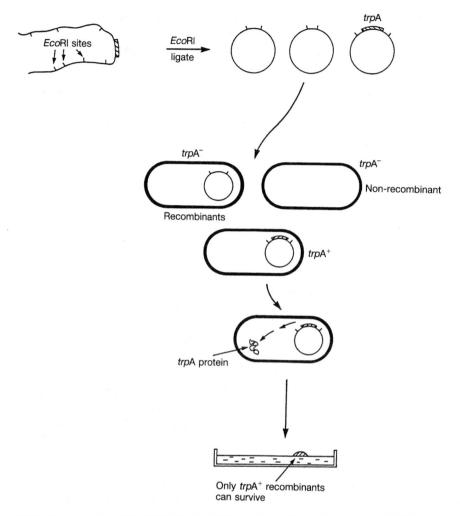

Figure 1 An example of the direct selection strategy: cloning the *trpA* gene of *E. coli*. DNA from a *trpA*⁺ strain of *E. coli* is restricted and the fragments ligated into a cloning vector. The recombinant DNA molecules are then used to transform an auxotrophic *trpA*⁻ strain and the transformants plated on minimal medium. Auxotrophs cannot grow on minimal medium so the only colonies that appear will be recombinants that contain the cloned *trpA* gene. Adapted with permission from ref. (1).

are inappropriate for the gene being sought (as is usually the case) it will be necessary to prepare a clone library and identify the desired gene by a second stage of experimentation.

Two types of clone library are commonly used:

- a genomic library, which is prepared from the total DNA of the organism under study, and which consists of a sufficiently large number of clones for there to be a statistically-high chance of the gene of interest being present (preparation of genomic libraries is described in Chapter 2)

- a complementary or copy DNA (cDNA) library, which is derived from the mRNA present in an individual tissue or cell type and which, therefore, contains copies of only those genes that are actively expressed in the tissue or cell type (preparation of cDNA libraries is covered in Chapter 3)

A genomic library may have to contain an immense number of clones in order to represent the whole genome. Exactly how many are required depends on the size of the fragments being cloned, the total size of the genome and the required probability of an individual gene being present (see Chapter 2, Section 3). If the probability is set at 95% then a library representing the entire *E. coli* genome, cloned as 17 kb fragments in a λ vector, must contain a total of 820 clones. Similarly, a yeast library would contain 3225 clones, and a human one 564 000 clones. Use of a cloning vector with a larger size capacity (e.g. a cosmid) reduces this figure, but for the larger genomes the library still consists of several tens or hundreds of thousands of clones. Searching through these for the gene of interest may be a tedious and difficult task.

In contrast, cDNA libraries can be quite small. Generally, a cDNA library is used if the gene of interest is known to be expressed at a relatively high level in a particular tissue. If the gene contributes 1% of the total mRNA then 1% of the clones in the cDNA library prepared from the tissue ought to contain the gene and fewer than 500 clones would be sufficient to have a good chance of finding the desired gene. In practice, there may be a difference in the efficiency of converting individual mRNA sequences into cDNA, so 1% of the mRNA may be converted into more or less than 1% of the cDNA but, generally, the strategy has been successful for highly or moderately expressed genes. Problems arise only if the gene contributes less than 0.1% of the total mRNA, as truly representative cDNA libraries of high numbers of clones may be difficult to obtain.

If a cDNA library is being used two factors must be kept in mind:

- degradation of the mRNA during preparation and conversion to cDNA leads to incomplete clones that represent only parts of the transcripts they are derived from
- a cDNA is a copy of the mRNA, not the gene itself, so neither introns nor upstream promoters are present

These two factors limit the amount of information that can be obtained about a gene from its cDNA. Beyond an understanding of which genes are expressed in a particular tissue, and the nucleotide sequences of the coding regions of the genes, very little can be discovered without the equivalent genomic clone. cDNA clones do, however, have considerable value, not least as intermediates in isolation of the desired genomic sequences. This point will be returned to in Section 2.2.1.

2.2 Identifying a gene in a gene library

If your gene can be obtained by cloning an isolated restriction fragment or by direct selection during plating-out then your task is straightforward. You need not recourse to specialized techniques and will be able to follow protocols in Volume I to obtain your desired clone. If, on the other hand, your strategy for gene isolation involves preparation of a clone library then you will need a means of identifying the gene you are interested in from the library. This may be self-evident, but many projects fail because insufficient thought has been put into how the correct clone will be distinguished from all the rest.

In principle, gene identification can be achieved by either of two approaches:

- the gene itself is identified

- the translation product of the gene is identified

The latter is a specialized procedure that requires first, that the gene is being transcribed and translated in the host cell, and second, that an antibody specific for the translation product is available. The desired clone is identified through crossreaction between its protein lysate and the antibody. For practical details the reader should consult refs. (2) and (3). Here, I will cover only the more general and commonly-used approach, where the gene itself is identified. This involves the important and powerful technique called hybridization analysis.

2.2.1 Hybridization analysis (see Chapter 5)

Two polynucleotides, whether DNA or RNA, will base-pair if all or part of their sequences are complementary, with the stability of the hybrid depending on the extent of base-pairing that occurs. Nucleic acid hybridization is used in molecular biology in a number of ways, including gene identification from clone libraries. In outline, the procedure is as follows (*Figure 2*):

1. The individual members of the clone library are placed on a nitrocellulose or nylon membrane. This can be achieved either by spotting on a sample of each clone (appropriate for small cDNA libraries, where the members are often stored as individual clones in microtitre trays) or, as shown in *Figure 2*, by transferring clones in bulk from the surface of an agar culture.

2. The cellular material is digested, leaving DNA, which is denatured and attached to the membrane.

3. The membrane is placed in a solution containing the hybridization probe, which is a polynucleotide that will base-pair with the gene being sought and no other. The probe is labelled with either a radioactive or non-radioactive marker, as described in Chapter 4. The membrane is then washed so that non-hybridized probe molecules are removed.

4. Areas of specific hybridization are detected by an appropriate technique, such as autoradiography if a radioactive marker has been used. Hybridization signals indicate clones that contain DNA molecules complementary to the probe.

Clearly, the key is having a probe that hybridizes only to the desired gene. Fortunately, there are a number of ways of obtaining a suitable probe; these are described below.

(a) Transfer colonies to nitrocellulose or nylon

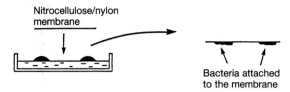

Nitrocellulose/nylon membrane

Bacteria attached to the membrane

(b) Degrade cells, purify DNA

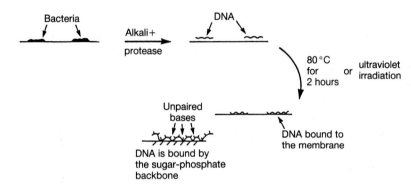

Bacteria

Alkali+ protease

DNA

80 °C for 2 hours   or   ultraviolet irradiation

Unpaired bases

DNA is bound by the sugar-phosphate backbone

DNA bound to the membrane

(c) Probe with labelled DNA

Wash

Apply X-ray film

X-ray film

Non-specific binding

Specific binding

(d) The resulting autoradiograph

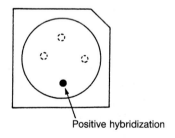

Positive hybridization

Figure 2 An example of hybridization analysis: identifying a recombinant of interest from a clone library by hybridization with a radioactively-labelled probe. Adapted with permission from ref. (1).

2.2.1.1 Probing a cDNA library for an abundant clone

Often, the aim of the project is to clone a gene known to be highly active in a particular tissue. If this is the case then transcripts of the desired gene will make up a relatively large component of the total mRNA of the tissue and clones representing the gene should be relatively abundant in a cDNA library prepared from this mRNA. If individual clones are chosen at random and used to probe the rest of the library then the family of abundant clones will quickly be identified.

2.2.1.2 Heterologous gene probing

Frequently, a gene being sought is known to have a similar nucleotide sequence to a second gene that is already available as an identified clone. This second gene could be the equivalent gene from a different organism, or possibly another member of a gene family that includes the desired gene. The second gene can be used as a heterologous probe for the gene being sought.

2.2.1.3 Oligonucleotide probes

If the amino acid sequence of the translation product is known, then it may be possible to design an oligonucleotide specific for the required gene. In most cases, this oligonucleotide need be no longer than 16–20 nucleotides (called a 16–20-mer) to act as a gene-specific probe. Because of the degeneracy of the genetic code a single oligonucleotide can be predicted only in exceptional circumstances (i.e. when the equivalent amino acid sequence consists entirely of methionine and/or tryptophan residues, the only amino acids to be specified by just one codon each) but a fairly small pool of oligonucleotides covering all possibilities can usually be synthesized. The size of the pool is minimized by choosing a region of the translation product that has a high proportion of single- or two-codon amino acids, and can be reduced further if the organism in question has a strong codon bias. For details of the use of oligonucleotide probes in hybridization analysis see Chapter 5, Section 3.11.

2.2.1.4 Using an identified cDNA to locate a genomic clone

Whatever the nature of the hybridization probe, the chance of success is improved by using as small a library as possible. All hybridization probes give at least some false-positive signals through hybridization to related but uninteresting sequences (possibly those displaying only short regions of similarity with the probe). For this reason, a cDNA library is preferable for the initial gene isolation, assuming that the gene being sought is expressed at a sufficiently high level in a particular tissue or cell type. The cDNA clone can then be used as a probe for the gene sequence in the genomic library. Because of its length (possibly more than 1 kb) and 100% complementarity with the gene sequence (except for the absence of introns) a cDNA is a highly-specific probe and the hybridization can be carried out under conditions that result in destabilization of all but the most perfectly-matched hybrids, resulting in a very low occurrence of false-positive signals.

2.2.2 Confirmation of clone identifications

You have probed your clone library, obtained some clear hybridization signals, and identified the clone containing the gene you have been looking for. Congratulations? No, not yet: before getting excited you must confirm that you do indeed have the correct clone. However clear-cut the result may appear it is always advisable to carry out a second, independent set of experiments to check the results of your original hybridization analysis. You have several choices:

- you could carry out another hybridization analysis with a second, unrelated probe. This may not always be possible but with oligonucleotide probing a second oligonucleotide predicted from a second region of the translation product could be used for confirmation

- if you know something about the expression profile of the gene you are trying to isolate, then you could hybridize the cloned sequence to samples of RNA prepared from different tissues and/or developmental stages to determine whether the RNA samples that the clone hybridizes to are those that are expected to contain transcripts of the gene. This technique is called RNA dot-blot analysis and is described in Chapter 5, Section 2.4.3. With cDNA clones, RNA dot-blotting is straightforward, but if you have isolated a genomic clone then you should bear in mind that the cloned DNA may contain more than one gene and that the hybridization results might be a composite of the expression profiles of two or more genes. Subclones should be prepared from the insert DNA and one carrying a relatively small fragment (< 1.5 kb), that hybridizes with the original probe, used for analysis of the expression profile

- if you know, or can predict, the sequence of all or part of the gene then the quickest way to check the identity of a clone is by PCR (see Chapter 7). The best strategy is to design a pair of primers that will give a reasonably small PCR product (200–300 bp) and then determine if these give a product of the expected size after PCR with the cloned DNA as template. If the PCR product is small then it can easily be sequenced to provide further confirmation

- you could sequence the entire cloned DNA fragment (see Chapter 6).

If you obtain a positive result with one or more of these confirmatory tests than you can generally assume that you have isolated the correct clone, but beware—there are still pitfalls, especially if you are working with a genomic library. Very few eukaryotic genes are unique within their genome: most are members of gene families, where the members all have similar sequences and include some pseudogenes that are not expressed. Final confirmation that you have cloned the gene you want and not some other member of its gene family requires a direct comparison between the gene sequence and the sequence of the translation product, or between the gene and a cDNA that is known to represent the transcript of the gene. If, as is often the case, this kind of confirmation is not possible then in future studies of the clone you should keep in mind the possibility that it is the wrong gene.

3 Studying gene structure and expression

Once a confirmed identification has been made and the desired gene obtained from a clone library, or by PCR, the route is open to analysis of the structure, function and expression of the gene, making use of the various molecular biology techniques that have been developed over the last 25 years. Exactly which techniques are used depends on the objectives of the research programme. This book does not aim to give the reader a detailed guide to all the techniques that can be used to study genes, as there are many large molecular biology manuals that do this (e.g. ref. 4). Instead, the objective is to explain the types of procedure that are common to all projects directed at understanding gene structure and activity. These procedures—most of which have already been mentioned above since they are also important in gene isolation—are summarized in the following sections.

3.1 Nucleic acid labelling (Chapter 4)

Nucleic acid labelling is central to many molecular biology procedures. In general terms, labelling is used for three purposes:

- to enable the position of a particular DNA molecule to be determined, as in hybridization analysis where the label is used to locate the position of a probe on a membrane
- to enable quantities of DNA or RNA that are too small to be visualized by ethidium bromide staining to be detected after electrophoresis (labelling is used in this way during the 'manual' procedures for DNA sequencing; see Chapter 6)
- to follow, through the course of an experiment, a single selected nucleic acid molecule against a background of many other molecules

Single- or double-stranded DNA and RNA molecules, as well as short oligo-nucleotides, can be labelled to high specific activity by any one of a variety of methods. These techniques were originally devised for use with radioactive markers, and radiolabelling is still widely used today because it routinely allows high specific activity (and hence low detection level) to be achieved. Radio-labelling does, however, have attendant health and disposal problems and these have prompted the development of non-radioactive systems, such as colorimetric and fluorescent markers, which now offer the same flexibility and, in many cases, the same sensitivity as radiolabelling.

3.2 Hybridization analysis (Chapter 5)

Hybridization analysis and the associated methods for the immobilization of nucleic acid molecules on membranes have widespread applications in molecular biology experiments. The use of the techniques in clone identification and determination of expression profiles have been outlined above. Other applications include:

- Southern hybridization (Chapter 5, Section 2.6), the first stage of which involves transfer of DNA molecules from an electrophoresis gel to a membrane, is used when the aim of the hybridization experiment is to detect a particular restriction fragment. Specific applications include:

  - identification of a restriction fragment, containing a sequence of interest, from within a larger, cloned DNA molecule

  - determination of the size of a restriction fragment containing a sequence of interest in a digest of total genomic DNA

  - RFLP analysis, in which differences in the lengths of the same restriction fragment in different genomic DNA preparations are used to infer genetic information about those genomes, such as the presence/absence of a disease marker in human DNA

- Northern hybridization (Chapter 5, Section 2.7), involves transfer of RNA molecules from gel to membrane and is used to determine the size(s) of the mRNA(s) transcribed from a gene.

- *In situ* hybridization is used to locate the positions of transcripts within cells (5) and to map genes on to chromosomes. A popular modification of the mapping procedure is FISH, in which two or more genes are labelled with different fluorescent markers so their relative positions on a chromosome can be distinguished (6).

3.3 DNA sequencing (Chapter 6)

The DNA sequence is virtually always looked upon as the primary objective once a gene has been cloned. Although a sequence on its own is of limited value, it is the necessary stepping-stone to more informative analyses of the cloned gene. Since 1977 there have been two efficient methods for DNA sequencing: the chain-termination procedure described by Sanger *et al.* (7) and the chemical cleavage method of Maxam and Gilbert (8). Although the latter was initially the more popular of the two procedures, the chain-termination method is now the one more routinely used, mainly because it enables longer sequences to be obtained and has proven to be amenable to automation. Non-automated or 'manual' DNA sequencing can be performed easily on the laboratory bench and is still frequently carried out, especially if the length of the DNA molecule to be sequenced is relatively small. Larger-scale sequencing projects, such at those directed at entire genomes, are entirely dependent on automated, high-throughput, procedures.

3.4 PCR (Chapter 7)

The polymerase chain reaction has such a vast range of applications that there is hardly a molecular biology procedure that cannot be carried out by a PCR-based technique. In addition to the standard technique, in which a DNA template is amplified by a pair of oligonucleotide primers, a number of modified versions of PCR have been devised. These include:

- RT-PCR (Chapter 7, Section 2.2.10) is used to amplify a DNA fragment from an RNA template. It can therefore be used to study RNA transcripts of genes, for example to determine in which tissues a gene is expressed
- RACE (9) is used to determine the sequence at the extreme 5′ or 3′ end of a cDNA molecule. This enables the exact initiation and termination positions for transcription of a gene to be identified within a DNA sequence
- *in situ* PCR (10), like *in situ* hybridization, can be used to locate transcripts within cells but, being more sensitive, it can detect rare transcripts as well as those that are highly abundant
- RAPD analysis is one of several techniques used in molecular phylogenetics, where the aim is to determine the evolutionary relationships between genomes. In RAPD analysis, the primers are not specific for a single position within a genome, but instead anneal at many positions and so give a complex banding pattern when the products are examined by agarose gel electrophoresis. Differences in the banding patterns when two related genomes are analysed result from the sequence variations between those genomes. RAPD analysis enables those sequence variations to be typed and the resulting data indicate the degree of evolutionary similarity between the genomes.

3.5 Studies of gene expression

Many techniques are available for analysis of the mode and regulation of transcription of a gene. These include a group of techniques that, together, are referred to as 'transcript mapping' and which enable the precise start- and end-points of a transcript to be mapped on to a DNA sequence. These are described in Chapter 9. Other techniques are available for direct sequencing of RNA molecules (11) and studies of interactions between DNA molecules and proteins (12–14).

4 Interpretation of molecular biology experiments

In any scientific research programme it is essential to be critical of experimental data and circumspect in their interpretation. A broad consideration of scientific method is inappropriate in a book like this, but it is appropriate that the reader be warned about the caution needed in interpreting the primary data obtained from molecular biology experiments. This point has already been stressed with reference to the confirmation of clone identification when screening a cDNA or genomic library: it is equally applicable to other techniques in molecular biology.

In some respects, the types of data that result from molecular biology experiments are not conducive to rigorous scientific analysis. In most experimental sciences, a statistical analysis is possible because the data obtained are numerical.

In molecular biology, this is rarely the case as frequently the data are obtained as bands on an autoradiograph or in some similar non-numerical form. Statistical analysis by conventional means is inappropriate. Repeating the experiment may identify clear artefacts but cannot provide data from which confidence limits and degrees of error can be calculated. For these reasons, a full understanding of the technical details of the procedure is required, along with sufficient critical appraisal to assign an empirical confidence limit to a particular result. This is one of the reasons why the use of pre-packaged kits complete with 'foolproof' instructions is looked on by some as dangerous to the development of molecular biology, or at least of molecular biologists.

Fundamental scientific technique demands that experiments are carried out in duplicate, that negative and positive controls are included, and that a result is accepted only when confirmed by at least two independent approaches. So, a DNA sequence should be obtained from both cDNA and genomic clones, or at least from two different clones of the same type, and the sequencing reactions should be checked by sequencing control DNA, whose sequence is known, in each experiment. This last point is rarely observed but it does allow a formal degree of confidence to be assigned to ambiguous positions in the unknown sequences. Scientific technique also demands that results are not predicted in advance, and that those which do not fit hypotheses are not simply ignored. This means that you must assign a meaning to every band that appears on a gel or autoradiograph. It is not good enough to say 'There is the one I want, the rest are unimportant'.

There is no reason why molecular biology should be exempt from the scientific tenets that apply to other disciplines. The techniques are not foolproof and the phenomena are, as always, studied in an indirect manner. Do not waste your time obtaining 'results' that have no value.

References

1. Brown, T. A. (2001) *Gene cloning and DNA analysis: an introduction*, 4th edn. Blackwell Scientific, Oxford.

2. Huynh, T. V., Young, R. A., and Davis, R. W. (1985). In *DNA cloning: a practical approach* (ed. D. M. Glover). Vol. I, p. 49. IRL Press at Oxford University Press, Oxford.

3. St John, T. P. (1990). In *Current protocols in molecular biology* (ed. F. M. Ausubel, R. Brent, R. E. Kingston, D. D. Moore, J. G. Seidman, J. A. Smith, and K. Struhl), Unit 6.7. John Wiley, New York.

4. Ausubel, F. M., Brent, R., Kingston, R. E., Moore, D. D., Seidman, J. G., Smith, J. A., and Struhl, K. (ed.) (2000). *Current protocols in molecular biology*. John Wiley, New York.

5. Zeller, R. and Rogers, M. (1989). In *Current protocols in molecular biology* (ed. F. M. Ausubel, R. Brent, R. E. Kingston, D. D. Moore, J. G. Seidman, J. A. Smith, and K. Struhl), Unit 14.3. John Wiley, New York.

6. Heiskanen, M., Peltonen, L., and Palotie, A. (1996). *Trends Genet.*, **12**, 379.

7. Sanger, F., Nicklen, S., and Coulson, A. R. (1977). *Proc. Natl Acad. Sci. USA* **74**, 5463.

8. Maxam, A. M. and Gilbert, W. (1977). *Proc. Natl Acad. Sci. USA* **74**, 560.

9. Frohman, M. A., Dush, M. K., and Martin, G. R. (1988). *Proc. Natl Acad. Sci. USA* **85**, 8998.

10. Bagasra, O., Seshamma, T., Pomerantz, R., and Hanson, J. (1996). In *Current protocols in molecular biology* (ed. F. M. Ausubel, R. Brent, R. E. Kingston, D. D. Moore, J. G. Seidman, J. A. Smith, and K. Struhl), Unit 14.8. John Wiley, New York.

11. Beier, H. and Gross, H. J. (1991). In *Essential molecular biology: a practical approach* (ed. T. A. Brown). Vol. II, p. 221. IRL Press at Oxford University Press, Oxford.

12. Galas, D. and Schmitz, A. (1978). *Nucl. Acids Res.* **5**, 3157.

13. Garner, M. M. and Revzin, A. (1981). *Nucl. Acids Res.* **9**, 3047.

14. Ausubel, F. M., Brent, R., Kingston, R. E., Moore, D. D., Seidman, J. G., Smith, J. A., and Struhl, K. (ed.) (2000). *Current protocols in molecular biology*, Chapter 12. John Wiley, New York.

Chapter 2

Construction of genomic libraries in λ and cosmid vectors

Anna-Maria Frischauf

Universität Salzburg, Institut für Genetik und Allgemeine Biologie,
Hellbrunnerstrasse 34, A-5020 Salzburg, Austria

1 Introduction

A genomic DNA library is a set of cloned fragments of genomic DNA. For most organisms no prior information about the genome is required for library construction (though special methods for DNA isolation may be required). The first genomic libraries of complex genomes were constructed in phage λ vectors cloned in *Escherichia coli*. This was followed by the introduction of cosmid vectors, plasmids which also use the phage λ packaging system (see Volume I, Chapter 9, Section 3.6). Packaging of recombinant DNA molecules into phage λ heads provides an excellent selection for a minimum insert size. It also specifies a maximum insert size of about 22 kb for a λ vector and 40 kb for a cosmid. More recently, several additional systems for the production of genomic libraries have become available. P1 vectors, bacterial artificial chromosomes (BACs) and P1 artificial chromosomes (PACs) (1) are also propagated in *E. coli*, while yeast artificial chromosomes (YACs) are grown in *Saccharomyces cerevisiae* (2, 3). All the new vectors offer significant increases in the size of inserts that can be accommodated—greater than 100 kb for P1, up to 300 kb for BACs and PACs, and more than 1 Mb for YACs. These large-insert vectors have the advantage that libraries containing fewer clones are needed in order to cover an entire genome. This greatly facilitates the construction of ordered clone maps and the physical linking of genetic markers. There are, however, some disadvantages inherent to propagating clones of very large DNA fragments in either yeast or *E. coli*. Not only is there a greater tendency for the DNA to undergo deletions or rearrangements, but the copy number of the cloned DNA also has to be quite low, leading to a relatively low yield of the rather complex insert DNA. Another important practical problem with such libraries is that very high-quality, high-molecular-weight insert DNA, of consistent size, must be prepared prior to cloning. Library construction is therefore laborious and technically demanding.

Ideally, different kinds of libraries complementing each other are used in the characterization of a large genome. This is possible and widely done with the human genome. YAC, PAC, and BAC clones are available from various public sources, as well as commercially, and can be used in conjunction with home-made or commercial cosmid libraries, including chromosome-specific libraries. Similar resources can also be obtained for other popular model organisms or commercially important species, where genome projects are often publicly funded. There still remains, however, a need for the construction of genomic libraries in individual laboratories. Examples include all organisms for which no libraries are yet available and genomes displaying special features (e.g. a chromosome translocation) from members of well-studied species such as humans. It is also frequently necessary to construct subgenomic libraries of YAC and other high-capacity clones in order to isolate a single desired gene as a reasonably small DNA fragment.

The widespread use of cosmid and λ vectors in *E. coli* over many years has led to straightforward, standard protocols that allow fast and efficient construction of genomic libraries. Host strains have been developed to maximize the yield and stability of recombinants and knowledge of the limitations and difficulties of the system has accumulated. Most necessary components, such as vector arms and packaging extracts, can either be obtained commercially as 'ready for use' reagents or can be prepared very cheaply in large quantities. The molecular weight of the starting DNA is not too critical, particularly for λ libraries, and a routine DNA preparation yields clonable DNA. Unless there are strong reasons for choosing long-insert systems, genomic libraries in cosmid vectors are the most easily accessible source of clones to cover most genomes.

Once the library has been constructed, pools of clones can be stored for repeated plating of the amplified library. Clones containing specific sequences can then be identified directly by hybridization analysis of libraries transferred onto filters, followed by picking of the colonies, rescreening and characterization. Alternatively, for projects involving the characterization of large numbers of clones, colonies from the primary, unamplified library can be transferred into microtitre plates for later 'gridding' into ordered arrays on filters or pooling for PCR screening.

2 Overview of library construction

Cloning in λ and cosmid vectors depends on the λ packaging system to introduce DNA with high efficiency into bacterial cells. In order to be packaged into λ heads DNA must contain two *cos* sequences in the correct orientation at a distance that corresponds to 78–105% of the wild-type λ genome (4). If the recombinant DNA is to be propagated as a phage, all the genes that are necessary for lytic growth of λ have to be included in the vector. This leads to 7–22 kb insert capacity for a λ replacement vector such as λEMBL3 (ref. 5; see also Volume I, Chapter 9, Section 4.2.2). Cosmids are plasmids containing the *cos* sites necessary for packaging into λ heads but no other λ-related components (Volume I, Chapter 9, Section 3.6). A cosmid vector can therefore accommodate inserts up to 42 kb (6).

To be cloned, the genomic DNA has to be cut into fragments of a size that will permit propagation in the λ or cosmid vector. This is usually done with *Mbo*I or *Sau*3A (restriction enzymes that leave overhanging ends that can be ligated into *Bam*HI restriction sites of the vector). Before ligation to the vector, the digested genomic DNA has to be either size-fractionated or treated with phosphatase (or both) to prevent the cloning of DNA fragments that are created by random co-ligation of DNA pieces not contiguous in the genome. Size-fractionation eliminates these artefactual products because they exceed the capacity of the packaging system (7). Phosphatase treatment removes the 5'-phosphate groups from the insert DNA and therefore prevents ligation of the molecules to each other, but not to vector DNA which retains its 5'-phosphate groups (8). A further alternative is partial filling-in of the overhanging ends created by different enzymes to give ends compatible only with vector-to-insert ligation (9). Vectors are digested with *Xho*I or *Sal*I and partly filled-in with dCTP and dTTP. Insert DNA is digested as usual with *Mbo*I or *Sau*3A and filled in with dATP and dGTP. This creates overhanging ends that do not allow insert-to-insert or vector-to-vector ligation.

After ligation of insert and vector DNA the mixture is packaged *in vitro* using extracts from *E. coli* strains that are lysogenic for defective λ phages. The packaged recombinant phage or cosmid is then introduced into the appropriate *E. coli* host by infection. Plaques or colonies can either be screened directly as a primary library or pooled and replated as needed for screening. Colonies can also be picked into microtitre plates for storage.

3 Representation

What should one wish for in the ideal genomic library? First, randomness: all sequences should be represented with equal probability. Unfortunately, this condition is never completely fulfilled in a real library. It is important to keep this limitation in mind and to apply strategies that minimize the problems caused by the unequal representation of different sequences in a genomic library.

A complete genomic library is usually prepared by cutting the DNA as randomly as possible to a size suitable for cloning in the chosen vector. In the ideal case the probability of the presence of any given sequence in a library of N clones with an insert size of f (as a fraction of the genome) is given by the following equations (10):

$$P = 1 - (1 - f)^N$$

or

$$N = \frac{\ln(1 - P)}{\ln(1 - f)}$$

This means that in a human genomic library of 2×10^5 clones with 15 kb inserts corresponding to the DNA from one haploid genome, a given unique sequence will be present with a probability of 63%; if there are 4×10^5 clones then the probability is 86% and if there are 10^6 clones the probability is 99%. Any deviation from randomness in the library will change this probability and usually decrease it.

Historically, *Eco*RI was the first enzyme extensively used to partially digest DNA to clonable size. This was not based on any special advantage of this enzyme but rather on the absence of suitable λ vectors with other cloning sites. Since the average size of an *Eco*RI fragment is 3.2 kb it follows that there is a significant number of *Eco*RI fragments that are greater than 22 kb (the maximum insert size of a λ vector). Such fragments were therefore not clonable. If not all sites are cut at the same rate and the total number of sites is relatively small, some fragments will be under-represented, even though they can be cloned in a complete *Eco*RI digest. Since λ vectors containing *Bam*HI cloning sites have become available, most libraries are now made using partial digestion with either *Mbo*I or *Sau*3A, a pair of isoschizomers recognizing the sequence 5'-GATC-3' and resulting in overhanging ends that can be cloned into *Bam*HI sites.

All the problems mentioned for partial digestion with *Eco*RI in principle also exist for *Mbo*I and *Sau*3A digestion. However, they are less severe, since the total number of sites for enzymes with a 4-bp recognition sequence is much larger. There are fewer long stretches completely devoid of sites and, even with significant differences in rates of cutting for specific sites, the presence of many alternative sites should result in a more random fragmentation.

As an alternative to partial digestion with an enzyme that will produce ends compatible with the cloning site of the vector, one can shear the DNA by physical means (e.g. sonication), fill in the overhanging ends and ligate with adapters, using the techniques described in Volume I, Chapter 7, Section 4.2. This approach avoids the difficulties arising from the uneven distribution of restriction sites in genomic DNA, but does not ensure completely random fragmentation since differences in secondary structure resulting from base sequence and composition may well create preferred or under-represented breakpoints. The main disadvantage of the procedure is the increased number of steps in the construction of the library, which leads to lower efficiencies, more work and the greater likelihood of introducing artefacts due to random co-ligation of DNA fragments. Despite these drawbacks, such libraries may be helpful in cloning gaps left by sequences not represented in standard *Sau*3A libraries. Libraries prepared in yeast may be even more useful since they also change the bias caused by the cloning host.

Fragmentation of the DNA is not the only step that introduces a bias into a library. Host restriction systems may deplete the DNA in sequences that are attacked by the respective system. Cloning hosts are generally deficient in the EcoK restriction-modification system. Some strains used for making packaging extracts and some cloning hosts, however, do have active McrA and McrB systems, which can restrict cytosine-methylated DNA (ref. 9; see also Appendix 3). Recombination systems also play an important role in the stability of cloned sequences. For cosmid libraries it is essential to have a very tight *recA*⁻ allele, otherwise many cosmids are quite unstable. For many types of λ vectors (e.g. EMBL3) *rec*⁺ or *recBC* hosts are required (11). It has been shown that *sbcC* hosts enhance the propagation of palindromic sequences (12). Since in the lytic propagation of phage λ there is a minimum size requirement for packaging into phage particles, phages with deletions of the insert are less likely to be found, so sequences unstable under these conditions will be missing from the library.

Non-random representation of DNA sequences means that some sequences are over-represented and others are either under-represented or completely lost. The first effect is generally less important and can be minimized by avoiding amplification of the library or by keeping pools of amplified recombinants separate (see Section 6.9). Complete loss of the desired sequences from a library requires a drastic change in cloning strategy, such as changing the enzymes used for creating inserts, changing vectors, or switching to cloning in yeast instead of *E. coli*. Before resorting to such drastic measures easier alternatives can be tried. If a sequence is severely under-represented in a cosmid library, a switch to a λ library can sometimes solve the problem. Occasionally, 'forced cloning' is an efficient alternative to a complete change in techniques. Here, a restriction map representing the genomic equivalent of a cDNA, or in the vicinity of the last probe in a chromosome walk, is produced by hybridizing the cDNA or probe to digests of genomic DNA. A restriction fragment is then selected for cloning which should be of a size as far removed as possible from the peak of the size distribution for that particular enzyme (e.g. a 10 kb *Sau*3A fragment, a 15 kb *Bgl*II fragment, etc.). Genomic DNA is then digested with this enzyme and size-fractionated on an agarose gel. The fraction of the desired size is cloned into the appropriate vector (for the two examples above, a *Bam*HI λ replacement vector such as EMBL3). If the required sequence is at all clonable in the particular vector–host system it can usually be obtained in this way. This approach is, however, laborious.

4 The choice between λ and cosmid libraries

What are the relative advantages of λ and cosmid libraries? If genes of unknown size are to be isolated from one or a small number of libraries, possibly with some chromosome walking, then cosmid cloning is the system of choice because each recombinant clone contains a longer DNA insert and the total number of recombinants that have to be screened is therefore smaller. This is especially true if the genome is complex and there is an abundant supply of DNA for cloning. λ libraries are easier to make and screening of λ plaques is somewhat simpler than screening of cosmid colonies. More importantly, the yield of recombinants per microgram of starting DNA is higher with λ vectors and the requirement that the starting DNA be high molecular weight is less stringent. These two advantages make λ cloning useful when the supply of DNA is very limited and its quality not good, as is the case with sorted chromosomes and YAC clones that have been prepared by elution from an electrophoresis gel.

5 Vectors

When a library is constructed, the choice of vector should be carefully considered. Beyond the actual cloning properties of the vector, it is useful to take into account the kinds of probes the library is to be screened with and whether clones from the library are to be used as probes to screen other libraries (e.g. cDNA libraries). Homologies between vector sequences can be an irritating problem in the counter-screening of different libraries. Different replication origins and antibiotic resistances can also be helpful in alternating between different libraries, as is the change between phage and plasmid vectors.

6 Library construction

Table 1 gives a summary of buffers (ref. 13, and others cited in this chapter), and other materials used throughout the library constructions.

6.1 Phage λ vector preparation

EMBL3 (16) is a λ replacement vector (*Figure 1*) with short polylinker sequences containing sites for *Sal*I, *Bam*HI and *Eco*RI flanking the middle fragment. A very similar vector, λEMBL3cos (11) offers the additional advantage of very easy restriction mapping by partial digestion and is used in exactly the same way as EMBL3. The presence of the *red* and *gam* functions on the middle fragment allows genetic selection against non-recombinant vector phages, by plating the library mixture on a P2 lysogen, which allows only recombinant phages to grow (the Spi$^-$ phenotype, described in Volume I, Chapter 9, Section 4.1). The background in this procedure is very low ($< 10^{-4}$). To reach such a low level of non-recombinant vector in the library it is necessary to check the Spi$^-$ background by self-ligation of the vector preparation, followed by test packaging and plating (see Sections 6.8 and 6.9), before the library is made. When EMBL3 is used as a *Bam*HI (or *Sal*I) vector, it is also possible to select against non-recombinants by excluding the middle fragment from the ligation reaction by using *Eco*RI restriction to remove the *Bam*HI ends. The small linker fragments can then be removed by isopropanol precipitation, as described in *Protocol 1*. The resulting ligation is as efficient as can be obtained with purified vector arms. If this is done in the absence of genetic selection it leads to a significantly higher background (10^{-1}–10^{-2}), but

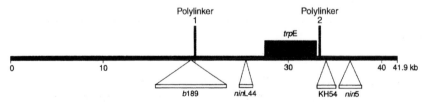

Figure 1 Structures of the λEMBL3 and λEMBL4 replacement vectors. Inserts of 7–20 kb can be cloned in the *Bam*HI, *Eco*RI or *Sal*I restriction sites. Recombinants are identified by virtue of their Spi$^-$ phenotype. The two vectors differ in the relative positions of the restriction sites in the polylinkers. Solid boxes above the map indicate DNA sequences from other sources that were inserted into these vectors during their construction and the open boxes below the map are deletions of the wild-type λ genome. Reprinted from Brown, T. A., *Molecular Biology Labfax*, Vol. 1, 2nd edn, p. 351, (1998), with permission of the publisher, Academic Press.

Table 1 Buffers and other materials needed for construction of genomic libraries

1. Buffers

TE	10 mM Tris-HCl, pH 7.6, 1 mM EDTA
Hi	10 mM Tris-HCl, pH 7.6, 10 mM MgCl$_2$, 100 mM NaCl
Med	10 mM Tris-HCl, pH 7.6, 10 mM MgCl$_2$, 50 mM NaCl, 1 mM DTT
Lig	40 mM Tris-HCl, pH 7.6, 10 mM MgCl$_2$, 1 mM DTT, 0.5 mM ATP
M1	0.55 ml water, 5 μl 2-mercaptoethanol, 30 μl 0.5 M Tris-HCl, pH 7.6, 1.5 ml 0.05 M spermidine plus 0.1 M putrescine, pH 7.0, 45 μl 1 M MgCl$_2$, 375 μl 0.1 M ATP. Add together in the order given
L Dil	10 mM Tris-HCl, pH 7.6, 10 mM MgSO$_4$, 1 mM EDTA
10 × HMFM	6.3 g K$_2$HPO$_4$, 1.8 g KH$_2$PO$_4$, 0.45 g trisodium citrate, 0.09 g MgSO$_4$.7H$_2$O, 0.9 g (NH$_4$)$_2$SO$_4$, 44 g glycerol, water to 100 ml
Phage freezing medium (5×)[a]	0.5 M NaCl, 0.05 M MgSO$_4$, 0.25 M Tris-HCl, pH 7.5, 0.05% gelatin, 50% glycerol
10× TAE	50 mM sodium acetate, 10 mM EDTA, 400 mM Tris-base, adjusted to pH 7.8 with acetic acid
56% CsCl	56 g CsCl made up to 100 ml total volume; same procedure for 42% and 31% CsCl

2. Materials

Nitrilotriacetic acid	Titriplex 1, Merck
T4 DNA ligase, restriction enzymes	New England Biolabs
Alkaline phosphatase (calf intestine)	Roche
Agarase	Calbiochem

3. *Escherichia coli* strains

K802	*supE44 hsdR galK2 galT22 metB1 mcrA mcrB1* (14)
NM646	*supE44 hsdR galK2 galT22 metB1 mcrA mcrB1* (P2*cox*3) (11)
BHB2688	N205 *recA* (λ *imm*434 *c*Its *b*2 *red*3 *E*am4 *S*am7)/λ (15)
BHB2690	N205 *recA* (λ *imm*434 *c*Its *b*2 *red*3 *D*am15 *S*am7)/λ (15)
DH5	*supE44 hsdR17 recA1 endA1 gyrA96 thi-1 relA1* (13)

4. Solvents

Phenol:chloroform:isoamyl alcohol[b]	50:48:2 (v/v/v)

[a] A. Craig, personal communication.

[b] See Appendix 4, *Protocol 6* for preparation of phenol solutions.

one that is still acceptable in most libraries. The *Eco*RI restriction cannot be checked on an agarose gel as the bands do not change mobility, therefore ligation, packaging and plating is used as an assay. It is possible to remove the middle fragment by gradient centrifugation or gel purification. The fast, convenient, but more expensive alternative is to buy purified vector arms that are available from several commercial suppliers.

The preparation of double-restricted EMBL3 that can be used with or without genetic selection is described in *Protocol 1*. If no genetic selection is used it should be remembered the digestion of the vector and the removal of the smaller linker fragments has to be very nearly complete, and that on an agarose gel 1–5% of uncut vector may not be visible but will result in a very significant background, especially when the efficiency of library construction is low.

Protocol 1

Preparation of λ vector

The example used is the preparation of EMBL3 for a *BamHI-MboI* library

Reagents

- λEMBL3 vector DNA
- *BamHI* and *EcoRI* restriction enzymes
- Hi buffer (see *Table 1*)
- 0.5 M EDTA, pH 7.0
- Phenol–choroform–isoamyl alcohol, 50:48:2 (v/v/v) (see *Table 1*)

- TE buffer (see *Table 1*)
- TE buffer containing 100 mM NaCl
- 3 M sodium acetate, pH 6.0, prepared by adding glacial acetic acid to 3 M sodium acetate until this pH is obtained

Method

1 Digest 10 μg EMBL3 DNA with 30 units *BamHI* in 50 μl Hi buffer for 1 h at 37°C and then put on ice. Test an aliquot (0.5 μg) for completeness of digestion on an agarose gel. It is important to heat the sample for 2 min at 68°C, then chill, before applying to the gel. The *cos* sites are then denatured and it is easy to decide whether the digestion is complete.

2 If the digestion is not complete, add 30 units *BamHI* and continue the incubation for 30 min, then check again. If the DNA has not been digested to completion, repurify the vector DNA and start again.

3 If digestion is complete, take a 0.1 μg aliquot, heat for 10 min at 68°C and freeze for later use as a control for ligation and packaging (see Section 6.9). To the remainder add 100 units *EcoRI* (the volume of enzyme solution, which contains glycerol, should be less than 10% of the total reaction volume) and continue the digestion for 30 min at 37°C.

4 Add 0.5 M EDTA, pH 7.0, to 15 mM final concentration, heat at 68°C for 10 min, and extract once with phenol–chloroform–isoamyl alcohol. Re-extract the organic phase with TE buffer containing 100 mM NaCl, adjust the salt concentration of the combined aqueous phases to 0.45 M with 3 M sodium acetate, pH 6.0, add 0.6 volumes of isopropanol and leave on ice for 10 min.

5 Centrifuge for 10 min at 16000 **g** in the microfuge, carefully take off the supernatant, then wash the pellet once with 100 μl 0.45 M sodium acetate plus 60 μl isopropanol and once with 70% ethanol. Leave the tube open for 10 min at room temperature to dry off the ethanol.

6 Resuspend the pellet in 18 μl TE to a final concentration of 0.5 μg DNA/μl.

6.2 Cosmid vector preparation

When preparing a cosmid library it is important to avoid the inclusion of more than one vector molecule per recombinant. Additional vector molecules reduce the possible size of the insert and create confusion in the restriction mapping, especially if this is carried out by linearization at the *cos* site and partial digestion (17).

With vectors that have appropriate restriction sites, the polymerization of vector molecules can be prevented by cutting with different combinations of restriction enzymes to create two different arms that are prepared separately (8). Alternatively, if a vector with two *cos* sites and a unique restriction site between them is used, it is possible to restrict between the *cos* sites, treat with alkaline phosphatase, then cut with the cloning enzyme and use the vector without physical separation of arms (18). In either case there should be no linear vector molecules present that can polymerize and ligate to insert DNA.

The vector used in the example described in *Protocol 2* is pcos6, a vector with two *cos* sites and a unique *Pvu*II site between them (19). The structure of pcos6 is given in *Figure 2*. The protocol includes a control reaction to verify that there is extensive dephosphorylation of the *Pvu*II site to avoid inclusion of more than one vector in the recombinant cosmids. Vectors with two *cos* sites are also available from commercial suppliers (e.g. superCos from Stratagene).

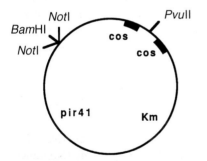

Figure 2 Schematic representation of the cosmid vector pcos6 (19). Abbreviations: cos, *cos* site; Km, kanamycin resistance gene; pir41, replication origin.

Protocol 2

Preparation of cosmid vector

Reagents

- pcos6 vector DNA
- Med buffer (see *Table 1*)
- *Pvu*II and *Bam*HI restriction enzymes
- Alkaline phosphatase (calf intestine)
- 0.5 M EDTA, pH 7.0
- 3 M sodium acetate, pH 6.0, prepared by adding glacial acetic acid to 3 M potassium acetate until this pH is obtained
- 100 mM nitrilotriacetic acid, pH 7.0 (see *Table 1*)
- TE buffer (see *Table 1*)
- Lig buffer (see *Table 1*)
- T4 DNA ligase
- Hi buffer (see *Table 1*)
- Phenol–choroform–isoamyl alcohol, 50:48:2 (v/v/v) (see *Table 1*)

Protocol 2 continued

Method

1   Digest 10 μg pcos6 DNA in 50 μl Med buffer with 20 units *Pvu*II for 1 h at 37 °C. Take an aliquot (0.1 μg) for a ligation control (see Section 6.9) and check the digestion by running 0.5 μg on a gel. Add 2 units alkaline phosphatase and continue incubation for 30 min at 37 °C.

2   If the restriction is complete, add 100 mM nitrilotriacetic acid, pH 7.0, and 0.5 M EDTA, pH 7.0, to 10 mM final concentration each, and incubate for 15 min at 68 °C.

3   Add 0.1 volumes of 3 M sodium acetate, pH 6.0, plus 2.5 volumes of ethanol, place at −20 °C for 1 h, then spin for 10 min at 16 000 **g** in the microfuge. Resuspend in 18 μl TE and take an aliquot (0.5 μg) to check dephosphorylation.

4   To check dephosphorylation, ligate separate aliquots (0.5 μg of each) taken before and after phosphatase treatment in 10 μl Lig buffer with 200 units T4 DNA ligase at 15 °C overnight. Compare the two ligated samples to cut, unligated vector on an agarose gel.

5   If dephosphorylation is complete, digest 8 μg of restricted, phosphatased pcos6 DNA in 20 μl Hi buffer with 20 units *Bam*HI for 1 h at 37 °C and check digestion on a gel.

6   If the digestion is complete, add 0.5 M EDTA, pH 7.0, to 15 mM final concentration, incubate for 10 min at 68 °C, then extract once with phenol–chloroform–isoamyl alcohol, twice with ether, and precipitate with ethanol, as described in step 3.

7   After centrifugation at 16 000 **g** wash the pellet once with 70% ethanol, air-dry and resuspend in 15 μl TE to a final concentration of 0.5 μg/μl.

6.3 Partial digestion of DNA

Protocol 3 describes the partial digestion of genomic DNA with *Mbo*I. It is carried out in the same way for cosmid and λ cloning. Identical, or at least very similar partial digestion products can be used for cloning into both types of vector.

Protocol 3

Preparation of partially digested DNA and phosphatase treatment

Reagents

- *Mbo*I restriction enzyme
- 0.5 M EDTA, pH 7.0
- Med buffer (see *Table 1*)
- 100 mM nitrilotriacetic acid, pH 7.0 (see *Table 1*)
- Phenol–choroform–isoamyl alcohol, 50:48:2 (v/v/v) (see *Table 1*)

- Alkaline phosphatase (calf intestine)
- 10 mM Tris-HCl, pH 7.6, 100 mM NaCl, 1 mM EDTA
- 3 M sodium acetate, pH 6.0, prepared by adding glacial acetic acid to 3 M potassium acetate until this pH is obtained
- TE buffer (see *Table 1*)

Protocol 3 continued

Method

1   Carry out the following analytical test to estimate the appropriate restriction conditions to produce fragments of the desired size:

(a) Dissolve 1 μg of high-molecular-weight DNA in 30 μl Med buffer;

(b) Take a 5-μl aliquot (0 min). Add 0.2 units of *Mbo*I and incubate at 37 °C;

(c) Take 5-μl aliquots at 5, 10, 20, 40, and 80 min, add to 1 μl 0.5 M EDTA, pH 7.0, and immediately incubate for 10 min at 68 °C;

(d) When all the aliquots have been collected, load them on to a 0.35% agarose gel with uncut λ DNA and *Hind*III-λ as size-markers. Run the gel at a very low voltage overnight (0.5 V/m);

(e) If necessary, repeat the experiment with more or less restriction enzyme.

2   Estimate the optimal enzyme concentration and digestion time from the results of the analytical experiment (step 1), bearing in mind the comments in the text immediately following this Protocol.

3   Carry out the preparative digestion by mixing 10 μg DNA (more if the DNA is to be size-fractionated on gradients) in 150 μl Med buffer at 37 °C with the calculated amount of *Mbo*I. Before adding enzyme, take a 3-μl aliquot to run on the gel. Take 70-μl aliquots at the calculated digestion times, add to 10 μl 0.1 M EDTA, pH 7.0, and incubate for 15 min at 68°C. Take a 3-μl aliquot from each time-point and run on a gel as described for the analytical digestions. If the DNA is to be size-fractionated but not phosphatased move directly to *Protocol 5* for gradient fractionation or to *Protocol 6* for gel fractionation.

4   To the rest of each aliquot add 1.5 units alkaline phosphatase and incubate for 30 min at 37 °C. Add 8 μl 100 mM nitrilotriacetic acid, pH 7.0, and incubate for 15 min at 68 °C.

5   Extract three times with phenol–chloroform–isoamyl alcohol and re-extract the organic phase with 10 mM Tris-HCl pH 7.6, 100 mM NaCl, 1 mM EDTA. Extract twice with ether and precipitate with ethanol (adjust salt to 0.3 M with sodium acetate, taking into account the NaCl already present, and add 2.1 volumes of ethanol). Leave on ice for 15 min, microfuge for 10 min at 16 000 **g**, take off the supernatant very carefully, wash the pellet once with 70% ethanol, and air-dry for 15 min.

6   Resuspend in TE to 0.5 μg/μl taking into account the losses during purification. Check the concentration by electrophoresing 1 μl on a gel with a standard of known concentration. If the DNA is to be size fractionated continue with *Protocols 5* or *6*.

When checking a partial digestion product on a gel it should be kept in mind that ethidium bromide binds in proportion to the amount of DNA. Long DNA molecules are therefore proportionately more stained than short ones. Since for cloning the number-average rather than the mass-average is the relevant variable, the partial digestion that is maximally stained at twice the molecular weight of the optimum size is the one that contains the maximum number of correctly-sized molecules (20). This means a sample showing maximum fluorescence at

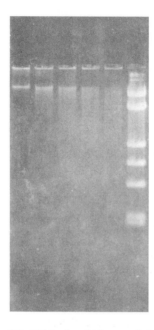

Figure 3 Analytical digestion products run on a 0.35% agarose gel. From left to right: 0, 5, 10, 20, 40 min restriction with *Mbo*I; the lane at the extreme right is a mixture of uncut and *Hind*III-λ. In this marker lane, the fragment sizes in kb are (from the bottom upwards): 2.0, 2.3, 4.4, 6.7, 9.4, 23.1 and 48.5.

20–30 kb should be used for λ cloning and maximum fluorescence at 60–80 kb for cosmid cloning.

Figure 3 shows an example of an analytical digest reaction as described in *Protocol 3*, step 1. The optimal digestion time would be about 10 min for cosmid cloning and less than 20 min for λ cloning. The extent of the optimal digestion depends somewhat on the starting material and on whether the DNA is to be size-fractionated or not. If the molecular weight of the starting DNA is high and if the partial digest is to be treated with alkaline phosphatase, with no size-fractionation, then it is preferable to digest the DNA to a lesser extent than would be optimally efficient for cloning, in order to have a larger average size of the DNA inserts. The reduced efficiency is more than compensated by avoiding the loss of material that occurs during size-fractionation. If the partially digested DNA is size-fractionated on a gradient or gel the exact size of the DNA in the digestion reaction is critical only for yield. In any case, it is useful to mix several samples that have been digested to slightly different extents to take into account the different rates at which individual restriction sites are cut and also to allow for some error in the size-determination on the gel. If the original DNA preparation is rather degraded it is better to digest to a smaller average size than usual ensure that a large number of molecules have two ligatable ends. In most manipulations, DNA molecules 20–40 kb long do not require special treatment to avoid shearing.

6.4 The choice between size-fractionation and alkaline phosphatase treatment

When preparing a genomic library it is very important to avoid random co-ligation of insert DNA fragments, since this leads to cloned sequences that have no counterpart in the genome. These artefacts can be recognized on genomic Southern blots (see Chapter 5, Section 2.6) by differences between genomic and cloned restriction fragments and by mapping of different parts of one clone to different chromosomal loci. The analysis of such clones wastes an enormous amount of time.

There are two ways of preventing such ligation events: size-fractionation and alkaline phosphatase treatment of the insert DNA. Since phage λ has a strict size requirement for packaging DNA into phage heads, only molecules less than 105% of the length of wild-type phage can be packaged. If the insert DNA has been size-fractionated on a gradient or a gel so that the ligation of two insert molecules and the required vector components produces a recombinant phage or cosmid molecule that exceeds the λ packaging capacity, then these molecules will not be packaged and therefore not transferred into bacterial cells. Alternatively, insert DNA can be treated with alkaline phosphatase to remove 5'-phosphate groups so that the molecules are unable to ligate to each other but are still able to ligate to vector DNA. It is essential that dephosphorylation should be as complete as possible to exclude co-ligation of DNA fragments not contiguous in the genome.

Both procedures work well when carried out properly. The advantage of size-fractionation on a gel is that the size of the insert can be more tightly controlled. Alkaline phosphatase treatment combined with size-fractionation reduces the level of random co-ligation products even further. Phosphatase treatment alone has the advantage of taking much less time and resulting in no loss of material. The way to obtain relatively large inserts when doing phosphatase treatment is to digest less than required to give optimal efficiency. This results, on average, in 15 kb λ clones and 40 kb cosmid clones (assuming the size of the cosmid vector is 6 kb). The number of inserts resulting from random co-ligation of fragments can be equally low in both procedures (much less than 5%).

6.5 Alkaline phosphatase treatment

Protocol 4 describes a set of control reactions to verify the dephosphorylation of the insert DNA (carried out in *Protocol 3*, steps 4 and 5). If the results of the test (see step 2) are unambiguous, the final library will not contain a significant amount of incorrectly-ligated clones. A further measure to reduce the risk is to work with a large excess of vector (see Section 6.7). In *Protocol 4*, an additional control reaction is included to check for the presence of a nuclease or residual *Mbo*I activity in the DNA preparation. If one of these are present then the DNA will be further digested during the ligation, producing new 5'-phosphate termini that can ligate to other insert molecules. This will also lead to clones containing more than one DNA fragment.

Protocol 4

Assessment of insert DNA dephosphorylation

Reagents

- Lig buffer (see *Table 1*)
- T4 DNA ligase

Method

1   Take 0.25 μg samples of the DNA before and after alkaline phosphatase treatment and incubate each overnight at 15°C in Lig buffer with 200 units T4 DNA ligase. Incubate an additional sample of phosphatased DNA in Lig buffer without adding ligase.

2   Compare these three samples to phosphatased, unligated DNA on a 0.35% agarose gel. The DNA that has not been dephosphorylated should have increased in molecular weight on ligation, the other three samples should look identical. If the phosphatased DNA has become ligated to itself, start again at *Protocol 3*, step 4, with the phosphatase treatment. If the sample incubated without ligase has been partially degraded, start again at *Protocol 3*, step 5, with additional phenol extractions.

6.6 Size-fractionation

Size-fractionation is the standard procedure for preventing the packaging of DNA fragments that are not contiguous in genomic DNA but have arisen during the ligation of insert to vector DNA. In all procedures the main aim is to remove the small fragments, because fragments that are too long to be packaged are unimportant. It is possible to fractionate by sucrose gradient centrifugation, NaCl gradient centrifugation (*Protocol 5*) and gel electrophoresis (*Protocol 6*). Gel electrophoresis gives the best separation, but lower yield. Careful treatment of the DNA is necessary to avoid degradation of the clonable ends. There are many possible protocols for purifying DNA after gel electrophoresis (see Volume I, Chapter 6) and *Protocol 5* gives just one example.

Protocol 5

Size-fractionation on a gradient

Equipment and reagents

- SW40 rotor (Beckman)
- 3 M sodium acetate, pH 6.0, prepared by adding glacial acetic acid to 3 M potassium acetate until this pH is obtained
- TE buffer (see *Table 1*)
- Phenol–choroform–isoamyl alcohol, 50:48:2 (v/v/v) (see *Table 1*)
- Linear 5–25% NaCl gradient prepared in TE buffer[a]

Protocol 5 continued

Method

1   More starting material has to be used if the DNA is to be size-fractionated on a gradient. Start with 100–200 μg DNA.

2   Take the DNA samples after heat inactivation of *Mbo*I (after *Protocol 3*, step 3), extract once with phenol–chloroform–isoamyl alcohol, and precipitate by adding 0.1 volumes of 3 M sodium acetate, pH 6.0, plus 2.5 volumes of ethanol and placing at −20 °C for 1 h. Spin for 10 min in the microfuge at 16 000 **g** and resuspend the dried pellets overnight in TE. After checking the samples on a 0.35% agarose gel pool the appropriate ones.

3   Prepare a linear 5–25% NaCl gradient in TE. Apply 50–100 μg DNA per gradient. Spin in the SW40 rotor (Beckman) for 4 h at 260 000**g** at 18 °C.

4   Collect 20 fractions, check aliquots on a gel, pool the appropriate fractions, dilute with TE, ethanol precipitate as in step 2, centrifuge at 3500 **g** and dissolve in TE to 0.5 μg/μl.

ᵃ A gradient can be prepared with a gradient-forming device or manually by placing a 20% NaCl solution in TE in a centrifuge tube, and then freezing and thawing once at −70 °C (−20 °C is insufficiently low a temperature).

Protocol 6

Size-fractionation in a gel

Reagents

- Phenol–choroform–isoamyl alcohol, 50:48:2 (v/v/v) (see *Table 1*)
- 3 M sodium acetate, pH 6.0, prepared by adding glacial acetic acid to 3 M potassium acetate until this pH is obtained

- TE buffer (see *Table 1*)
- TE buffer containing 100 mM NaCl
- TAE buffer (see *Table 1*)

Method

1   Take the DNA samples after heat inactivation of *Mbo*I (after *Protocol 3*, step 3), extract once with phenol–chloroform–isoamyl alcohol, and precipitate with ethanol as described in *Protocol 5*, step 2. Resuspend the pellets overnight in TE. After checking the samples on a 0.35% agarose gel, pool the appropriate ones.

2   Pour a 0.6% low-melting-point agarose gel in TAE buffer. Load size markers on both sides and leave empty slots between the sample and the size markers. Apply the DNA and run overnight at a low voltage (0.6 V/cm). Cut off the lanes with the size markers and a small strip of the genomic DNA and stain with ethidium bromide in electrophoresis buffer.

Protocol 6 continued

3   If the DNA has run far enough cut out the desired size-fraction from the unstained part of the gel. (If it has to be run further, destain the strips in electrophoresis buffer, put them back into the chamber and continue the run. The DNA does not need to be stained again.)

4   Electroelute the DNA into a small volume of TAE buffer.[a] Ethanol precipitate, centrifuge and take up in TE, 100 mM NaCl to 0.1 µg/µl.

5   Heat to 65 °C for 5 min, add 10 units agarase, incubate at 37 °C for 2 h, extract once with phenol–chloroform–isoamyl alcohol, ethanol precipitate and wash the pellet twice with 70% ethanol. Take up in TE buffer to 0.5 µg/µl.

[a] Carry out the electroelution as follows. Transfer the gel slice into a dialysis tube which has been rinsed in distilled water. The tubing is prepared by boiling for 10 min in 50 mM sodium carbonate, 1 mM EDTA, pH 8.0, and can be stored at 4 °C for several weeks. Place 50 volumes of TAE per volume of gel slice into the tube and clamp the ends. Place the tube in the electrophoresis tank such that the long dimension of the agarose block is parallel with the dialysis tube walls and to the electrodes of the tank. Electrophorese for 30 min using the conditions under which the original separation was performed. If UV-transparent apparatus is available the electroelution can be monitored every few minutes. The fluorescing band of DNA will be seen to transfer from the gel. Without disturbing the position of the tube in the tank, reverse the current to enable the DNA to detach from the dialysis tube and enter the buffer solution; 60 s is sufficient, otherwise the DNA may bind to the opposite wall of the tube. Remove the buffer containing the DNA from the dialysis tube and rinse the tube with 0.5–1.0 ml TAE, then combine the two solutions and ethanol precipitate as described in step 4.

6.7 Ligation of insert DNA to the vector

The best yield of recombinant clones for a given amount of DNA is obtained in the presence of a large excess of vector. In practice, background from non-recombinant phages or inclusion of linear cosmid vector molecules limits the vector excess that is desirable. For a phage vector with a good genetic selection a tenfold molar excess can be used for precious DNA samples. For routine use a two- to three-fold excess is adequate.

   To determine the appropriate amount of vector it is necessary to estimate the molar amount of DNA. This is not too difficult for size-fractionated DNA, but very inaccurate for a partial digest reaction that has not been size-fractionated. However, since a wide range of molar ratios greater than 2:1 (vector to insert DNA) will give acceptable results this is not a critical variable. Empirically, the weight of insert DNA multiplied by 5.0 for λ and by 0.5 for cosmid cloning gives a reasonable amount of vector DNA to be used for slightly under-digested DNA (see Section 6.3). Examples are given in *Protocol 7*. Too much vector is better than too little because an excess of vector also reduces the probability of insert–insert ligation if the dephosphorylation is incomplete. This is, of course, no replacement for alkaline phosphatase treatment but provides an additional safeguard against random co-ligation products.

Protocol 7

Ligation of vector to insert DNA

Reagents

- Lig buffer (see *Table 1*)

Method

1 Set up the appropriate ligation.

   (a) For λ libraries, with under-digested, phosphatase-treated but not size-fractionated DNA: ligate 5 μg EMBL3 DNA restricted with *Bam*HI and *Eco*RI to 2 μg dephosphorylated insert DNA in 25 μl Lig buffer overnight at 15 °C.

   (b) For λ libraries, with size-fractionated DNA: ligate 5 μg EMBL3 DNA restricted with *Bam*HI and *Eco*RI to 1 μg insert DNA in 25 μl Lig buffer overnight at 15 °C.

   (c) For cosmid libraries: ligate 2 μg cosmid vector prepared as described in *Protocol 2* to 2.5 μg insert DNA in 20 μl Lig buffer overnight at 15 °C.

2 Take 1 μl of the ligation mixture for an analytical packaging reaction. The ligation can then be continued at 4 °C for 3 days. After that, it can either be packaged or frozen in liquid nitrogen and kept at −70 °C.

For optimal ligation efficiency the reaction mix should be as concentrated as possible. Furthermore, commercial packaging extracts are expensive. To obtain the specified efficiency, the packaging mix must not be diluted too much with ligation solution. Occasionally, the presence of substances that inhibit ligation in the component solutions can make it impossible to achieve a good ligation at a very high concentration. In that case a more dilute reaction mixture may still give efficient ligation and a reasonable yield of recombinants. If home-made packaging extract is used the packaging reaction can be scaled-up to fit the ligation volume.

6.8 *In vitro* packaging

The preparation of high-efficiency packaging extract requires some experience and time spent on optimizing conditions. Packaging extracts are also available commercially (e.g. from Stratagene). In most cases, it is convenient and economical to use commercial extracts.

In vitro packaging is carried out with extracts from either a single or two complementing strains of *E. coli* lysogenic for defective λ phages. One component systems use a *cos⁻* lysogen (21). The most widely used system consists of extracts from two strains with different mutations in structural proteins of the phage and defects in DNA excision (15, 22). In *Protocol 8* (23) the strains BHB2688 and BHB2690 are used. They give extracts of reproducible and very high efficiency on thawing of every aliquot of the extract.

Protocol 8

Packaging extracts

Equipment and reagents

- Water baths or incubators at 30°C, 38°C, 42°C and 43°C
- Ti50 or Ti75 rotor (Beckman)
- Sonicator with a microtip device
- Sorvall SS-34 rotor
- *E. coli* BHB2688
- *E. coli* BHB2690
- LB agar plates (10 g Bacto-tryptone, 5 g Bacto-yeast extract, 10 g NaCl, 15 g Bacto-agar per 1 l). Check the pH and adjust to 7.0–7.2 with NaOH. Sterilize by autoclaving at 121°C, 103.5 kPa (15 lbf/in²), for 20 min

- LB medium (10 g Bacto-tryptone, 5 g Bacto-yeast extract, 10 g NaCl per 1 l). Check the pH and adjust to 7.0–7.2 with NaOH. Sterilize by autoclaving at 121°C, 103.5 kPa (15 lbf/in²), for 20 min
- BHB2688 resuspension buffer (50 mM Tris-HCl, pH 7.6, 10% sucrose)
- Lysozyme solution (2 mg/ml in BHB2688 resuspension solution)
- M1 buffer (see *Table 1*)
- BHB2690 resuspension buffer (20 mM Tris-HCl, pH 8.0, 3 mM MgCl₂, 10 mM β-mercaptoethanol, 1 mM EDTA)

Method

A. Preparation of a freeze-thaw lysate of *E. coli* BHB2688

1. Remove a single colony of *E. coli* BHB2688 from a stock culture and touch on to the surface of an LB agar plate that has been prewarmed to 42°C. Streak onto an LB agar plate at room temperature (see Volume I, Chapter 2, *Protocol 2* for the streak plate method). Incubate the first plate at 42°C and the second at 30°C. The strain should not grow at 42°C: if it does, then check another colony and obtain a new stock if necessary.

2. Remove the cells from the 30°C plate and resuspend in a minimal volume of LB medium prewarmed to 30°C. Transfer the entire resuspension, in equal aliquots, to 6 × 400 ml of prewarmed LB medium in 2-l flasks. Incubate at 30°C with shaking until the absorbance at 600 nm reaches 0.3. Transfer the flasks to a 43°C water bath or incubator and shake gently for 20 min to induce the λ prophage, then transfer to 38°C and shake vigorously for 3 h.[a]

3. Chill the cultures in ice-water, then centrifuge for 10 min at 2000 **g** at 4°C. Decant the supernatants and resuspend each pellet in 1.2 ml cold BHB2688 resuspension buffer.[b]

4. Add 120 μl lysozyme solution to each resuspended pellet, mix, then place the tube in liquid nitrogen. Either proceed to step 5 or store the extract at −70°C.

5. Thaw each sample slowly on ice (this should take about 1 h). Add 0.5 ml M1 buffer to each one, shake gently, and then centrifuge at 2°C for 30 min at 150000 **g** in a precooled Ti50 or Ti75 rotor.

6. Pool the supernatants, then distribute 50, 100, and 250 μl aliquots into cold tubes, freeze in liquid nitrogen, and store at −70°C.

Protocol 8 continued

B. Preparation of a sonic extract of *E. coli* BHB2690

1   Test the strain as described in Part A, step 1.

2   Grow a single 400-ml culture and induce the λ prophage as described in Part A, step 2. After induction, shake vigorously for 2 h at 38 °C.

3   Chill the culture in ice-water, then centrifuge for 10 min at 2000 **g** at 2 °C. Decant the supernatant and resuspend the pellet in 1.5 ml cold BHB2690 resuspension buffer.[c]

4   Transfer the suspension to a clear plastic tube. Sonicate while cooling within an ice–salt mixture, approximately 15 times for 5 s at full power using a microtip sonicator. The solution has to be ice-cold at all times and there should be no foaming. Sonication is completed when the solution is translucent.

5   Centrifuge the solution for 10 min at 3500 **g** in a Sorvall SS-34 rotor. There should be a very small pellet: if not the sonication has been insufficient. Add 0.6 ml M1 buffer to the supernatant, distribute aliquots of 20, 50 and 100 μl into cold tubes, freeze in liquid nitrogen, and store at −70 °C.

[a] These incubations induce the prophage version of the defective λ genome carried by the *E. coli* strain, resulting in synthesis of the λ proteins.

[b] The cell suspension must be kept cold at all times, so carry out these manipulations on ice.

The McrA and McrB restriction systems have been shown to be active during packaging (11). This should lead to under-representation of sequences containing methylcytosines. New packaging strains which do not contain this restriction system have been developed (N. Murray, personal communication) and may be used. It should be remembered, however, that when packaging extract is prepared from a different set of strains it is essential to optimize the protocol to give extracts of comparable efficiency. Some of the commercially available packaging extracts also do not have McrA and McrB restriction activity (e.g. Stratagene Gigapack Gold).

The preparation of packaging extract requires the basic precautions involved in handling proteins: the extracts should be prepared quickly, kept cold at all times and frozen by dropping into liquid nitrogen. Foaming should be avoided. *Protocol 8* can be scaled up or down without any differences in the quality of the extracts. Packaging extracts are stable for a long time if kept at −70 °C and it is more efficient to make a large quantity each time an extract is prepared.

The efficiency of a set of packaging extracts depends not only on the extracts but also on the DNA concentration, the individual DNA preparation, the order of addition of the extracts and the duration of the packaging reaction. For good extracts, the number of p.f.u. rises linearly for more than 4 h. This parameter should be tested for every new extract.

Protocol 9

Packaging

Reagents

- L Dil buffer (see *Table 1*)
- BBL top agar (180 mg BBL trypticase peptone, 50 mg NaCl, 60 mg Bacto-agar per 10 ml). Check the pH and adjust to 7.0–7.2 with NaOH. Sterilize by autoclaving at 121 °C, 103.5 kPa (15 lbf/in^2), for 20 min
- BBL agar plates (18 g BBL trypticase peptone, 5 g NaCl, 6 g Bacto-agar per 1 l). Check the pH and adjust to 7.0–7.2 with NaOH. Sterilize by autoclaving at 121 °C, 103.5 kPa (15 lbf/in^2), for 20 min

- LB medium (see *Protocol 8*)
- LB agar plates (see *Protocol 8*). Add the appropriate antibiotic for selection of the cosmid (see Appendix 4 for preparation and use of antibiotic solutions)

Method

1   For commercial packaging extracts follow the manufacturer's protocol.

2   For 'home-made' packaging extract take 1–2 μl ligation mixture or phage DNA (not more than 0.1 μg) and bring to room temperature.

3   Add 2 μl sonic extract, then 6 μl freeze–thaw lysate, both freshly thawed. Mix gently and incubate at room temperature for 4 h, or as determined from the time-course of the packaging reaction.

4   Dilute with 200 μl L Dil. Make appropriate dilutions in L Dil for titration. To the remainder add 2 μl chloroform, shake, leave for 5 min at room temperature, then store at 4 °C.

5   To 100 μl plating cells (see *Protocol 10*) add the calculated amount of packaged cosmid or phage. Incubate for 15 min at 37 °C.

6   (a) For λ, add cells with adsorbed phages to 3 ml BBL top agar at 42 °C. Plate on to a BBL agar plate.

   (b) For cosmids, add cells with packaged, adsorbed cosmids to 1 ml LB medium, incubate for 1 h at 37 °C, centrifuge for 30 s in a bench-top centrifuge at 16 000 **g**, resuspend the cells in a drop of LB medium and plate on the appropriate antibiotic-selection medium.

6.9 Plating and amplification

The *in vitro* packaged cosmid or λ DNA is introduced into *E. coli* cells by infection. If genetic selection with a λ vector such as EMBL3 is used, the library must be plated on the selective host, a P2 lysogen (e.g. *E. coli* NM646). In addition, re-ligated vector without insert should be checked for background on the P2 lysogen. To check the efficiency of the ligation an aliquot of the sample before and after ligation can be plated on a host that allows the growth of both vector and recombinant phage (e.g. *E. coli* K802). This permits a ligation failure caused by damaged ends of the insert DNA to be distinguished from inhibition of the ligation reaction by contaminants, such as EDTA or residual phenol. This control can only be used if complete vector rather than purified arms has been used and the small *Bam*HI–*Eco*RI fragment was not removed from the middle region. Cosmids are plated on a *recA⁻* strain such as DH5, which gives good plating efficiencies.

Before finally plating the library, the titre of the packaged material is determined using the batch of plating cells that will be used for the preparative plating. The efficiency of plating with *in vitro* packaged material can depend strongly on the plating cells. *Protocol 10* describes a simple method in which saturated overnight cultures that have been well-aerated up to the point of harvesting are used. Different batches of plating cells are tested using *in vitro* packaged material since *in vivo* packaged phage shows the same efficiency on all plating cells. The cells retain their efficiency for at least 2 weeks when stored at 4 °C.

Protocol 10
Preparation of plating cells

Reagents
- LB agar plates (see *Protocol 8*)
- LB medium (see *Protocol 8*)

Method

1   Streak out the appropriate *E. coli* strain (see the text above) on an LB agar plate.

2   Pick a single colony into LB medium. Grow overnight at the appropriate temperature with vigorous aeration (380 r.p.m.) in a shaker.

3   Take the saturated culture, chill, centrifuge for 10 min at 2000 **g**, pour off the supernatant, and resuspend the cells immediately in 10 mM MgSO$_4$. Test the plating efficiency by plating an aliquot of the *in vitro* packaged material that is to be used for library construction (see *Protocol 9*, steps 5 and 6). The cells will keep for at least 2 weeks in the refrigerator.

For a given size of plate, the appropriate amount of titrated, packaged mixture is added to a given number of plating cells. It is not possible to add unlimited amounts of packaged material to a given amount of plating cells. This is not problem if the library construction was efficient. If, however, a dilute ligation mixture was used and then packaged in a large volume of packaging extract it may be necessary to concentrate and purify the packaged material by centrifugation through a CsCl step gradient (*Protocol 11*).

Protocol 11

CsCl step gradient

Equipment and reagents

- Polyallomer tubes
- L Dil buffer (see *Table 1*)

Method

1 Layer into polyallomer tubes 0.5 ml each of 56%, 42% and 31% CsCl solutions in L Dil, apply packaged DNA on top, and fill up the tubes with L Dil.

2 Centrifuge for 3 h at 126 000 g at 20 °C.

3 Collect fractions from the top with a Pasteur pipette. The phage band is underneath the visible protein band. Titrate and pool the cosmid or phage fractions and dialyse for 6 h against two changes of L Dil.

After collecting and dialysing, large volumes of the library can be added to the cells and the library can be plated to the desired density of colonies for screening (*Protocols 12–15*). For 23 × 23 cm plates approximately 100 000–150 000 colonies or plaques represent a good compromise between the need for having independently-growing recombinants and the advantage of having the smallest possible number of plates to screen.

The λ and cosmid libraries can either be amplified or screened directly. The advantage of amplification is that it provides an almost inexhaustible supply of library that is viable for very long periods of time, if properly stored. The main disadvantage of amplification is that it severely distorts the concentration of each independent recombinant in the mixture because of differential amplification of different clones. Therefore, as a matter of principle, it is always preferable to use a primary library. Plated primary cosmid libraries can be stored at −70°C on filters soaked with the appropriate freezing medium (HMFM for cosmids, phage freezing medium for phages: see *Table 1*) (24) for very long periods.

For genome analysis, when it is expected that the library will be screened with a large number of probes and that the majority of the clones will be characterized and should be easily accessible, primary colonies are picked into microtitre plates. From there they can be mechanically spotted onto filters for repeated hybridization screening, or identified by their plate and well location and recovered.

Phage libraries are stable without further preservation in the cold-room on well-covered plates for many months. Alternatively, there are several protocols available for the freezing of phage plaques (e.g. see ref. 25), although these procedures are not as widely used as storage of frozen colonies. The limiting factor in the use of primary libraries is often the quality of the filters that are being hybridized over and over again with eventual loss of signal. It may, therefore, be more convenient to amplify a library and to plate it out whenever it has to be screened with a set of probes. If a large library is amplified it is very important to divide it into several fractions that are amplified independently, for example by keeping the product from different 23×23 cm plates separate. This prevents over-amplified recombinant clones from appearing at a high frequency in the whole library, restricts different over-amplified clones to different fractions, and ensures that independent clones are isolated if another fraction is screened.

Protocol 12

Test-packaging of phage libraries

Reagents

- L Dil buffer (see *Table 1*)
- *E. coli* K802 and NM646 (see *Table 1*)

Method

1 Package each of the following: 10 ng EMBL3; 10 ng EMBL3 restricted with *Bam*HI and re-ligated; 10 ng EMBL3 restricted with *Bam*HI + *Eco*RI; 10 ng EMBL3 restricted with *Bam*HI + *Eco*RI and re-ligated; 1 μl aliquot of the ligation mixture. The packaging is carried out according to the manufacturer's protocol (scaling the reaction mixture down) or as described in *Protocol 9*. If packaging extract is limiting, only EMBL3 restricted with *Bam*HI + *Eco*RI and re-ligated, and the ligation mixture aliquot, are packaged.

2 The packaged phage are diluted in L Dil to a concentration where approximately 100 plaques are expected from plating 10 μl on the non-selective host (*E. coli* K802). The reduction of plating efficiency on restriction is usually about 1000-fold. Re-ligation should not increase the plating efficiency for the double-restricted vector by more than 10- to 50-fold. On the selective host (*E. coli* NM646) a 10^4-fold greater amount of vector should be used. The ligation mixture is plated under the assumption of efficiencies of 10^4, 10^5 and 10^6 recombinants/μg insert DNA.

3 Part of the analytical packaging reaction can be used to test the plating efficiency of the different batches of plating cells prepared previously (*Protocol 10*).

Protocol 13

Test-packaging of cosmid libraries

Reagents

- L Dil buffer (see *Table 1*)

Method

1 Package 1 μl of ligation mixture according to the manufacturer's protocol or as described in *Protocol 9*. Dilute with L Dil and plate three concentrations that differ by factors of 10, using several different batches of plating cells. Compare with plating cells to which no cosmids have been added.

Protocol 14

Plating of phage libraries

Equipment and reagents

- BBL top agar and top agarose (see *Protocol 9*; for top agarose replace the Bacto-agar with agarose). See Appendix 4 for the preparation and use of MgSO$_4$ supplements
- 22 × 22 cm nylon filters (see Chapter 5, Section 2.3.3)
- L Dil buffer (see *Table 1*)
- BBL agar plates (see *Protocol 9*)

Method

1   Take an aliquot of titrated, packaged ligation mixture containing approximately 100 000–150 000 p.f.u. in no more than 2 ml L Dil, add 2 ml plating cells (*Protocol 10*), adsorb for 15 min at 37 °C, add 30 ml BBL top agarose (for direct screening) or top agar (for amplification), each with 10 mM MgSO$_4$ and cooled to 42 °C.

2   Plate on prewarmed 23 × 23 cm BBL plates. Incubate overnight or until the plaques are still not confluent at 37 °C.

3   Prepare plaque lifts on 22 × 22 cm nylon filters (see Chapter 5, Section 2.5) if the primary library is to be screened directly.

4   For amplification, scrape off the top agar into a beaker, add 30 ml L Dil + 0.5 ml chloroform, stir gently for 15 min and centrifuge at 5000 **g**. Collect the supernatant, re-extract the agarose with 10 ml L Dil, 0.1 ml chloroform for 10 min, centrifuge as before, collect the supernatant and pool with the first extraction. Do not pool recombinant phages from different plates.

5   The library can be stored at 4 °C for several years with slow loss of titre. Alternatively, it can be frozen. It is advisable to split it into two aliquots and use both means of storage.

Protocol 15

Plating of cosmid libraries

Equipment and reagents

- 22 × 22 cm nylon filters (see Chapter 5, Section 2.3.3)
- Whatman 3MM paper
- Sterile plastic sheet
- Transparent foil
- LB medium (see *Protocol 9*)
- LB agar plates (see *Protocol 9*)
- 10 × HMFM buffer (see *Table 1*)

Method

1   Take an aliquot of packaged, titrated ligation mixture containing ≈ 100 000–150 000 cosmids and adsorb to 1.5 ml of plating cells (*Protocol 10*) for 15 min at 37 °C. Add

Protocol 15 continued

15 ml LB medium and shake gently at 37 °C for 1 h. Plate a 15-μl aliquot directly on a small LB agar plate for counting. Centrifuge the rest for 5 min at 500 **g**, take off the supernatant and resuspend the pellet in 1.5 ml LB medium.

2   Plate on a 22 × 22 cm nylon filter (see Chapter 5, Section 2.5) for direct screening or on a 23 × 23 cm LB agar plate for amplification. Grow overnight at 37 °C—longer if colonies are still too small or shorter if they are too big (this is very host- and vector-dependent).

3   For amplification, add 20 ml LB medium and scrape off colonies. Pour into a 50 ml sterile tube and rinse the plate with 10 ml LB medium. Do not mix recombinants from different plates. Add a one-tenth volume of 10 × HMFM, aliquot into small tubes and drop into liquid nitrogen. Store at −70 °C.

4   To store the primary library (24):

(a) Make two filter replicas and prepare them for hybridization. It is possible to make a third replica filter and to scrape the colonies off for library amplification. All filters are marked by making 10 holes in identical positions. Take the regrown master filter and place it for 10 min on Whatman 3MM paper soaked with LB medium containing 10% 10 × HMFM. The paper must not be too wet;

(b) Place the filter on a piece of sterile plastic sheet cut to 23 × 23 cm, and cover with a transparent foil with a suitable grid (obtained by photocopying graph paper on to transparent foil). Mark the holes in the filter on to the transparent graph paper. Put a second sheet on top and transfer the markings;

(c) Keep the top sheet as a record to compare with the autoradiograph. Leave the bottom sheet on the colonies and put at −70 °C for 20 min. Put a precooled plastic square, cut to size, on top and clamp with clips. Store at −70 °C;

(d) If a positive colony has been identified by hybridization, mark its position on the record sheet, retrieve the frozen library plate, open while cooled on dry ice, match the graph papers and take a scalpel and cut out a 2 × 2 mm square through both the transparent foils and the filter below. Pick up the filter piece and put it into 0.2 ml LB medium containing 10% 10 × HMFM. Plate bacteria for rescreening from a dilution of this stock and freeze the labelled stock for further use.

References

1. Osoegawa, K., Woon, P. Y., Zhao, B., Frengen, E., Tateno, M., Catanese, J. J., and de Jong, P. J. (1998). *Genomics*, **52**, 1.
2. Burke, D. T., Carle, G. F., and Olson, M. V. (1987). *Science*, **236**, 806.
3. Larin, Z., Monaco, A. P., and Lehrach, H. (1991). *Proc. Natl Acad. Sci. USA*, **88**, 4123.
4. Feiss, M. and Becker, A. (1983). In *Lambda II* (ed. R. W. Hendrix, J. W. Roberts, F. Stahl, and R. A. Weisberg), p. 305. Cold Spring Harbor Laboratory Press, Cold Spring Harbor.
5. Murray, N. E. (1983). In *Lambda II* (ed. R. W. Hendrix, J. W. Roberts, F. Stahl, and R. A. Weisberg), p. 395. Cold Spring Harbor Laboratory Press, Cold Spring Harbor.

6. Collins, J. and Hohn, B. (1979). *Proc. Natl Acad. Sci. USA*, **75**, 4242.

7. Maniatis, T., Hardison, R. C., Lacy, E., Lauer, J., O'Connell, C., Quon, D., Sim, G. K., and Efstratiadis, A. (1978). *Cell*, **15**, 687.

8. Ish-Horowicz, D. and Burke, J. F. (1981). *Nucl. Acids Res.*, **9**, 2989.

9. Zabarovsky, E. R. and Allikmets, R. L. (1986). *Gene*, **42**, 119.

10. Clarke, L. and Carbon, J. (1976). *Cell*, **9**, 91.

11. Whittaker, P. A., Campbell, A. J., Southern, E. M., and Murray, N. E. (1988). *Nucl. Acids Res.*, **16**, 6725.

12. Chalker, A. F., Leach, D. R. F., and Lloyd, R. G. (1988). *Gene*, **71**, 201.

13. Sambrook, J., Fritsch, E. F., and Maniatis, T. (ed.) (1989). *Molecular cloning: a laboratory manual*, 2nd edn. Cold Spring Harbor Laboratory Press, Cold Spring Harbor.

14. Wood, W. B. (1966). *J. Mol. Biol.*, **138**, 179.

15. Hohn, B. and Murray, K. (1977). *Proc. Natl Acad. Sci. USA*, **74**, 3259.

16. Frischauf, A.-M., Lehrach, H., Poustka, A., and Murray, N. (1983). *J. Mol. Biol.*, **170**, 827.

17. Rackwitz, H. R., Zehetner, G., Frischauf, A.-M., and Lehrach, H. (1984). *Gene*, **30**, 195.

18. Poustka, A., Rackwitz, H.-R., Firschauf, A.-M., Hohn, B., and Lehrach, H. (1984). *Proc. Natl. Acad. Sci. USA*, **81**, 4129.

19. Ehrich, E., Craig, A., Poustka, A., Frischauf, A.-M., and Lehrach, H. (1987). *Gene*, **57**, 229.

20. Seed, B., Parker, R. C., and Davidson, N. (1982). *Gene*, **19**, 201.

21. Rosenberg, S. M., Stahl, M. M., Kobayashi, I., and Stahl, F. W. (1985). *Gene*, **38**, 165.

22. Sternberg, N., Tiemeier, D., and Enquist, L. (1977). *Gene*, **1**, 255.

23. Scherer, G., Telford, J., Baldari, C., and Pirrotta, V. (1981). *Dev. Biol.*, **86**, 438.

24. Herrmann, B. G., Barlow, D. P., and Lehrach, H. (1987). *Cell*, **48**, 813.

25. Whittaker, P. A. and Lavender, F. L. (1989). *Nucl. Acids Res.*, **17**, 4406.

Chapter 3

Construction of a cDNA library

Madan K. Bhattacharyya

G303 Agronomy Hall, Iowa State University, Ames, Iowa 50011-1010, USA

1 Introduction

This chapter primarily describes a protocol for cDNA library construction that has worked routinely in our laboratory. This protocol includes methods for RNA and poly(A$^+$) RNA isolation, complementary DNA (cDNA) synthesis and cDNA cloning. In writing this protocol, construction of a cDNA library is considered for etiolated hypocotyls of dark-grown soybean seedlings as an example. This protocol is for unidirectional cloning of cDNA in the λ-based vector Uni-ZAP XR (Stratagene). However, this general protocol, with suitable modifications, should be applicable for construction of cDNA libraries in any other cloning vector. Reverse transcriptase, the enzyme that synthesizes a cDNA molecule using RNA as a template, frequently fails fully to transcribe RNA molecules, especially if they are very long. Therefore, full-length molecules become only a minor portion of the total cDNA. A new technique, based on addition of a biotin group to the cap structure of mRNA, has been developed for the construction of cDNA libraries in which more than 90% of the clones are full length (ref. 1, and references there in). This technique, or others like it, should be considered if the transcripts to be cloned are especially long.

In constructing a cDNA library, several factors should be considered. They include:

(a) The type of tissues where the mRNA of interest is most abundant. This can be a difficult task, especially for a gene which is yet to be isolated. Some traits are regulated at the post-translational level, while others are under developmental regulation at the transcriptional level. Some are constitutively expressed. Usually, mRNA from the stage at which the trait is expressed and from the early developmental stages of the organism are good starting points for constructing a library for cloning a novel cDNA. It is much simpler to select tissues for genes that are transcriptionally activated by environmental stimuli.

(b) If possible, start the protocol with a large amount of high-quality poly(A$^+$) RNA.

(c) Selection of vector and cloning strategy can also be critical, especially if the transcript of a gene is unusually long (e.g. over 4 kb). Unidirectional cloning

Table 1 Titre of soybean cDNA libraries

| Library | Volume plated | | | | | | Average | |
| | 1 µl | | 0.1 µl | | 0.01 µl | | Recombinant (p.f.u./µg)[a] | % Wild type |
	Wild type	Recombinant	Wild type	Recombinant	Wild type	Recombinant		
L1	2	–[b]	0	162	0	11	3.4×10^6	0.1
L2	6	–	0	362	0	28	8.0×10^6	0.2
L3	7	–	1	337	0	31	8.1×10^6	0.2

[a] From a total of 6 µg vector DNA, ligated to cDNA fraction pools L1, L2 and L3 (obtained from 5 µg poly(A⁺) RNA), we can package approximately 3.9×10^7 p.f.u.
[b] Not counted.

using the mRNA poly(A$^+$) tail as the priming site for reverse transcriptase, as described in this protocol, may not be suitable for cloning longer transcripts. Use of random hexamers as primers in the cDNA synthesis allows cloning of different parts of a long transcript, and thereby cloning of the whole transcript. One can also try new cDNA cloning technologies for cloning longer transcripts (1). The strategy that will be used to identify the desired clone once the library has been constructed (e.g. antibody screening or nucleic acid hybridization) also influences the choice of cDNA cloning vector.

(d) Construction of large unamplified libraries can be critical in cloning rare transcripts. Current cDNA cloning protocols, as described here, allowed us to synthesize a library of over 3.9×10^7 p.f.u. from 5 μg poly (A$^+$) RNA (*Table 1*), which should be suitable for most cloning experiments.

The protocol used in the construction of a soybean etiolated hypocotyl cDNA library comprises the following steps.

- isolation of total RNA
- preparation of poly(A$^+$) RNA
- cDNA synthesis
- preparation of cDNA for cloning
- establishment of the cDNA library
- characterization of the cDNA library
- amplification and storage of the cDNA library

2 Isolation of total RNA

The method given in *Protocol 1* is based on the selective precipitation of RNA with lithium chloride and works well with most plant tissues. See Volume I, Chapter 4 for other RNA isolation methods and for general information regarding work with RNA.

Protocol 1
Isolation of total RNA

Equipment and reagents
- Mortar and pestle
- Liquid nitrogen
- RNA extraction buffer (50 mM Tris-HCl, pH 9.0, 150 mM LiCl, 5 mM EDTA, 5% SDS)
- 1:1 phenol–chloroform (see Appendix 4, *Protocol 6*)
- 8 M LiCl$_2$; sterilize by autoclaving at 121°C, 103.5 kPa (15 lbf/in^2), for 30 min
- Milli-Q water; sterilize by autoclaving at 121°C, 103.5 kPa (15 lbf/in^2), for 30 min
- Refrigerated bench-top centrifuge (e.g. Sorvall RT6000B)

Protocol 1 continued

Method

1 Harvest the tissues and quickly freeze in liquid nitrogen. Tissues can be stored at $-80\,°C$ until use.

2 Cool a large mortar and pestle by pouring liquid N_2 into it twice.

3 Grind the tissue samples to a fine powder. If necessary add liquid N_2 during the grinding process.

4 Transfer the powder to a red-top centrifuge tube (Sarsted) containing 20 ml RNA extraction buffer. Use approximately 5 g tissue per 20 ml buffer. Dissolve the powder by vigorous shaking for 2 min.

5 Add 20 ml phenol–chloroform to the tissue extract, and shake vigorously for 2–3 min. Spin in a refrigerated bench-top centrifuge at 2000 **g** for 10 min.

6 Transfer the upper phase to a new red-top tube, add an equal volume of phenol:chloroform and shake vigorously for 2 min. Centrifuge as in step 5.

7 Repeat step 6.

8 Transfer the upper phase to a new tube and add an equal volume of chloroform. Shake vigorously and centrifuge for 10 min, as in the previous steps.

9 Transfer the upper phase to a new tube and add a one-third volume of 8 M $LiCl_2$ (final concentration 2 M $LiCl_2$). Mix the sample and store overnight at $4\,°C$.

10 The following morning, spin the tubes in a refrigerated centrifuge at 2000 **g** for 10 min. Carefully collect the supernatant in a new tube. This supernatant contains DNA that can be precipitated by adding two volumes of ethanol.

11 Resuspend the RNA pellet in 6 ml autoclaved Milli-Q water. Add 2 ml 8 M $LiCl_2$ and store overnight at $4\,°C$. Carry out this resuspension step on ice.

12 Recentrifuge the RNA samples at 2000 **g** for 10 min. Resuspend the RNA pellet in 1 ml autoclaved Milli-Q water, and store at $-80\,°C$.

3 Preparation of poly(A$^+$) RNA

Magnetic beads with attached oligo(dT) primers have been shown to work satisfactorily in isolating poly(A$^+$) RNA from plant tissues (2). We routinely prepare poly(A$^+$) RNA by a similar method as described in *Protocol 2*. Poly(A$^+$) RNA is first annealed to biotinylated oligo(dT) which is then mixed with magnetic particles coated with streptavidin. Streptavidin binds to the biotin–RNA complex and is subsequently separated from the poly(A$^-$) RNA using a magnetic stand. The protocol is based on one provided by Promega, from whom the materials can be obtained. See also Volume I, Chapter 4, Section 7 for alternative methods for poly(A$^+$) RNA preparation.

Protocol 2

Preparation of poly(A$^+$) RNA

Equipment and reagents

- Magnetic separation stand (Promega)
- Streptavidin-magnetic beads (Promega)
- RNase-free water (see Volume I, Chapter 4, Section 2 for preparation of ribonuclease-free reagents)
- Biotinylated oligo(dT) probe (50 pmol/μl; Promega)
- 20× SSC (3 M NaCl, 0.3 M trisodium citrate, pH 7.0)

Method

1 Mix 5 mg of total RNA with RNase-free water to a final volume of 2.43 ml and incubate the mixture at 65°C for 10 min.

2 Add 500 pmol of biotinylated-oligo(dT) and 60 μl of 20× SSC to the RNA. Mix gently. Slowly cool the tube at room temperature so that the oligo(dT) anneals to the poly(A) tails of the RNA. This process takes about 30 min.

3 Equilibrate the streptavidin-magnetic bead mix in 0.5× SSC as follows:

   (a) Resuspend the streptavidin-magnetic bead mix by gently flicking the tube. Place the tube in a magnetic stand to collect all the streptavidin–magnetic bead particles (SA-MB) at the side of the tube close to the magnetic stand. Remove the liquid without disturbing the brown beads.

   (b) Resuspend the SA-MB particles in 1.5 ml 0.5× SSC by gently flicking the tube. Place the tube in the stand and remove the supernatant after all the SA-MB particles have accumulated at the side of the tube close to the magnet.

   (c) Repeat step (b) twice.

   (d) Resuspend the equilibrated SA-MB in 0.5 ml 0.5× SSC.

4 Once the temperature of the RNA and oligo(dT) mix is around 25°C, add the equilibrated SA-MB mix to the oligo(dT)-annealed RNA sample and incubate at room temperature for 10 min. Gently invert the tube every 1–2 min so that a thorough mixing occurs between the SA-MB particles and the oligo(dT)-annealed RNA.

5 Place the tube in the magnetic stand, and collect the SA-MB-RNA complex at the side of the tube. Transfer the supernatant to a tube labelled 'poly(A$^-$), round I'.

6 Equilibrate the SA-MB-RNA complex in 0.1× SSC by washing four times (1.5 ml SSC per wash). After the final wash, remove as much of the supernatant as possible without disturbing the SA-MB–RNA complex.

7 Dissociate the poly(A$^+$) RNA from the oligo(dT) by gently mixing 1 ml RNase-free water into the SA-MB–RNA complex.

8 Capture the SA-MB–oligo(dT) complex as before by placing the tube in the magnetic stand. Transfer the poly(A$^+$) RNA-containing supernatant to a tube labelled 'poly(A$^+$), round I'.

Protocol 2 continued

9 Repurify the poly(A$^+$) RNA by repeating steps 1–8 for the poly(A$^+$) RNA obtained in step 8. Poly(A$^-$) RNA obtained from this poly(A$^+$) RNA sample is labelled 'poly(A$^-$), round II', and the poly(A$^+$) RNA obtained in this second purification step is labelled 'poly(A$^+$), round II'.

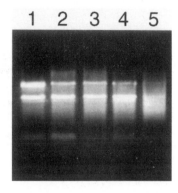

Figure 1 Isolation of poly(A$^+$) RNA from soybean etiolated hypocotyls. Lanes: 1, total RNA; 2, poly(A$^-$) RNA from one round of poly(A$^+$) RNA separation; 3, poly(A$^+$) RNA from one round of separation; 4, poly(A$^-$) RNA from two rounds of poly(A$^+$) RNA separation; 5, poly(A$^+$) RNA from two rounds of separation. RNA samples were separated in a neutral gel.

We typically obtain 200 µg poly(A$^+$) RNA from 10 mg total RNA isolated from soybean hypocotyls in round I purification. From this, 18 µg poly(A$^+$) RNA is isolated in round II separation. It is important to monitor the quality of the poly(A$^+$) RNA samples in a neutral gel before proceeding to the cDNA synthesis and cloning steps (*Figure 1*).

4 cDNA synthesis

Molecular analysis of RNA transcripts became much simpler with the discovery of reverse transcriptase. This enzyme synthesizes a cDNA molecule using an RNA molecule as the template. Each cDNA molecule, which is initially single stranded, can subsequently be used as a template to synthesize the second strand by *Escherichia coli* DNA polymerase I. This double-stranded DNA molecule can be cloned into vectors developed from plasmids or phages. The clones are generally maintained in an *E. coli* host as a library of cDNA molecules representing transcripts of, for example, a specific tissue or organ. Only a subset of all the genes in a genome are transcribed in a particular tissue, so the complexity of a cDNA library is much lower than that of a genomic library, especially with higher eukaryotic organisms. cDNAs for most transcripts can be isolated by screening less than 10^6 plaques or colonies.

When preparing a cDNA library in the Uni-ZAP XR vector, poly(A$^+$) RNA is reverse transcribed from an oligonucleotide primer that has an 18-base poly(dT) sequence at the 3′-end. At the 5′-end of the poly(dT) sequence there is an *Xho*I restriction enzyme recognition site preceded by a 26-base 'GAGA' sequence. After the second-strand cDNA synthesis (by *E. coli* DNA polymerase I) and polishing of the ends of the double-stranded cDNA (by *Pfu* DNA polymerase), *Eco*RI adapters are ligated to the blunt ends. The cDNAs are digested completely with *Xho*I and cloned into the *Eco*RI–*Xho*I sites of the Uni-ZAP XR vector. Cloning efficiency in this vector is monitored by the blue-and-clear selection system based on the expression of β-galactosidase (see Volume I, Chapter 8, Section 3.2). In recombinant molecules, β-galactosidase expression is disrupted and, therefore, recombinant plaques are clear instead of the blue of wild-type plaques.

Protocol 3 (based on the Stratagene instruction manual) gives the steps for cDNA synthesis prior to cloning in Uni-ZAP XR.

Protocol 3

cDNA synthesis

Reagents

- RNase-free water (see *Protocol 2*)
- Linker primer (1.4 µg/µl) (sequence 5′-GA-GAGAGAGAGAGAGAGAGAGAGACTCG AGTTTTTTTTTTTTTTTTTTTT-3′; Stratagene)
- 10× first-strand buffer (0.5 M Tris-HCl, pH 8.3, 0.75 M KCl, 0.3 M MgCl$_2$)
- dNTP mix (10 mM dATP, 10 mM dGTP, 10 mM dTTP, 5 mM 5-methyl dCTP)
- RNase Block ribonuclease inhibitor (40 U/µl) (Stratagene)

- Moloney murine leukaemia virus reverse transcriptase (MMLV-RT; 50 U/µl)
- [α-^{32}P]dATP (800 Ci/mmol, 10 µCi/µl)
- 10× second-strand synthesis buffer (188 mM Tris-HCl, pH 8.3, 906 mM KCl, 46 mM MgCl$_2$)
- RNase H (1.5 U/µl)
- DNA polymerase I (9.0 U/µl)

Methods

A. First-strand synthesis

1   Mix 5 µg poly(A$^+$) RNA and RNase-free water to a final volume of 37.5 µl.

2   Add 2 µl linker primer to this RNA solution and incubate at 65 °C for 10 min. Cool slowly to 25 °C.

3   Spin the tube at 16 000 **g** for 10 s to mix and place on ice. Add the following components, in order: 5 µl 10× first-strand buffer, 3 µl dNTP mix, 1 µl RNase Block ribonuclease inhibitor and 1.5 µl MMLV-RT.

4   Mix the reaction gently. Remove 5 µl of the mixture and transfer to a microfuge tube that contains 0.5 µl (5 µCi) of [α-^{32}P]dATP. Mix gently. This is the first-strand synthesis control reaction.

5   Incubate the remainder of the first-strand synthesis reaction, and the radioactive control reaction, at 37 °C for 1 h.

6   Place the first-strand synthesis reaction on ice and the control reaction in the freezer (−20 °C) until analysis by electrophoresis on an alkaline agarose gel (3).

Protocol 3 continued

B. Second-strand synthesis[a]

1 Add the following components, in order, to the non-radioactive first-strand synthesis reaction tube on ice: 20 μl 10× second-strand synthesis buffer, 6 μl dNTP mix, 114 μl RNase-free water, 2 μl (20 μCi) [α-^{32}P]dATP, 2 μl RNase H and 11 μl DNA polymerase I.

2 Gently mix the reaction and spin the tube in a microfuge at 16000 **g** for 10 s. Incubate for exactly 2.5 h at 16 °C.

3 After the incubation place the reaction tube on ice.

[a] In this step, RNA–cDNA hybrids produced following reverse transcription are treated with *E. coli* RNase H to generate small RNA fragments that remain associated with the first-strand cDNA. These RNA fragments are primers for synthesis of the second strand by *E. coli* DNA polymerase I.

5 Preparation of cDNA for cloning

Double-stranded cDNA molecules are processed through several steps before ligating to the cloning vector. The double-stranded cDNAs carry uneven ends. To facilitate the blunt-end adapter ligation, cDNA molecules are 'nibbled back' or filled-in with *Pfu* DNA polymerase. Following adapter ligation, the *Xho*I site of the linker at the 3′-end of cDNA molecule is completely digested. Unligated adapters

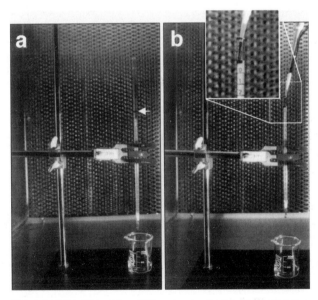

Figure 2 Separation of the cDNA molecules from adapters. (a) Preparation of a Sepharose CL-2B gel filtration column in a 2-ml pipette. (b) Loading of a cDNA sample at the surface of the Sepharose column. Refer to protocol 8 on page 51.

and the digested portion of the ligated adapter are removed by passing the re-action mixture through a Sepharose CL-2B column. The cDNA fractions from the column are subsequently cloned into Uni-ZAP XR vector to establish the cDNA library.

The preparation of cDNA for cloning therefore involves the following steps:

- creation of blunt ends
- ligation to *Eco*RI adapters
- phosphorylation of *Eco*RI adapters
- digestion with *Xho*I
- size fractionation of cDNA molecules
- processing of cDNA fractions
- quantification of cDNA fractions

Protocol 4

Creation of blunt ends

Reagents

- dNTP mix (2.5 mM dATP, 2.5 mM dCTP, 2.5 mM dGTP, 2.5 mM dTTP)
- 1:1 phenol–chloroform (see Appendix 4, *Protocol 6*)
- *Pfu* DNA polymerase (2.5 U/μl)
- 3 M sodium acetate, pH 4.6, prepared by adding glacial acetic acid to 3 M sodium acetate until this pH is obtained

Method

1 Add the following components to the second-stranded synthesis reaction: 23 μl dNTP mix and 2 μl *Pfu* polymerase.

2 Mix the reaction and centrifuge the tube at 16 000 **g** for 10 s. Incubate at 72 °C for exactly 30 min. Place the tube on ice after completion of the incubation period.

3 Add 200 μl 1:1 phenol–chloroform to the reaction mix and wrap the lid with a small piece of Parafilm to prevent leakage (also wrap the lid with Parafilm in subsequent steps before centrifugation as a precautionary measure). Vortex the tube vigorously for 30 s and spin at 16 000 **g** in a microfuge for 2 min at room temperature.

4 Transfer the upper aqueous layer to a new microfuge tube without disturbing the interface. Add an equal volume of chloroform to the aqueous phase and vortex vigorously for 30 s.

5 Spin the reaction mix for 2 min at 16 000 **g** in a microfuge at room temperature.

6 Transfer the upper aqueous phase to a new tube and add 20 μl of 3 M sodium acetate, pH 4.6, and 400 μl absolute ethanol.

7 Precipitate the cDNA overnight at −20 °C.

8 Spin the tube in a microfuge at 16 000 **g** for 1 h at 4 °C.

Protocol 4 continued

9 Carefully transfer the supernatant to a new tube. Gently wash the white cDNA pellet with 500 μl 70% ethanol and spin at 16 000 **g** for 2 min at room temperature.

10 Carefully remove the ethanol and dry the pellet under vacuum.

Protocol 5

Ligation to *Eco*RI adapters

Reagents

- *Eco*RI adapter solution (0.4 μg/μl)
- 10× ligase buffer (0.5 M Tris-HCl, pH 7.5, 70 mM MgCl$_2$, 10 mM DTT)
- 10 mM ATP
- T4 DNA ligase (4 U/μl)

Method

1 Resuspend the cDNA pellet obtained in step 10 of *Protocol 4* in 8 μl of the *Eco*RI adapter solution and incubate on ice for at least 30 min.

2 Save 1 μl of this double-stranded cDNA/adapter suspension in a separate tube at −20 °C for analysis.

3 To the remainder of the double-stranded cDNA/adapter suspension add the following components, in order: 1 μl 10× ligase buffer, 1 μl 10 mM ATP and 1 μl T4 DNA ligase

4 Mix the components gently and centrifuge at 16 000 **g** for 10 s at 4 °C.

5 Incubate the ligation reaction overnight at 8 °C, or 2 days at 4 °C.

6 Stop the ligation reaction by incubating for 30 min at 70 °C. Spin the tube at room temperature and place on ice.

Protocol 6

Phosphorylation of *Eco*RI adapters

Reagents

- 10× ligase buffer (see *Protocol 5*)
- 10 mM ATP
- T4 polynucleotide kinase (10 U/μl)

Method

1 Add the following components to the adapter-ligated, double-stranded cDNA from step 6 of *Protocol 5*: 1 μl 10× ligase buffer, 2 μl 10 mM ATP, 6 μl autoclaved water and 1 μl T4 polynucleotide kinase.

2   Mix the components and spin at 16 000 **g** for 10 s at room temperature.

3   Incubate the reaction mixture at 37 °C for 30 min.

4   Incubate the reaction mixture at 70 °C for 30 min to inactivate the kinase.

5   Spin the mixture for 2 min at room temperature and place the tube on ice.

Protocol 7

Digestion with *Xho*I

Reagents

- *Xho*I buffer supplement (187 mM NaCl, 13 mM MgCl$_2$)
- *Xho*I restriction enzyme (40 U/μl)
- 10× STE buffer (0.2 M Tris-HCl, pH 7.5, 1 M NaCl, 0.1 M EDTA)

Method

1   Add the following components, in order, to the kinased reaction mix from step 5 of *Protocol 6*: 28 μl *Xho*I buffer supplement and 3 μl *Xho*I.

2   Mix the reaction by pipetting, then incubate at 37 °C for 1.5 h.

3   Add 5 μl of 10× STE buffer and 125 μl of absolute ethanol to the reaction.

4   Mix the reaction and incubate overnight at −20 °C.

5   Spin the tube in a microfuge at 16 000 **g** at 4 °C for 60 min.

6   Transfer the supernatant to a new tube and dry the cDNA pellet under vacuum.

7   Resuspend the pellet in 14 μl 1× STE buffer at room temperature for 60 min.

Protocol 8

Size fractionation of cDNA molecules[a]

Equipment and reagents

- Sepharose CL-2B (Sigma)
- 1× STE buffer (see *Protocol 7*)
- TBE buffer (for 5× TBE, mix 54.0 g Tris base, 27.5 g boric acid, 3.72 g Na$_2$ EDTA, 2H$_2$O and add distilled water to 1l)
- Sterile 2-ml plastic pipette
- 30-ml plastic syringe
- Clamps and a stand to support the drip column
- Loading dye (mix 50 μl glycerol, 40 μl saturated bromophenol blue solution and 1 μl 10× STE)

Method

1   Transfer 10 ml Sepharose CL-2B suspension in 20% ethanol to a 50-ml red-top tube (Sarsted). Add 20 ml 1× STE buffer and mix well by inverting the tube several times.

2 Spin the Sepharose CL-2B gel suspension at 875 **g** in a bench-top centrifuge at room temperature for 2 min.

3 Discard the supernatant and resuspend the gel pellet in 30 ml 1× STE buffer by inverting the tube.

4 Spin the suspension as in step 2.

5 Repeat steps 3 and 4 in succession for at least three more times.

6 After five or six washes with 1× STE buffer, resuspend the gel pellet in 5 ml 1× STE buffer. The gel suspension is now ready for loading into the column.

7 Partially pull out the cotton plug from a 2-ml sterile plastic pipette using a sterile needle that is bent at the end. Cut off the external portion, a little more than half of the cotton plug, using a sharp clean razor blade. Push the remaining part of the plug into the pipette using a sterile needle.

8 Connect the cotton-plug end of the pipette to a 30-ml plastic syringe and push the cotton plug to the narrow or lower end of the pipette with the help of a rapid and forceful pushing action of the plunger into the 30-ml syringe. This can also be achieved by connecting the cotton plug-end of the pipette to a source of pressurized air.

9 Load the pipette with gel suspension. Hold the 2-ml pipette in a near-horizontal position to slow down the movement of gel suspension through the column. Do not trap any air bubbles during the gel loading procedure. Load nearly 2 ml of gel into the pipette. Do not let the column dry out at any stage.

10 Set the pipette in a vertical position with the help of clamps and a stand (*Figure 2a*).

11 Continue loading the gel suspension until it reaches close to '0' mark on the pipette.

12 Wash the gel with 5 ml 1× STE buffer. Mark a ring on the column approximately 2 cm above the interface, as shown by the arrow in *Figure 2a* on page 48.

13 Place the column horizontally and use a razor blade to score the pipette surface along the marked ring. Carefully break the column at this position.

14 Place the column in a vertical position. Add 1× STE buffer to the cut end, and mix carefully with the gel interface to obtain a smooth interface (*Figure 2b*). Add 1× STE buffer five more times to re-establish the column.

15 Add 3.5 μl loading dye to 14 μl of the double-stranded cDNA solution in 1× STE buffer.

16 Allow the last drop of 1× STE buffer to soak into the gel interface and quickly place the dye/cDNA solution onto the gel surface as shown in *Figure 2b*. Let the cDNA enter into the gel completely. Rinse tube that contained the dye/cDNA with 10 μl 1× STE buffer and immediately place onto the gel surface. Once this cDNA wash has entered completely into the gel, fill the column with 1× STE buffer.

17 Collect single-drop fractions in microfuge tubes. Each fraction is ≈ 50 μl in volume.

18 Continue collecting fractions until the blue dye front starts to elute from the column.

19 Measure the radioactivity in 2 μl of each fraction (*Figure 3a*). The first peak of radio-activity contains the cDNA. The second peak (not shown in *Figure 3a*) represents mostly free nucleotides and adapter molecules.

20 Run 3 μl of each fraction that contains double-stranded cDNA molecules in a 0.8% neutral gel prepared with 0.5× TBE buffer (*Figure 3b*).

21 Divide fractions representing the first peak into three classes. For example: class I comprised of fractions 8, 9, 10, 11 and 12, class II comprised of fractions 13, 14 and 15, and class III comprised of fractions 16, 17, 18, and 19 (as shown in *Figure 3a*).

22 Pool class I fractions (L1 in *Figure 3a*). Similarly, pool class II and III fractions (L2 and L3, respectively, in *Figure 3a*).

ᵃ The success of a high-quality cDNA library construction depends upon the complete separation of adapter molecules from double-stranded cDNA. This is achieved by fractionating the double-stranded cDNA sample through a drip column containing Sepharose CL-2B gel.

Protocol 9

Processing of cDNA fractions

Equipment and reagents

- Handheld Geiger counter
- 1× STE buffer (see *Protocol 7*)
- 1:1 phenol–chloroform (see Appendix 4, *Protocol 6*)

Method

1 Add 1× STE buffer to each fraction pool (L1, L2 and L3) to a final volume of 300 μl.

2 Add 300 μl 1:1 phenol–chloroform.

3 Vortex vigorously for 30 s and spin at room temperature in a microfuge at 16 000 **g** for 2 min.

4 Transfer the upper aqueous layers to new microfuge tubes and add an equal volume of chloroform.

5 Vortex and spin as in step 3.

6 Transfer the aqueous phase to new tubes, add two volumes of absolute ethanol and mix.

7 Precipitate the cDNA overnight at −20 °C.

8 Spin the samples in a microfuge at 16 000 **g** for 60 min at 4 °C. Transfer the supernatant to new tubes. Spin the tubes for 2 min at room temperature and carefully remove the excess supernatant without disturbing the cDNA pellet. This supernatant should not carry any radioactivity. Monitor with a handheld Geiger counter. Carefully wash the pellets with 200 μl 80% ethanol. Avoid disturbing the pellets. Spin the tubes in a microfuge at 16 000 **g** for 2 min at room temperature.

Protocol 9 continued

9 Carefully remove the ethanol and save along with the supernatant. Make sure that the ethanol does not carry any radioactive cDNA molecules.

10 Dry the cDNA pellet under vacuum initially for 1–2 min. Over-drying of the cDNA pellets may interfere with the cDNA solubilization process.

11 Add 5 μl sterile water to each cDNA pellet. Place the tubes on ice for 1 h so that the cDNA molecules dissolve completely. To determine if the cDNA has completely solubilized, hold the solution in a pipette tip and test the empty tube in which the cDNA was solubilized against the Geiger counter. Most of the radioactivity should be in the cDNA solution rather than in the nearly empty microfuge tube.

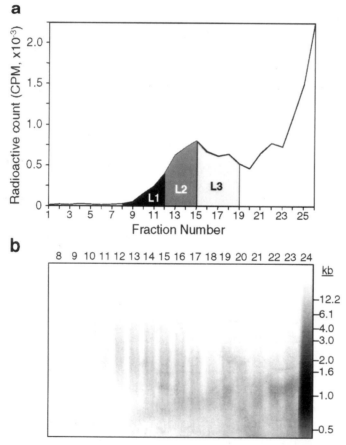

Figure 3 Size fractionation of the cDNA molecules. (a) [α-^{32}P]dATP labelled cDNA fractions collected from a drip Sepharose column. Fraction pools: L1, cDNA molecules pooled from fractions 8 to 12; L2, cDNA molecules pooled from fractions 13–15; L3, cDNA molecules pooled from fractions 16–19. (b) Neutral gel electrophoresis of cDNA fractions 8–23, with [α-^{32}P]dATP-labelled total double-stranded cDNA in lane 24. The dried agarose gel was exposed to X-ray film overnight.

Protocol 10

Quantification of cDNA fractions[a]

Equipment and reagents

- Petri dishes (100 mm diameter)
- UV transilluminator
- Agarose (high melting point, DNA grade)
- 50× TAE buffer (mix 242 g Tris base, 57.1 ml glacial acetic acid, 18.61 g Na$_2$EDTA.2H$_2$O, and add distilled water to 1 l)
- Ethidium bromide stock solution (10 mg/ml)
- Standard DNA solution of known concentration (e.g. 100 ng/μl phage λ DNA in 100 mM EDTA)

Method

1 Prepare 100 ml 0.8% agarose in TAE buffer. Cool the molten agarose solution to 50°C, add 10 μl ethidium bromide stock solution and swirl to mix. Pour 10 ml of the agarose solution into individual Petri dishes. Proceed to the next step once the agarose has solidified and the plates are dry.

2 Prepare a dilution series of the standard DNA solution in 100 mM EDTA, pH 8.0. Suitable concentrations are 100 ng/μl, 50 ng/μl, 25 ng/μl, 12.5 ng/μl, 6.25 ng/μl and 3.12 ng/μl.

3 Spot 0.5 μl of each dilution onto the surface of an agarose plate.

4 Immediately spot 0.5 μl of each cDNA fraction pool onto the plate. Allow 10–15 min at room temperature for all samples to absorb into the agarose.

5 Remove the lid of the Petri dish, and visualize the plate under UV transillumination (*Figure 4*).

6 Estimate the concentrations of the cDNA samples by comparing with the standards.

[a] Quantification of double-stranded cDNA molecules is very critical before cloning.

6 Establishment of the cDNA library

Vectors based on λ phage are widely used in many cloning experiments (see Volume I, Chapter 9, Section 4). With the development of very high-quality λ packaging extracts, a cloning efficiency of up to 10^7 recombinant p.f.u./μg vector DNA can be obtained. The cloning efficiency of a plasmid vector is 10- to 100-fold less than that of a λ vector, but with the development of electroporation systems for *E. coli* (Volume I, Chapter 8, Section 2.2), one can construct a cDNA library in a plasmid vector with less difficulty. Examples of cDNA libraries in plasmid vectors include those used in yeast two-hybrid systems (3). Several λ vectors are currently available, such as λgt10, which is suitable for screening with a nucleic acid probe, and λgt11, which is suitable for screening with an antibody. An expression library also enables clones for DNA-binding proteins to be identified by probing with oligonucleotides (4).

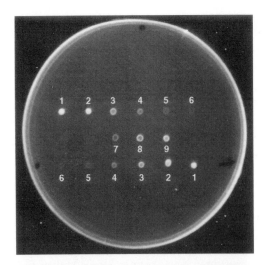

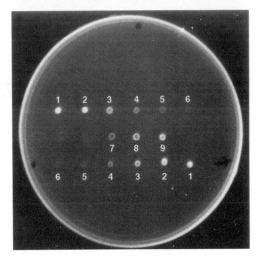

Figure 4 Quantification of pooled cDNA fractions. Half a microlitre of DNA solution was spotted onto an agarose plate containing ethidium bromide and DNA samples were visualized on a UV transilluminator. Spots: 1, 50 ng DNA; 2, 25 ng; 3, 12.5 ng; 4, 6.25 ng; 5, 3.12 ng; 6, 1.55 ng; 7, cDNA fraction pool L1 (see *Figure 3a*); 8, cDNA fraction pool L2; 9, cDNA fraction pool L3.

The λZAP vector system is particularly convenient for construction of cDNA libraries. It combines both the high cloning efficiency of the λ vectors and the convenience of a plasmid system for molecular analysis of the cDNA clones (5). The cDNA insert in a λZAP clone can be excised as a double-stranded pBluescript phagemid and as single-stranded DNA for sequencing or site-directed muta-genesis. Other advantages include the presence of T3 and T7 promoters, enabling RNA copies of the cDNA to be produced for use as hybridization probes (see Chapter 4, Section 2.3.3), and the ability to synthesize the protein coded by the cDNA as a β-galactosidase fusion. A variation of the λZAP system, the Uni-ZAP XR vector, is used in the protocols given below. This vector allows directional cloning so the resulting cDNA expression library requires screening for only three instead of the six possible open reading frames.

Ligation of 50–100 ng of cDNA per μg of λ vector arms gives consistently high efficiency in cDNA cloning. Following ligation, packaging of the recombinant DNA molecules results in viable phage particles that infect selected *E. coli* strains. Using blue-clear selection, based on the β-galactosidase gene, one can determine the ratio between recombinant p.f.u. which are clear (loss of β-galactosidase function) and blue, wild-type p.f.u. With Uni-ZAP XR, the number of recombinant p.f.u. should be 10- to 100 fold higher than wild-type p.f.u. The recombinant p.f.u. can be screened by hybridization probing or antibody testing. Isolated Uni-ZAP clones are then excised with the assistance of helper phage and a suitable *E. coli* strain to obtain pBluescript phagemids (5), eliminating the need to subclone the cDNA insert from λ vector to plasmid.

Establishment of the cDNA library therefore consists of the following steps:

- ligation of the cDNA into the λ vector
- packaging of the cDNA library
- plating and titre determination

Protocol 11

Ligation of the cDNA into the λ vector

Reagents

- Uni-ZAP XR vector double digested with *Eco*RI and *Xho*I and dephosphorylated (1 μg/μl) (Stratagene)
- 10× ligase buffer (see *Protocol 5*)
- 10 mM ATP
- T4 DNA ligase (4 U/μl)

Method

1 Add the following components in order: 1 μl Uni-ZAP XR vector, approximately 100 ng cDNA, 0.5 μl 10× ligase buffer, 0.5 μl ATP, water to a final volume of 4.5 μl and 0.5 μl of T4 DNA ligase.

2 Mix by gently flicking the microfuge tube.

3 Incubate overnight at 12 °C.

4 At the end of the ligation, store the reaction at 4 °C for a short period.

Protocol 12

Packaging of the cDNA library

Reagents

- Gigapack III gold packaging extract (Stratagene)
- SM buffer (50 mM Tris-HCl, pH 7.5, 100 mM NaCl, 8 mM MgSO$_4$, 0.01% gelatin)

Protocol 12 continued

Method

1 Place the packaging extracts on dry ice. Take one extract and partly thaw by holding the tube between your fingers.

2 Transfer 1 µl of the ligation mix from step 4 of *Protocol 11* into the partly thawed packaging extract and mix with a pipette tip.

3 Once the extract is completely thawed, place the tube in a water bath at 22 °C and incubate for exactly 2 h.

4 Add 0.5 ml SM buffer and a drop of chloroform to each packaging reaction and store at 4 °C for a short period.

Protocol 13
Plating and titre determination

Reagents

- *E coli* XL1-Blue MRF′ (see Appendix 3)
- LB-tetracycline agar plates. For LB agar, mix 10 g Bacto-tryptone, 5 g Bacto-yeast extract, 10 g NaCl, 15 g Bacto-agar per 1 l. Check the pH and adjust to 7.0–7.2 with NaOH. Sterilize by autoclaving at 121°C, 103.5 kPa (15 lbf/in^2), for 20 min. (see Appendix 4 for preparation and addition of tetracycline)
- LB medium (10 g bacto-tryptone, 5 g bacto-yeast extract, 10 g NaCl per 1 l). Check the pH and adjust to 7.0–7.2 with NaOH. Sterilize by autoclaving at 121°C, 103.5 kPa (15 lbf/in^2), for 20 min
- 10 mM and 1 M MgSO$_4$ (see Appendix 4)
- 20% maltose (see Appendix 4)
- SM buffer (see *Protocol 12*)
- NZY agar plates (10 g NZ amine, 5 g Bacto-yeast extract, 2 g MgSO$_4$, 15 g Bacto-agar per 1 l). Check the pH and adjust to 7.0–7.2 with NaOH. Sterilize by autoclaving at 121°C, 103.5 kPa (15 lbf/in^2), for 20 min
- NZY top agar (as for NZY agar but with 7 g agarose/l instead of Bacto-agar)
- 0.5 M IPTG
- X-gal (20 mg/ml in DMSO)

Method

1 Streak a sample of *E. coli* XL1-Blue MRF′ onto a LB-tetracycline agar plate and incubate overnight at 37 °C.

2 Inoculate 50 ml LB broth containing 10 mM MgSO$_4$ and 0.2% maltose with a single colony from the LB-tetracycline plate.

3 Grow the bacteria overnight at 30 °C, shaking at 200 r.p.m.

4 Spin the cells at 1000 **g** for 10 min at 4 °C in a bench-top centrifuge.

5 Resuspend the cells in 25 ml ice-cold 10 mM MgSO$_4$ and then add more 10 mM MgSO$_4$ until an OD$_{600}$ of 0.5 is reached.

6 Make 10-fold and 100-fold dilutions of the packaged libraries in SM buffer. Mix 1 µl of the undiluted packaged library with 200 µl of cell suspension from step 5. Repeat from the 10-fold and 100-fold dilutions.

7 Incubate at 37 °C for 15 min to allow the phage to adhere to the bacterial surfaces.

8 Add the following components to 3 ml of NZY top agar, at 46 °C, just before pouring over an NZY agar plate (100 mm diameter) equilibrated to 37 °C: 15 µl IPTG, 300 µl X-gal, 200 µl bacteria and phage mix (from step 7).

9 Leave the plates at room temperature for 10 min, then incubate overnight at 37 °C.

10 Count the number of blue and clear p.f.u. and calculate the packaging efficiency as the number of p.f.u./µg vector (see *Table 1*).

7 Characterization of the cDNA library

Determination of the average insert size for a cDNA library provides information related to both the quality and quantity of the library. Randomly selected cDNAs can be converted to double-stranded phagemids and their inserts sized by digestion with restriction enzymes, but it is preferable to use the method described in *Protocol 14*, in which inserts in the λ vector are directly sized by PCR (see also Chapter 7).

Protocol 14

Characterization of the cDNA library

Equipment and reagents

- Thermal cycler
- SM buffer (see *Protocol 12*)
- 10× PCR buffer (use the buffer provided with the *Taq* polymerase enzyme, or refer to Chapter 7, Section 2.2 for details of how to devise a PCR buffer)
- dNTP mix (10 mM dATP, 10 mM dCTP, 10 mM dGTP, 10 mM dTTP)
- Agarose (see *Protocol 10*)

- Forward and reverse primers (M13 sequencing primers, 10 µM) (Sigma)
- *Taq* DNA polymerase mix (*Taq* DNA polymerase [Sigma] and *Pfu* DNA polymerase [Stratagene] mixed at a 20:1 ratio)
- 5× TBE (see *Protocol 8*)

Method

1 Pick a single, clear recombinant plaque and mix with 50 µl SM buffer and boil for 5 min. Spin the tubes for 1 min in a microfuge.

2 Use 2 µl of this boiled plaque extract as a DNA template in a PCR. Mix the following components in order: 4 µl 10× PCR buffer, 0.8 µl dNTP mix, 2 µl forward primer, 2 µl reverse primer, 2 µl DNA template, 29 µl water and 0.2 µl *Taq* DNA polymerase mix.

3 Carry out the PCR as follows:

Step 1: 94 °C for 2 min

Step 2: 94 °C for 30 s

Protocol 14 continued

Step 3: 55 °C for 30 s

Step 4: 72 °C for 7 min

Step 5: Repeat steps 2, 3 and 4 for 29 cycles

Step 6: 72 °C for 10 min

Step 7: 25 °C for 1 min

4   Electrophorese 10 μl of each PCR product in a 0.8% agarose gel prepared with 0.5 × TBE buffer (*Figure 5*).

5   From the insert sizes of at least 24 recombinant plaques, calculate the statistics given in *Table 2*.

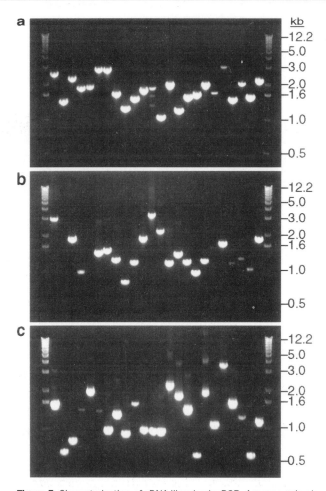

Figure 5 Characterization of cDNA libraries by PCR. Agarose gels showing PCR amplified cDNA inserts of 24 randomly picked plaques from (a) library L1, (b) library L2, and (c) library L3. The L1, L2 and L3 libraries were constructed from cDNA fraction pools L1, L2, and L3, respectively (see *Figure 3a*).

Table 2 Characterization of soybean cDNA libraries

Library[a]	Size of cDNA (bp)				
	Minimum	Maximum	Mean	Median	Standard deviation
L1	1065	3078	1969	1824	553
L2	823	3656	1675	1362	742
L3	593	3717	1440	1228	711

[a] See *Figure 3*.

8 Amplification and storage of the cDNA library

During preparation of the primary library, the cDNA is labelled with radioactive dATP. Therefore, it could be unstable and hence an amplification step is desirable shortly after the establishment of the library. A single amplification of the library will not result in under-representation of many of the clones. However, if the transcript of interest is very rare, screening of an unamplified library is highly recommended.

Protocol 15

Amplification and storage of the cDNA library

Reagents

- *E. coli* XL1-Blue MRF′ (see Appendix 3)
- SM buffer (see *Protocol 12*)
- NZY top agar and agar plates (150 mm diameter) (see *Protocol 13*)

Method

1 Prepare fresh bacterial cells as in *Protocol 13*, steps 1–5.

2 Mix approximately 5×10^4 p.f.u. of the cDNA library with 600 μl of cells and incubate for 15 min at 37 °C. Prepare 20 such mixes so that a total of 1×10^6 p.f.u. are amplified per library.

3 Mix 600 μl of phage-infected cells with 6.5 ml NZY top agar at 46 °C and quickly pour uniformly onto a NZY plate that has been preincubated at 37 °C for 1 h. After 10–15 min, transfer the plates to an incubator at 37 °C for 6–8 h.

4 When clear plaques have formed, add 8–10 ml SM buffer to each plate. Store the plates overnight at 4 °C, with gentle rocking.

5 Transfer the phage suspension from each plate and pool into sterile plastic containers. Rinse the plates with an additional 2 ml SM buffer and add to the pool. Add chloroform to 5% final concentration and mix well. Incubate the phage suspension for 15 min at room temperature.

6 Remove cell debris by centrifugation at 875 **g** for 10 min.

Protocol 15 continued

7 Add chloroform to 0.3% final concentration to the supernatant (carrying the phage particles) and store at 4 °C. For long-term storage add DMSO to the library to a final concentration of 7% and store at −80 °C.

Acknowledgements

I thank Robert A. Gonzales for comments on the manuscript, Allyson D. Willkins for preparing the manuscript, and Darla F. Boydston, Cuc K. Ly, and Broderick N. Stearns for photographs and illustrations. The work was supported by the Samuel Roberts Noble Foundation.

References

1. Seki, M., Carninci, P., Nishiyama, Y., Hayashizaki, Y., and Shinozaki, K. (1998). *Plant J.*, **15**, 707.

2. Jakobsen, K. S., Breivold, E., and Hornes, E. (1990). *Nucl. Acids Res.*, **18**, 3669.

3. Chien, C-T., Bartel, P. L., Sternglanz, R., and Fields, S. (1991). *Proc. Natl Acad. Sci. USA*, **88**, 9578.

4. Vinson, C. R., LaMarco, K. L., Johnson, P. F., Landschulz, W. H., and McKnight, S. L. (1988). *Genes Dev.*, **2**, 801.

5. Short, J. M., Fernandez, J. M., Sorge, J. A., and Huse, W. D. (1988). *Nucl. Acids Res.*, **16**, 7583.

Chapter 4
Nucleic acid labelling and detection

C. R. Mundy
Amersham Laboratories, Forest Farm, Whitchurch, Cardiff CF4 7YT, UK

M. W. Cunningham and C. A. Read
Amersham Laboratories, White Lion Road, Amersham, Buckinghamshire HP7 9LL, UK

Additional material provided by T. A. Brown

1 Introduction

The recent explosive progress in our understanding of genome structure and function has depended upon methods for detecting small amounts of nucleic acid. As described in Chapter 5, specific fragments of genomic DNA can be detected in complex mixtures by hybridization with labelled complementary DNA or RNA probes. A 1 kb fragment of a single-copy human gene is present at the level of only 300 fg/μg of total DNA. Thus, a very high level of sensitivity is required, and for several applications a high degree of spatial resolution is also important. The purpose of this chapter is to describe the available labels and labelling methods and to define as far as practicable the optimum combinations for a range of key applications.

1.1 Applications of nucleic acid labelling

The isolation, structural analysis and functional analysis of genes all involve the use and detection of labelled nucleic acid. The applications listed in *Table 1* fall into three groups. These are:

- the use of labelled nucleic acid as a hybridization probe to locate complementary sequences (e.g. filter hybridization and *in situ* hybridization)
- the processing of labelled nucleic acid for structural or functional analysis (e.g. mapping and sequencing studies)
- incorporation of label to monitor enzymatic reactions (e.g. during cDNA synthesis)

In some cases, label is uniformly distributed throughout the length of the nucleic acid molecules, allowing maximization of label density. For other techniques (e.g. nuclease S1 mapping), precise definition of one end of all labelled molecules

Table 1 Nucleic acid labelling and detection: applications and requirements

Application[a]	Required sensitivity	Required resolution	Required label position
Filter hybridization			
Single-copy mammalian DNA, Southern blot	1–5 pg	1 mm	end or uniform
Low copy mammalian DNA, Northern blot	100 fg–10 pg	1 mm	end or uniform
Single-copy *Drosophila* DNA, Southern blot	10–50 pg	1 mm	end or uniform
Single-copy *E. coli* DNA, Southern blot	50–250 pg	1 mm	end or uniform
Colony/plaque hybridization	50 pg–10 ng	1 mm	end or uniform
***In situ* hybridization**			
Fine chromosome mapping	1–10 ag	< 1 μm	end or uniform
Amplified DNA sequence or abundant transcript	1 fg–1 pg	1–10 μm	end or uniform
Mapping and sequencing			
Nuclease S1 mapping	10 pg–1 ng	200 μm	end
Primer extension mapping	10 pg–1 ng	200 μm	end or uniform
Restriction mapping	1–100 ng	500 μm	end or uniform
Gel retardation assays	1–100 ng	1 mm	end or uniform
DNA sequencing	10–50 pg	200 μm	end or uniform
Monitoring of reactions			
cDNA synthesis	1–10 ng	1 mm	uniform

[a] For details of the procedures see Chapter 3 (cDNA synthesis), Chapter 5 (filter hybridization), Chapter 6 (DNA sequencing) and Chapter 9 (nuclease S1 and primer extension mapping).

is achieved by end-labelling. Appropriate methods for all applications will be described in this chapter.

1.2 Sensitivity and resolution

Sensitivity levels are normally expressed in terms of the weight of detectable target nucleic acid. Required sensitivity levels often depend on the amount of the total sample that is available, or the amount that can be applied to a separation system such as an electrophoretic gel. The sensitivity levels listed in *Table 1* reflect these limitations.

The required spatial resolution can be defined in terms of the minimum distance between two features which must be resolved for usable data to be obtained. In the case of DNA sequencing, for example, the resolution required to enable reading of 300 bp of sequence after electrophoresis under defined conditions can be specified. Similarly, at the microscopic level, the required resolution can be set for various *in situ* hybridization applications. *Table 1* gives the required resolutions for each of the labelling applications.

1.3 Types of labels

The available labels have diverse properties which influence significantly their suitability for the various applications. Safety and disposal considerations have led to the increasing use of fluorescent, colorimetric and chemiluminescent labels in molecular biology applications. However, radioactive labels are still widely used, partly because they offer the greatest sensitivity and partly because the methodology required to detect radiolabelled nucleic acids is, in general, less complicated than the equivalent procedures for non-radioactive markers.

1.3.1 Radioactive labels

Radioactive labels have been used since the inception of recombinant DNA techniques, and without their prior availability these techniques may never have been developed. Certain elements exist in several isotopic forms. Their radioactive isotopes, called radionuclides, are unstable and decay with the co-incident emission of energetic particles and electromagnetic radiation. The degree of instability of a radionuclide is expressed as the time taken for 50% of a given sample of molecules to decay—the half-life (t½). The lower the half-life of a radionuclide, the greater the maximum specific activity (number of nuclear d.p.m. per mole) at which it can be obtained. The radionuclides used commonly in molecular biological applications are described in *Table 2*.

   Those radionuclides which are isotopic equivalents of natural component atoms of nucleic acids can be used for labelling. For example, tritium (^{3}H) can be substituted for hydrogen (^{1}H), or ^{32}P or ^{33}P for ^{31}P, by various procedures to produce labelled nucleoside triphosphates (*Figure 1*). Such labelled nucleotides may be used in a variety of enzyme-catalysed reactions to produce labelled DNA or RNA (*Table 3*). It is also possible to synthesize nucleotides labelled with ^{35}S by substituting this isotope for an oxygen atom of one of the phosphate groups (*Figure 1*). Similarly, non-natural cytidine nucleotides can be labelled with ^{125}I by modification of the base moiety.

   ^{3}H and ^{14}C may be substituted at any of several positions within the base and sugar of nucleotide molecules. Multiple substitution increases the maximum specific activity of the nucleotide and the labelled nucleic acid, because both the base and the sugar are incorporated into the nucleic acid molecule during enzymatic synthesis. Conversely, only the α-phosphate (see *Figure 1*) is incorporated during these reactions, so ^{32}P or ^{35}S must be substituted at this position. However,

Table 2 Physical characteristics of radionuclides used in molecular biology

Radionuclide	Half-life	Emission	
		Type	Maximum energy (MeV)
^{3}H	12.43 years	β⁻	0.018
^{32}P	14.3 days	β⁻	1.71
^{33}P	25.4 days	β⁻	0.249
^{35}S	87.4 days	β⁻	0.167
^{125}I	60 days	γ	0.035
		Auger e⁻	0.032

Table 3 Characteristics of the major radioactive labels used in molecular biology

Radionuclide	Specific activity of labelled nucleotides (Ci/mmol)	Labelling methods	Typical specific activity of probe (d.p.m./µg)	Detection limit[a] (d.p.m./cm²)	Achievable sensitivity[b] (pg target/cm²)	Achievable resolution[c] (µm)
^{32}P	400–6000	Nick translation	5×10^8	50[d]	2	200–500, 20–50
		Random priming	5×10^9		0.2	
		Phage RNA polymerase	1.3×10^9		0.6	
		End-labelling	5×10^6		200	
^{35}S	400–1500	Nick translation	1×10^8	400[e]	200	100–200, 10–20
		Random priming	7×10^8		50	
		Phage RNA polymerase	1.3×10^9		20	
^{125}I	1000–2000	Nick translation	1×10^8	100[d]	100	500, 1[f]
		Random priming	1.5×10^9		10	
		Phage RNA polymerase	1×10^9		10	
		Direct iodination	2×10^8		50	
^{3}H	25–100	Nick translation	5×10^7	8000[e]	3000	10, 0.5
		Random priming	1.5×10^8		1000	
		Phage RNA polymerase	5×10^7		3000	

[a] Represents an increase in film absorbance of 0.2 Å units after 24 h exposure, using as a source a 1.5 mm gel containing the isotope.

[b] Detected either from a filter (average band area 0.07 cm²) or by light microscopy for *in situ* signals (average cell area 7×10^{-7} cm²).

[c] The first figure(s) is the resolution achievable on filters, the second is the resolution *in situ*.

[d] Detection using an intensifying screen.

[e] Detection by fluorography.

[f] γ-Emissions are detected on film with an intensifying screen. Auger electrons are detected *in situ* on emulsion.

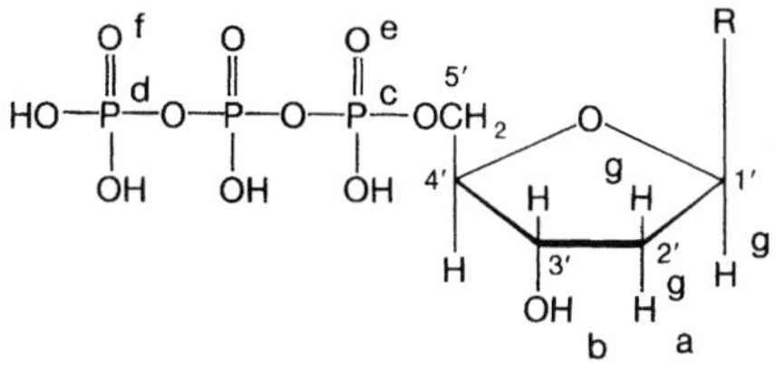

KEY

R : Purine base, e.g. adenine

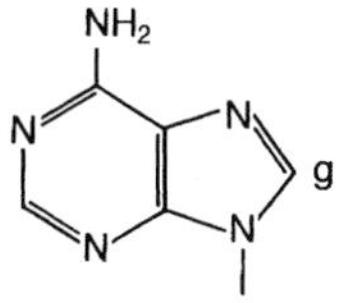

or Pyrimidine base, e.g. cytosine   uracil

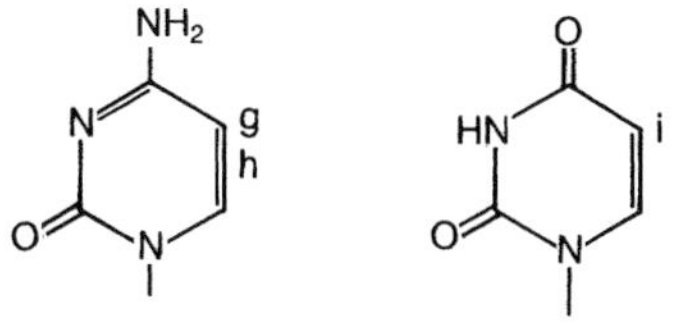

a : OH replaces H in ribonucleotides

b : H replaces OH in dideoxynucleotides

c : ^{32}P replaces ^{31}P in α labelled ^{32}P nucleotides

d : ^{32}P replaces ^{31}P in γ labelled ^{32}P nucleotides

e : ^{35}S replaces ^{16}O in α labelled ^{35}S nucleotides

f : ^{35}S replaces ^{16}O in γ labelled ^{35}S nucleotides

g : ^{3}H replaces ^{1}H in various ^{3}H nucleotides

h : ^{125}I replaces ^{1}H in iodocytidine 5'-triphosphate

i : biotin attached via allylamine linker arm in
    bio-(n)-dUTP or bio-(n)-UTP
    and
    digoxygenin attached via allylamine linker arm
    in dig-11-dUTP or dig-11-UTP

Figure 1 Labelling positions in nucleotides. ^{33}P replaces ^{31}P in exactly the same manner as shown for ^{32}P.

end-labelling by T4 polynucleotide kinase (Section 2.4.1) works on a different basis and requires ATP labelled at the γ-phosphate.

1.3.2 Non-radioactive labels

Non-radioactive labelling methods for DNA and RNA have become increasingly popular in recent years, largely because of the hazardous nature of radiolabelling, the environmental concerns regarding disposal of radioactive waste and the short shelf life of radioisotopes. The various types of non-radioactive label that are commonly used can be divided into two groups based on the methodology for signal detection:

- fluorescent labels, whose emissions are detected directly
- labels such as biotin or digoxygenin, which are detected indirectly by a 'detector complex' that generates a colorimetric or chemiluminescent signal

A range of fluorescent labels are available, most of these being derivatives of the fluorescein, rhodamine and cyanine dyes. In addition to their ease of use, they have the advantage that the fluorescent emissions of the different labels span the entire visible spectrum, which means that two or more labels can be detected from within a mixture. This feature has led to fluorescent labelling becoming the method of choice for certain applications such as automated DNA sequencing (Chapter 6, Section 6) and fluorescent *in situ* hybridization (Chapter 5, Section 4.2).

The advantage of indirect label detection is that this offers the possibility of amplifying the signal during the detection process, resulting in sensitivities that approach and sometimes exceed those possible by radioactive methods. The detection complex comprises one component that binds to the label and a second component that produces the signal. For biotin labels, the binding component is avidin or streptavidin—two proteins that use biotin as a cofactor and so attach tightly to it. Digoxygenin detector complexes make use of a monoclonal antibody that specifically binds the label. In both systems, the signal component is an enzyme, such as horseradish peroxidase or alkaline phosphatase, which can produce a coloured or chemiluminescent product when supplied with the appropriate substrate.

All of the non-radioactive markers are bulky molecules: biotin is a vitamin comprising two five-membered rings with attached side-chains, digoxygenin is a plant steroid, and the fluorescent dyes are multi-ringed structures. Surprisingly, nucleotides conjugated to these molecules still act as substrates for the polymerases and other enzymes used to label nucleic acids, although fluorescent markers are more commonly used as end-labels rather than in procedures that result in uniform labelling of the polynucleotide chain. A variety of means have been used to attach non-radioactive labels to nucleotides, one example for biotin and digoxygenin being illustrated in *Figure 1*.

The key difference between the use of radioactive and non-radioactive labels lies with their detection: similar labelling procedures can be used for both types of marker. Detailed consideration of the use of non-radioactive markers therefore appears in Section 3, after the procedures for radioactive labelling and detection have been described.

2 Radioactive labelling and detection

2.1 Labelling and detection strategies

2.1.1 Labelling methods

The various methods available for radioactive labelling of nucleic acids are summarized in *Table 4*. Labels are usually introduced into nucleic acids by *de novo* synthesis, using polymerase enzymes and labelled dNTPs or NTPs. Most polymerases synthesize DNA or RNA using a pre-existing DNA or RNA molecule as a template, the nucleotide monomers being incorporated into the newly-synthesized nucleic acid molecule in an order that reflects the base sequence of the template, following the pairing rules. The template molecule is usually single-stranded, although during the nick translation reaction a double-stranded, nicked template is made accessible by simultaneous polymerization and exonuclease digestion reactions (Section 2.3.1).

There are two special types of polymerization reaction which do not involve faithful synthesis of a nucleic acid molecule complementary to a template. The enzyme terminal deoxynucleotidyl transferase from calf intestine adds dNTPs to the 3′-end of DNA (or, less efficiently, RNA) molecules, and poly(A) polymerase generates a 'tail' of A residues at the 3′-end of an RNA molecule.

All the polymerases in common use require a free 3′-hydroxyl group to act as a priming site for synthesis. At each step of synthesis, a new phosphodiester bridge is formed between the 3′-carbon of the terminal nucleotide of the primer and the 5′-carbon of the incoming nucleotide. The α-phosphate of the newly-added nucleotide forms this bridge, and the β- and γ-phosphates are discarded. If a priming site is absent, a free 3′-hydroxyl group can be provided by one of several methods:

- by adding an oligonucleotide primer capable of base-pairing with part of the template molecule

- by producing recessed 3′-ends by digestion with a restriction endonuclease

- by internal nicking with an endonuclease such as DNase I

As well as *de novo* synthesis, labels can also be introduced into nucleic acids by modification reactions, the simplest of these being the addition of a labelled phosphate group to the 5′-end of a DNA or RNA molecule. This is accomplished by the enzyme T4 polynucleotide kinase (prepared from *Escherichia coli* cells infected with phage T4) which transfers the γ-phosphate from ATP (see Section 2.4.1).

2.1.2 Detection methods for radionuclides

The sensitivity and resolution achievable with the various radioactive labels are determined by their rates of decay and the energy and type of the radiation emitted. If film is used as the detection medium, it is important to optimize the autoradiographic conditions to suit the radionuclide. In general, maximum specific activity enables maximum sensitivity of detection, and minimum energy

Table 4 Properties of radiolabelling methods

Method	Template[a]	Product	Enzyme(s) required	Labelled precursor	Label position
Nick translation	dsDNA	dsDNA	DNase I + DNA polymerase I	dNTP	uniform
Random priming	ssDNA	dsDNA	DNA polymerase I or Klenow polymerase	dNTP	uniform
Single primer labelling	ssDNA	dsDNA	DNA polymerase I or Klenow polymerase	dNTP	uniform
				Oligonucleotide	5'-end
Phage polymerase	dsDNA	RNA	SP6, T3 or T7 RNA polymerase	NTP	uniform
				γ-labelled NTP	5'-end
End-filling	dsDNA	dsDNA	DNA polymerase I, Klenow polymerase or T4 DNA polymerase	dNTP	3'-end
Kinasing	ssDNA	ssDNA	T4 polynucleotide kinase	γ-labelled ATP	5'-end
	dsDNA	dsDNA			
	RNA	RNA			
Tailing	ssDNA	ssDNA	Terminal deoxynucleotidyl transferase	dNTP or ddNTP	3'-end
	dsDNA	dsDNA			
	RNA	RNA			
Poly(A) tailing	RNA	RNA	Poly(A) polymerase	NTP	3'-end
Ligation	ssDNA	ssDNA	T4 RNA ligase	pCp[b]	3'-end
	dsDNA	dsDNA			
	RNA	RNA			

[a] ssDNA can be produced by denaturing dsDNA. Methods for end-labelling will not generally work well at internal nicks.
[b] pCp is 5',3'-cytidine biphosphate. Label is usually incorporated in the 5'-phosphate.

of emission enables maximum resolution. Autoradiography is discussed in detail in Section 2.5.

The energy and energetic particles emitted during nuclear decay can also be detected and counted in various types of instrument. X-rays, such as those emitted by ^{125}I, can be detected directly at high efficiency by the photomultiplier tubes used in commercial scintillation counters, provided the machine is adjusted correctly. In contrast, β-particle (electron) emissions are non-penetrating, and cannot be detected directly. Collision of β-particles with hard materials such as metals and glass generates secondary bremsstrahlung X-rays. The amount of X-radiation produced depends on the energy of the β-particles and the nature of the target material. A radionuclide such as ^{32}P, which emits energetic β-particles, can be detected at moderately high efficiency in a scintillation counter by virtue of the bremsstrahlung X-rays generated within the instrument (this is called Cerenkov counting). β-particles generate light on collision with certain phosphors, by raising the phosphor molecule to an excited energy state which subsequently decays to the ground state with the emission of light. Thus immersion of the weaker β-emitters (e.g. ^{35}S and ^{3}H) within a phosphor solution (scintillation fluid) permits efficient detection by the photomultiplier tube of the scintillation counter. Conversion of the recorded c.p.m. to actual d.p.m. requires an estimate of counting efficiency.

Applications of commercially-available scintillation counters in molecular biology have been limited largely to monitoring of enzymatic reactions, because positional information across a two-dimensional surface cannot be obtained. Phosphor-screen light detectors and multiwire proportional counters are now available which feed information directly in digital form to a computer but, at present, film remains the most sensitive, cheap and convenient detection medium for radionuclides.

2.1.3 Choosing a labelling and detection strategy

The key criteria which must be applied when choosing a labelling and detection strategy are the required sensitivity and the required spatial resolution. These are described for the major applications in *Table 1*, and in *Table 3* the sensitivity levels and resolutions routinely achievable by various combinations of radiolabel and labelling method are compared. It is clear that there are many labelling options for those techniques (e.g. colony or plaque screening and Southern hybridization of prokaryotic or lower eukaryotic DNA; see Chapter 5) that require low sensitivity and resolution. The highest sensitivity requirements, such as those for Northern hybridization with blots of low copy number mRNA molecules, are best met by ^{32}P-labelling.

The detection of single-copy sequences by *in situ* hybridization requires the ultimate sensitivity and resolution. ^{32}P is not useful for high-resolution *in situ* hybridization, because the high-energy β-particles cause exposure of the emulsion over distances of several microns. ^{3}H does provide high resolution, but sensitivity within reasonable exposure times is limited by its low maximum specific activity. [125]I gives high resolution with high sensitivity after short exposure times because of its high maximum specific activity (1).

In many cases it is necessary to use an oligonucleotide probe, which restricts the available labelling options. Oligonucleotide probes can be labelled radioactively at either end, using T4 polynucleotide kinase or terminal deoxynucleotidyl transferase (Section 2.4).

2.2 Practical considerations

2.2.1 Determining the efficiency of radiolabelling

When establishing a labelling method or using a new preparation of DNA, it is often advisable to determine the efficiency of labelling. This is relatively simple when using a radioactive label, and various methods such as precipitation by TCA, TLC, absorption to DE81-paper or filtration through nitrocellulose can be used. *Protocol 1* gives a method for TCA precipitation that is widely applicable and *Protocol 2* gives a method for absorption to DE81-paper that is rapid, avoids the use of corrosive materials and is suitable for either multiple or single samples.

Protocol 1

Determination of incorporation of radiolabel by precipitation with TCA

Equipment and reagents

- Scintillation vials
- Cellulose acetate, cellulose nitrate or glass fibre filters (2.4 cm diameter)
- Filtration apparatus (e.g. Millipore)
- 8–10% TCA. **Caution: TCA is a corrosive acid: wear gloves and safety glasses at all times**

- Infrared lamp
- Scintillation fluid
- Scintillation counter
- 1 mg/ml carrier DNA or RNA (e.g. calf thymus or herring sperm DNA, or yeast RNA)

Method

1 For each sample take four scintillation vials. Total counts and incorporated (TCA-precipitable) counts should be measured in duplicate.

2 Place filters into two vials. Spot 2–10 μl (10^4–10^6 d.p.m.) of sample on to each. This will be used to give the total number of counts in the sample. For most labelling reactions an initial dilution of 2 μl of reaction mixture into 18 or 198 μl water or 0.2 M EDTA (pH 8.0) will be necessary to give counts in the required range.

3 Pipette an equal volume of the sample into two 5 ml tubes on ice. Half fill the tubes with 8–10% TCA solution. With a Pasteur pipette, add 4 drops of carrier DNA or RNA and fill the tubes with TCA solution. The carrier should produce a white cloudy precipitate on contact with the TCA solution. Leave for 30 min on ice.

Protocol 1 continued

4   Place two filters in position on a filtration apparatus. Apply a vacuum and wet the filters with approximately 2 ml 8–10% TCA. Ensure that the TCA passes through the filters readily; if not, check that the apparatus is sealed and connected to the vacuum correctly.

5   Pour the TCA precipitates into the wells. Put the tubes back on ice and refill with TCA to rinse. Pour the liquid into the wells once the initial solution has been drawn through.

6   Rinse the walls of the wells with at least 2 ml TCA. When all the liquid has been drawn through, dismantle the apparatus and transfer the filters to the remaining scintillation vials. These will be used to give the amount of precipitable radioactivity in the sample.

7   Dry all four filters (e.g. under an infrared lamp for approximately 30 min). Avoid overheating the filters. Allow the vials to cool and add 10 ml scintillation fluid. If ^{32}P has been used it is possible to count directly without scintillant (Cerenkov counting).

8   Cap the vials and determine the radioactivity by liquid scintillation counting.

9   Calculate the percentage incorporation as follows:

$$\% \text{ incorporated} = \frac{\text{precipitated counts} \times 100}{\text{total counts}}$$

Protocol 2

Determination of incorporation of radiolabel by absorption to DE81-paper

Equipment and reagents
- DE81-paper (Whatman)
- 3MM paper (Whatman)
- Scintillation fluid
- Scintillation counter

- 0.2 M EDTA, pH 8.0
- 2× SSC (0.3 M NaCl, 30 mM trisodium citrate)
- Absolute ethanol

Method [a]

1   Mark a 1×1 cm grid in pencil on a sheet of DE81. For each sample, four 1×1 cm squares will be required: two for determining total label in duplicate and two for determining the label incorporated into nucleic acid.

2   Dilute an aliquot of labelling reaction into 0.2 M EDTA to give between 10^4 and 10^6 d.p.m. in the final samples counted. Usually, a 2-μl aliquot diluted into either 18 or 198 μl 0.2 M EDTA is adequate.

3   Carefully spot 3-μl aliquots of diluted reaction mix on to the gridded DE81-paper, resting on a double sheet of 3MM paper. The aliquots for determining incorporated counts should be grouped together, separate from those for total counts. A row of

blank squares surrounding the squares for incorporated counts should be included to avoid damage during the subsequent washing stages.

4   Allow the paper to dry for ≈ 10 min. Wash the paper carrying incorporated counts for 2×10 min in $2\times$ SSC (approximately 200 ml $2\times$ SSC for a 10×5 cm sheet). Briefly rinse in water, then in absolute ethanol and allow to dry for 10–15 min.

5   Cut out all the squares and count by liquid scintillation. If ^{32}P has been used it is possible to count directly without scintillant (Cerenkov counting).

6   Calculate the average values for incorporated and total counts for each sample using the formula in *Protocol 1*, step 9.

7   If desired, the efficiency of washing can be checked by carrying out a labelling reaction without substrate and processing as above.

[a] The method described is suitable for multiple samples but 2.4 cm discs or individual squares of DE81-paper can be used for single samples. Labelled DNA is absorbed to the paper while nucleotides are washed off.

2.2.2 Removal of unincorporated label

If required, unincorporated radiolabel can be removed from the labelled nucleic acid by a variety of methods, including ethanol precipitation and Sephadex chromatography. When using a ^{32}P label, these methods can be used as a rough and ready guide to incorporation efficiency by monitoring the incorporated and unincorporated label with a β-monitor. More precise measurements can be obtained by β-scintillation counting, which is also applicable for ^{35}S and ^{3}H labels. Such measurements assume complete separation of incorporated and unincorporated label, which is not always achieved. *Protocols 3* and *4* give procedures for both ethanol precipitation and Sephadex chromatography. For oligonucleotide probes, ethanol precipitation can give poor recovery, and alternative methods such as purification by gel electrophoresis, TLC, HPLC or Sephadex G-25 chromatography are recommended. A procedure for gel purification of oligonucleotide probes was given in a early volume in the *Practical Approach* series (2).

Protocol 3

Removal of unincorporated nucleotides by ethanol precipitation

Equipment and reagents

- β-monitor
- 0.2 M EDTA pH 8.0
- 5 M NaCl or 3 M sodium acetate, pH 7.0 (the latter is prepared by adding glacial acetic acid to 3 M sodium acetate until this pH is obtained)
- Carrier DNA (calf thymus or herring sperm DNA at 1 mg/ml)
- Ice-cold absolute ethanol
- TE buffer, pH 8.0 (see *Appendix 4*, Section 2)

Protocol 3 continued

Method[a]

1 To a 50 μl labelling reaction, add the following: 20 μl 0.2 M EDTA, pH 8.0, 20 μl 5 M NaCl or 3 M sodium acetate, pH 7.0, 20 μl carrier DNA[b] and 400 μl ice-cold absolute ethanol.[c]

2 Leave to precipitate at −80 °C for 30 min or at −20 °C overnight.

3 Spin in a microcentrifuge for 15 min and carefully aspirate and dispose of the radioactive supernatant in a suitable manner.

4 Wash the pellet in ice-cold absolute ethanol, centrifuge for 5 min, pour off or aspirate the supernatant and dry the resulting pellet under vacuum.

5 Resuspend the labelled DNA pellet in an appropriate volume of TE buffer, pH 8.0, for use as a probe.

6 If desired, a small aliquot of the solution can be removed for scintillation counting to calculate the specific activity of the probe. With ^{32}P, it is frequently sufficient to check the remaining activity with a β-monitor.

[a] This method is for a labelling reaction of 50 μl; for other volumes the amounts of reagents added should be scaled up or down accordingly.

[b] Carrier DNA is optional, but improves the recovery of probe at low concentrations (e.g. during random primer labelling). Alternative carriers are total RNA (e.g. from yeast) or glycogen. As little as 1 μg may be used effectively.

[c] If the reaction is scaled up then it is adequate to use three volumes of ice-cold ethanol, to enable precipitation to be carried out in a single microfuge tube.

Protocol 4

Removal of unincorporated nucleotides by Sephadex G-50 spun column chromatography

Equipment and reagents

- 1 ml disposable syringe
- Sterile glass wool
- TE buffer, pH 8.0 (see *Appendix 4*, Section 2)
- β-monitor
- Sephadex G-50 (Pharmacia) equilibrated in TE buffer, pH 8.0

Method

1 Plug the bottom of a 1 ml disposable syringe with a small amount of sterile glass wool, preferably siliconized to prevent adsorption of nucleic acid.

2 In the syringe, prepare a column of Sephadex G-50 equilibrated in TE buffer, pH 8.0.

3 Place the syringe in a 10-ml conical tube, in which a decapped 1.5-ml conical polypropylene reaction tube has been inserted. Centrifuge at 1600 g for 4 min.

4 Continue to add Sephadex G-50 suspension until the packed volume is 0.9 ml.

5 Add 100 µl TE buffer, pH 8.0, and recentrifuge as in step 3.

6 Repeat step 5. The volume collected should now be approximately 100 µl. If the volume is greater than 150 µl, repeat step 5 again.

7 Apply the sample to the column in a total volume of 100 µl.

8 Recentrifuge as above. The unincorporated nucleotides should remain in the syringe, while the labelled probe is eluted in approximately 100 µl. To minimize the amount of high-level radioactive waste, the syringe can either be left to decay and disposed of as low-level waste, or it can be thoroughly washed and the slurry disposed of as liquid waste.

9 If desired, a small aliquot of the eluate can be removed for scintillation counting to calculate the specific activity of the probe. With ^{32}P, it is frequently sufficient to check the remaining activity with a β-monitor.

2.2.3 Removal of the labelling enzyme

Enzyme can be removed from the labelling reaction by a phenol extraction prior to removal of unincorporated label. However, for most filter hybridization applications, this is not necessary. Virtually all the enzymes used for labelling require a divalent metal ion as a cofactor, so the chelating agent EDTA is generally added to terminate the reaction before purification.

2.2.4 Determining the specific activity of the probe

The way in which the specific activity of the probe is calculated varies with the labelling method and depends on whether or not net synthesis of probe occurs during the reaction (as with random primer labelling and methods using RNA polymerases), and whether the original template participates in the subsequent hybridization reaction (as with nick translation and random primer labelling). Worked examples will be given for the major labelling methods.

2.3 Methods for uniform labelling of DNA and RNA

The highest specific activities are obtained by labelling methods that result in uniform incorporation of the radionuclide throughout the DNA or RNA molecule. The primary means of achieving this are by *de novo* DNA or RNA synthesis in the presence of radiolabelled dNTPs. Nick translation (Section 2.3.1) and random primer labelling (Section 2.3.2) are the appropriate methods for DNA, and highly-efficient phage RNA polymerases are used to synthesize labelled RNA (Section 2.3.3).

2.3.1 Nick translation

DNA polymerase I (DNA pol I) from *E. coli* has three distinct activities (3):

- a template-dependent DNA polymerase activity which acts in the 5′→3′ direction

- 5′→3′ exonuclease activity
- 3′→3′ exonuclease activity

The nick translation reaction (4) uses the ability of DNA pol I to couple the first two of these activities to add nucleotides sequentially to the 3′-end of a nick within a DNA duplex, while removing nucleotides from the adjacent 5′-terminus. Nicks are introduced in a random manner by the addition of pancreatic DNase I, so that the net effect is to produce a uniformly-labelled population of molecules. Nucleotides labelled at the α-phosphate position with ^{32}P, ^{33}P or ^{35}S can be incorporated, as can ^{3}H- or ^{125}I-labelled nucleotides. Any double-stranded DNA molecule (linear, supercoiled or nicked circular) can be used as substrate.

Although nick translation reactions have traditionally been carried out with relatively high concentrations of DNA substrate (0.5–1 μg), it is possible to label as little as 50 ng in a standard reaction. This is an advantage when labelling purified insert DNA. Using nucleotide labelled to high specific activity (> 3000 Ci/mmol) with ^{32}P, it is possible to obtain probes labelled to greater than 10^9 d.p.m./μg with low amounts of substrate. Under equivalent conditions, with 500 ng DNA, specific activities of around 2×10^8 d.p.m./μg will be obtained. A small amount of input DNA, while making more economical use of a valuable DNA preparation, also reduces the level of any contaminants added (which might otherwise lead to poor labelling efficiency). Thus, for example, DNA inserts in low-melting-point agarose can be labelled successfully (to $> 5 \times 10^8$ d.p.m./μg). An advantage of using labels of high specific activity is that short reaction times are required in subsequent experiments using the labelled molecule.

Under standard conditions (i.e. saturating nucleotide concentrations, and constant incubation time and DNA concentration), the size of the probe fragments is determined by the concentration of DNase I. We have found that optimal results in filter hybridization are obtained with probes of around 500 bp (5). Short probes (less than 100 bp) show reduced sensitivity, while longer ones (greater 1 kb) produce increased background. Prolonged incubation can also decrease the probe size because of the continuing action of DNase I. Finally, we have found that the incubation temperature may be varied in the range 15–21 °C without negative effects in subsequent hybridizations. Incubations at approximately 18 °C are convenient and may result in increased incorporation of label when using shorter (30–60 min) incubation times.

Protocol 5 generates a ^{32}P-labelled probe of an appropriate size for filter hybridization, and allows a wide range of substrate concentrations to be chosen. It uses labelled dCTP, but equivalent protocols can be devised for other nucleotides by changing the mix of unlabelled nucleotides, omitting the one that is to be used as the label. Similar protocols can also be used for ^{35}S- and ^{3}H-labelled nucleotides, but lower specific activities will be obtained. The maximum specific activity available with a ^{35}S-labelled thionucleotide is about 1000 Ci/mmol, so the specific activity of the probe will be somewhat lower and, in particular, low amounts of substrate will label less efficiently. ^{3}H-labelled nucleotides are available at a maximum specific activity of ≈ 100 Ci/mmol so that probe specific activities are tenfold lower than those possible with ^{35}S-nucleotides. Probes of higher specific activity are attainable using more than one labelled nucleotide in a reaction, which is particularly useful for ^{3}H-labelled nucleotides as the chemical concentration of the labelled species will be relatively high.

Protocol 5

Nick translation

Reagents

- TE buffer, pH 8.0 (see *Appendix 4*, Section 2)
- 5× nucleotide mix (100 µM dATP, 100 µM dGTP, 100 µM dTTP in TE, pH 8.0)
- 10× buffer (600 mM Tris-HCl, pH 7.8, 100 mM MgCl$_2$, 100 mM 2-mercaptoethanol)
- [α-^{32}P]dCTP at 3000 Ci/mmol
- Enzyme mix (0.006 units/ml DNase I and 500 units/ml DNA pol I)
- 0.5 M EDTA, pH 8.0

Method

1 Dilute the DNA to be labelled to a concentration of 5–50 µg/ml in either distilled water or TE pH 8.0.

2 Add the following, in the order given, to a microfuge tube on ice: 50–500 ng DNA solution, 10 µl 5× nucleotide mix, sufficient water to bring the final volume to 50 µl, 5 µl 10× buffer, 10 µl (100 µCi) [α-^{32}P]dCTP (3000 Ci/mmol), 5 µl enzyme mix (total volume = 50 µl). Mix gently by pipetting up and down once or twice, and cap the tube. If necessary, spin briefly (2 s) in a microcentrifuge to collect the contents at the bottom of the tube.

3 Incubate the reaction mix at 15 °C for 60 min. If desired, any incubation time between 30 min and 3 h may be used.

4 Terminate the reaction by addition of 2 µl 0.5 M EDTA, pH 8.0.

5 A small aliquot (1–2 µl) can be removed for determination of incorporation (see *Protocols 1* and *2*). For probes with specific activities $> 10^9$ d.p.m./µg or if incorporation is $< 50\%$, it is recommended that unincorporated nucleotides are removed by one of the methods described in *Protocols 3* and *4*.

6 Store at -20 °C in the presence of 20 mM EDTA, pH 8.0, before use in hybridization analysis. Prolonged storage can lead to substantial degradation and probes of high specific activity should be stored for no longer than 3 days.

7 Denature the probe by heating to 95–100 °C for 5 min immediately before use.

8 Calculate the specific activity of the probe:

   (a) During nick translation, nucleotides are effectively excised and replaced, and there is usually no net synthesis of DNA. Thus, the probe specific activity is calculated simply as:

$$\text{specific activity (d.p.m./µg)} = \frac{\text{total activity incorporated (d.p.m.)}}{\text{amount of substrate added (µg)}}$$

   (b) Fifty per cent incorporation in the above reaction would give a specific activity of 2×10^8 d.p.m./µg with 500 ng input DNA, and 2×10^9 d.p.m./µg with 50 ng ($1 \, \mu\text{Ci} = 2.2 \times 10^6$ d.p.m.).

2.3.2 Random primer labelling

The Klenow fragment of DNA polymerase I is produced by proteolytic cleavage of the enzyme or, as is now more common, from an overproducing recombinant *E. coli* strain expressing the appropriate portion of the gene (3). The enzyme lacks the $5' \rightarrow 3'$ exonuclease activity of the intact DNA pol I enzyme and can be used *in vitro* to copy a single-stranded DNA molecule starting from the 3'-end of a primer DNA annealed to the template. A mixture of hexanucleotide primers of random sequence, derived either from digestion of calf thymus DNA or by chemical synthesis, allows effectively any DNA template to be copied (6, 7). As with nick translation, nucleotides labelled with ^{32}P, ^{33}P, ^{35}S, ^{3}H or ^{125}I can be incorporated. The absence of the $5' \rightarrow 3'$ exonuclease activity prevents degradation of the primer from its 5'-end and also ensures that incorporated nucleotides are not subsequently excised. This increases the efficiency of the reaction and allows a range of reaction times and temperatures to be adopted. The amount of newly-synthesized DNA that is produced often exceeds the amount of input DNA, suggesting that the same region of template can be copied more than once, by synthesis primed from different hexanucleotides. These factors lead to a highly efficient reaction which gives good levels of incorporation (70–80%) even with small amounts of substrate (as little as 25 ng), The use of [^{32}P]-nucleotides labelled to high specific activity allows the production of probes of greater than 10^9 d.p.m./µg. The technique is relatively insensitive to the purity of the substrate DNA and both mini-preparations and fragments in low gelling-temperature agarose can be labelled. Random primer labelling is now probably the most frequently method used for labelling insert DNA, as opposed to intact recombinant vector molecules.

A single-stranded template is required, which is usually obtained by brief heat denaturation of a double-stranded molecule, unless a single-stranded clone (e.g. in an M13 vector) is available. Both linear and circular DNA can be used. Reactions are generally carried out at room temperature or 37°C. The higher temperature allows a more rapid reaction so that, under appropriate conditions, maximum incorporation can be obtained within 30 min. Reactions at room temperature can be left overnight if convenient. Klenow polymerase and DNA polymerase from T7 bacteriophage have also used in conjunction with random primers (6- or 9-mers) to allow very rapid (2–5 min) labelling.

The size of the probe is determined primarily by the ratio of primer to substrate and by the nucleotide concentration. *Protocol 6* should generate probes with sizes suitable for filter hybridization. As with *Protocol 5*, dCTP is used as the label, but the procedure can easily be modified for other labelled nucleotides. Again, a range of substrate concentrations is quoted, but amounts at the lower end of the range will be appropriate in most cases. With 25 ng DNA, a probe with a specific activity of about 2×10^9 d.p.m./µg should be obtained. This can be increased to around 5×10^9 d.p.m./µg by the use of label at 6000 Ci/mmol. Addition of more label will also increase the specific activity of the probe, but will result in reduced efficiency of utilization. Specific activities are concomitantly lower with ^{35}S and ^{3}H. The use of multiple labels gives no real advantage with ^{32}P but, with ^{3}H, specific activities above 10^8 d.p.m./μg can be obtained if probes are labelled with three or four nucleotides. It should be noted, however, that incorporation rates are reduced with multiple labels, and that reaction times of at least 5 h are recommended. The calculation of the final specific activity of the probe is complicated by the fact that, while net synthesis of DNA occurs during the reaction, the initial labelled substrate also remains. The worked example given in *Protocol 6* should make this calculation clearer.

Protocol 6

Random primer labelling

Reagents

- TE buffer, pH 8.0 (see Appendix 4, Section 2)
- Random hexanucleotide primers (20 OD units/ml in TE buffer, pH 8.0, containing 4 mg/ml bovine serum albumin), available from several companies (e.g. Roche product no. 1277081
- [α-^{32}P]dCTP at 3000 Ci/mmol

- 5$\times$ nucleotide mix (100 μM dATP, 100 μM dGTP, 100 μM dTTP in TE buffer, pH 8.0)
- 10$\times$ buffer (600 mM Tris-HCl, pH 7.8, 100 mM MgCl$_2$, 100 mM 2-mercaptoethanol)
- Klenow fragment of DNA polymerase I
- 0.5 M EDTA, pH 8.0

Method

1 Dilute the DNA to be labelled to a concentration of 2–25 μg/ml in distilled water or TE buffer, pH 8.0.

2 If double-stranded DNA is being labelled, denature by heating to 95–100 °C for 2–5 min in a boiling water bath, then chill on ice.

3 Add the following, in the order given, to a microfuge tube on ice: 25–250 ng DNA solution, 10 μl 5$\times$ nucleotide mix, sufficient water to bring the final volume to 50 μl, 5 μl 10$\times$ buffer, 5 μl random primers, 10 μl (100 μCi) [α-^{32}P]dCTP (3000 Ci/mmol), 2 units Klenow polymerase (total volume = 50 μl). Mix gently by pipetting up and down once or twice and cap the tube. If necessary, spin briefly (2 s) in a microcentrifuge to collect the contents at the bottom of the tube.

4 Incubate the reaction mix at room temperature for 3 h to overnight, or at 37 °C for 30 min to 3 h.

5 Terminate the reaction by adding 2 μl of 0.5 M EDTA, pH 8.0.

6 A small aliquot (1–2 μl) can be removed for determination of incorporation (see *Protocols 1* and *2*). For probes with specific activities > 10^9 d.p.m./μg, or if incorporation is < 50%, it is recommended that unincorporated nucleotides are removed by one of the methods described in *Protocols 3* and *4*.

7 Store at −20 °C in the presence of 20 mM EDTA, pH 8.0, before use in hybridization analysis. Prolonged storage can lead to substantial degradation and probes of high specific activity should be stored for no longer than 3 days.

8 Denature the probe by heating to 95–100 °C for 5 min immediately before use.

9 Calculation of the probe yield and specific activity:

(a) During random primer labelling there is net synthesis of DNA, while the initial substrate remains unlabelled. Both participate in the subsequent hybridization, so:

$$\text{probe yield} = \text{ng initial substrate DNA} + \text{ng DNA synthesized}$$

As the average molecular weight of a nucleoside monophosphate in DNA is 350, for a labelled nucleotide at $y \times 10^3$ Ci/mmol:

$$\text{ng DNA synthesized} = \frac{\mu\text{Ci incorporated} \times 0.35 \times 4}{y}$$

Note that a multiplication factor of 4 is included as there are four nucleotides, only one of which is labelled.

(b) Once the probe yield has been calculated, the specific activity can be determined:

$$\text{specific activity (d.p.m./}\mu\text{g)} = \frac{\text{total activity incorporated (d.p.m.)}}{\text{probe yield (}\mu\text{g)}}$$

(c) A worked example may help:

- Assume that in the reaction mix set up in step 3, 70% incorporation was obtained with 25 ng DNA
- This means that 70% of 100 μCi = 70 μCi was incorporated at 3000 Ci/mmol
- So the amount of nucleotide incorporated $= 70 \times \dfrac{0.35}{3} = 8$ ng
- The amount of DNA synthesized $= 8 \times 4 = 32$ ng
- The total DNA $= 32 + 25 = 57$ ng $= 0.057\ \mu$g
- Since 70 μCi $= 1.5 \times 10^8$ d.p.m.:

$$\text{probe specific activity} = \frac{1.5 \times 10^8}{0.057} = 2.6 \times 10^9\,\text{d.p.m./}\mu\text{g}$$

(d) Similarly, 70% of incorporation with 250 ng DNA gives a total DNA concentration of $250 + 32 = 282$ ng, and a probe specific activity of

$$\frac{1.5 \times 10^8}{0.282} = 5 \times 10^8\,\text{d.p.m./}\mu\text{g}.$$

2.3.3 Synthesis of labelled RNA with bacteriophage RNA polymerases

The RNA polymerases from a number of bacteriophages show a high affinity for their homologous promoters, and are capable of transcribing a downstream sequence very efficiently *in vitro* (3). This has allowed the development of a wide variety of cloning vectors containing one or more bacteriophage RNA polymerase promoters upstream of a cloning site or polylinker into which the sequence to be transcribed is cloned (see Volume I, Chapter 9, Section 3.4). RNA polymerases from SP6, T7, and T3 bacteriophages are now available commercially, frequently derived from over-producing recombinant clones. Under appropriate conditions milligram quantities of RNA can be generated from a cloned sequence, for a variety of applications (8, 9). By the use of a single labelled nucleotide species, it is possible to generate nanogram to microgram amounts of an RNA hybridization probe of high specific activity. As with other labelling reactions, nucleotides labelled at the α-position with ^{32}P, ^{33}P or ^{35}S, or labelled with ^{3}H or ^{125}I can be used, although it has been found that some individual nucleotides give only very low levels of incorporation (as discussed below).

Many of the features of the RNA polymerase reactions are quite different from those involved in DNA labelling. The most obvious is that the product is RNA rather than DNA, and RNA hybridization probes have a number of specific properties. As they are single-stranded, they do not require denaturation before hybridization and will not reanneal during the hybridization experiment. In the presence of formamide, RNA–RNA and RNA–DNA hybrids are more stable than DNA–DNA hybrids. When carrying out hybridizations either with or without formamide, we have found that optimal signal to background ratios are obtained at temperatures somewhat higher than those commonly used with DNA probes. When using an RNA probe for the first time, it is therefore advisable to determine the optimum hybridization and washing temperatures empirically. These points are discussed further in Chapter 5.

In some circumstances, the background in hybridization experiments with RNA probes can be reduced by treating the filter with RNase A, which is highly specific for single-stranded RNA (10) and degrades probe molecules that are bound non-specifically to the filter (see Chapter 5, Section 3.8). In addition, unlabelled template can be removed by digestion with RNase-free DNase I before hybridization. Some workers have reported improved sensitivity when comparing RNA probes with DNA probes (10, 11) and it is likely that several of the above factors have contributed to this observation. In general, however, we have found similar sensitivities with RNA and DNA probes labelled to equivalent specific activities in both standard filter hybridization and in a rapid hybridization system.

Standard labelling reactions with RNA polymerases use a relatively high concentration of substrate (1–2 µg) but, as this is usually intact vector rather than purified insert, there is less difficulty in obtaining significant quantities. We have had some success in labelling plasmid mini-preparations, but more consistent results are obtained with DNA purified through a CsCl gradient. It is advisable to include an inhibitor of ribonuclease activity such as RNasin (Promega) or HPRI (Amersham Pharmacia), particularly if substrate of low purity is used. *Protocol 7* recommends 4–10 units of enzyme, but the optimal amount varies to some extent with the purity of the substrate and the quality of the enzyme preparation. In some cases as little as one unit can be used.

Under conditions of saturating nucleotide concentration, full-length transcripts are generated, but when producing labelled probes it is usually necessary to reduce the concentration of the labelling nucleotide in order to maximize the specific activity of the probe. A concentration of 12.5 μM is usually adopted, which approximates to the K_m of CTP, GTP, and UTP with the SP6 polymerase (10). In a 20 μl reaction this is equivalent to 100 μCi of a nucleotide labelled at 400 Ci/mmol, or 200 μCi at 800 Ci/mmol. Full-length RNA yields of more than 80% have been reported with an SP6 transcript of 1850 nucleotides and either CTP, ATP or GTP at 10 μM (10); in fact, most of the standard RNA polymerase labelling protocols would be expected to produce a high proportion of full-length transcripts. Depending on insert size, this may result in probes longer than the optimum of around 500 bp, but this is less of a problem with RNA probes, particularly since RNase A treatment can be used to reduce the hybridization background. Shorter probes can be generated (e.g. for *in situ* hybridization) by using very low concentrations of the labelled nucleotide.

Protocol 7 uses [α-^{32}P]UTP as the label, but equivalent procedures can be derived for other nucleotides, while keeping the concentration of the labelling species at 12.5 μM. We have, in practice, obtained the most reproducible results with either UTP or CTP as label. ATP appears to be somewhat less efficient at 12.5 μM with SP6 polymerase, possibly because the enzyme has a higher K_m for this nucleotide. T7 RNA polymerase has given a good level of incorporation with ATP as the label, but poor incorporation with GTP. In particular, the nucleotide analogue [α-^{35}S]GTPαS shows virtually no incorporation at 12.5 μM with T7 polymerase.

Because the initial DNA substrate does not participate in the hybridization reaction, as the probe is either not denatured before use or is removed by DNase I treatment, the effective specific activity of the probe depends only on the specific activity of the labelled nucleotide. The addition of more nucleotide may increase the probe yield, but will not increase its specific activity. Thus, a labelled nucleotide at 400 Ci/mmol will give a probe specific activity of 6.7×10^8 d.p.m./μg and at 800 Ci/mmol will give 1.3×10^9 d.p.m./μg. *Protocol 7* should give an incorporation of 70–80% within 1 h—equivalent to the synthesis of about 250 ng of transcript. In theory, higher specific activities should be possible by the use of more than one labelled nucleotide species, but this reduces the efficiency of the reaction too much to be of practical advantage.

Protocol 7

Labelling with bacteriophage RNA polymerases

Reagents

- TE buffer, pH 8.0 (see Appendix 4, Section 2)
- 5× buffer (200 mM Tris-HCl, pH 7.5, 30 mM MgCl$_2$, 10 mM spermidine)
- Nucleotide mix (2.5 mM ATP, 2.5 mM CTP, 2.5 mM GTP in 20 mM Tris-HCl, pH 7.5)
- 200 mM DTT, freshly prepared

- HPRI (Amersham Pharmacia)
- [α-^{32}P]UTP at 800 Ci/mmol, 20 mCi/ml or 400 Ci/mmol, 10 mCi/ml
- SP6 or T7 RNA polymerase
- RNase-free DNase I
- 0.25 M and 20 mM EDTA, pH 8.0

Protocol 7 continued

Method

1 Linearize the DNA template downstream of the phage polymerase promoter with an appropriate restriction enzyme. Phenol extract, ethanol precipitate, and resuspend in TE buffer, pH 8.0, at a concentration of 1–2 mg/ml.

2 Add the following in the order given to a microfuge tube at room temperature (spermidine may cause precipitation of the substrate DNA on ice): 4 μl 5× buffer, 1.5 μl nucleotide mix, 1 μl 200 mM DTT, 20 units HPRI ($\approx$ 1 μl), sufficient water to bring the final volume to 20 μl, 2 μg linearized template DNA, 10 μl [α-$^{32'}$P]UTP (800 Ci/mmol) (total volume = 20 μl).

3 Mix at room temperature, then add 4–10 units of SP6 or T7 RNA polymerase.

4 Incubate for 1 h at 40°C with SP6 RNA polymerase, or 1 h at 37°C with T7 polymerase.

5 If the DNA template is to be removed, incubate with 2 units of RNase-free DNase I for 10 min at 37°C. In general, we have found this to be unnecessary for filter hybridization, and the step can usually be omitted.

6 Terminate the reaction by addition of 2 μl 0.25 M EDTA, pH 8.0.

7 A small aliquot (1–2 μl) can be removed for determination of percentage incorporation (see *Protocols 1* and *2*). If incorporation is less than 50%, it is recommended that unincorporated nucleotides are removed by one of the methods described in *Protocols 3* and *4*.

8 Store at −20°C in the presence of 20 mM EDTA, pH 8.0, before use in hybridization analysis. Prolonged storage can lead to substantial degradation, and probes of high specific activity should not be stored for longer than 3 days.

9 The probe should not be denatured before hybridization.

10 Calculate of probe yield and specific activity:

(a) Since unlabelled vector DNA does not participate in the hybridization reaction, the specific activity of the probe depends solely on the specific activity of the labelled nucleotide, while probe yield can be calculated from the percentage incorporation of label. The specific activity of the probe is independent of labelling efficiency.

(b) This is illustrated by a worked example:

- In a reaction containing 200 μCi of [α-^{32}P]UTP at 800 Ci/mmol (= 800 mCi/mmol) an incorporation of 70% is obtained. The average molecular weight of a nucleoside monophosphate in DNA is 350, and 1 mCi is equivalent to 2.2×10^9 d.p.m. The specific activity of the nucleotide can therefore be expressed as:

$$\frac{800 \times 2.2 \times 10^9}{350} = 5 \times 10^9 \, \text{d.p.m./μg}$$

- With a single labelled nucleotide species, the specific activity of the probe is:

$$\frac{5 \times 10^9}{350} = 1.3 \times 10^9 \, \text{d.p.m./μg DNA}$$

- If any unlabelled UTP is added to the reaction, the specific activity of the probe will be reduced, though the yield may be improved.
- To calculate probe yield:
  - The amount of labelled nucleotide incorporated = $200 \times 70\% = 140\,\mu Ci$.
  - This is equivalent to 0.175 nmol (as the label is present at 800 Ci/mmol), which is equivalent to $0.175 \times 350 = 61.25$ ng of incorporated label.
  - With a single labelled nucleotide species, this is equivalent to $61.25 \times 4 = 245$ ng of probe.
- Note that the amount of probe is significantly less than the amount of whole vector $(1–2\,\mu g)$ present in the reaction.

As described in Volume I, Chapter 9, Section 3.4, a multitude of vectors are now available containing phage RNA polymerase promoters, often in combination with other features such as single-strand replication origins, *lacZ'* selection, annealing sites for sequencing primers and inducible promoters. A common arrangement is to place two different RNA polymerase promoters, say those for SP6 and T7, in opposite orientations, separated by a multiple cloning site. Both sense and antisense transcripts can then be generated, providing a valuable control, for example in *in situ* hybridizations. Transcription from vector sequences is avoided by restriction of the vector immediately downstream of the insert, so that run-off transcripts are produced.

2.4 Methods for end-labelling

There are a variety of enzymes capable of labelling either the 3'- or 5'-ends of nucleic acids (3). The most commonly used are probably T4 polynucleotide kinase for 5'-end labelling of DNA and Klenow polymerase or terminal deoxynucleotidyl transferase for 3'-ends. Incorporation is usually limited to one, or at most several, labelled residues per end, so the specific activity of the probe is lower than with methods that result in uniform labelling. For this reason, ^{32}P is frequently used to maximize the available signal. It is important to calculate the relative amounts of DNA termini and labelled nucleotide in a reaction. These can then be adjusted, depending on whether efficient utilization of label or maximal labelling are the priority. In most cases, the emphasis is on labelling to maximum specific activity, and so the labelled nucleotide will be present in a molar excess. The following example shows how these amounts can be calculated.

An end-labelling reaction contains:

- 200 μCi of label at 3000 Ci/mmol
- 2 μg of a 1500 nucleotide DNA fragment (single- or double-stranded)

The amounts of both label and available ends can usually be expressed in picomoles for direct comparison. Thus:

- 3000 Ci/mmol = 3 μCi/pmol, therefore 200 μCi = 67 pmol

- $2\,\mu g\,DNA = \dfrac{2 \times 10^6}{350} = 5700\,pmol\,total\,nucleotide$

Each DNA strand is 1500 nucleotides, so pmol ends $= \dfrac{5700}{1500} = 3.8\,pmol\,ends.$

Thus, label has been included at molar excess over ends.

2.4.1 End-labelling with T4 polynucleotide kinase

Polynucleotide kinase from bacteriophage T4 catalyses two distinct labelling reactions which may be used with either single- or double-stranded DNA or with RNA (3). The first, termed the forward reaction, results in very efficient labelling of DNA containing a terminal 5'-hydroxyl group. This is particularly convenient for use with chemically synthesized oligonucleotides, which generally carry such groups, but most other nucleic acid species will require prior dephosphorylation at their 5'-ends with either bacterial or calf intestinal alkaline phosphatase. T4 polynucleotide kinase catalyses the transfer of the terminal γ-phosphate group from a ribonucleoside triphosphate donor to the 5'-terminus of the DNA molecule. The most popular donor is [γ-^{32}P]ATP, although other γ-labelled nucleotides can also be used. Both ^{3}H- and ^{35}S-labelled nucleotides can be used, although the reaction is somewhat less efficient with thionucleotides, and it may be necessary to increase both the reaction time and the enzyme concentration to obtain efficient labelling.

The second reaction catalysed by polynucleotide kinase is termed the exchange reaction (12). It is somewhat less efficient than the forward reaction, but is considerably more convenient for labelling DNA containing a 5'-phosphate group. The reaction is driven by the presence of excess ADP. The enzyme catalyses the transfer of the 5'-phosphate group of DNA to ADP, converting it to ATP, and then rephosphorylates the 5'-end using a ribonucleoside triphosphate donor, as in the forward reaction. Again, [γ-^{32}p]ATP is the most commonly-used donor.

Protocols 8 and 9 give procedures for both the forward and exchange reactions based on a number of published methods, together with details for 5'-dephosphorylation of DNA with CIAP. Both labelling reactions, in particular the exchange reaction, are most efficient with single-stranded molecules or with protruding 5'-ends. If the DNA molecules have blunt or recessed 5'-ends, then a short incubation at 70 °C followed by rapid chilling on ice may improve efficiency.

Protocol 8

5'-end labelling with T4 polynucleotide kinase: forward reaction

Reagents

- 10 mM Tris-HCl, pH 8.0
- CIAP
- 5 M NaCl
- Ice-cold absolute ethanol

Protocol 8 continued

- 10× CIAP buffer (0.5 M Tris-HCl, pH 9.0, 10 mM MgCl$_2$, 10 mM ZnCl$_2$, 10 mM spermidine)
- Buffer-saturated phenol (see Appendix 4, *Protocol 6*)
- 24:1 chloroform : isoamyl alcohol (see Appendix 4, *Protocol 6*), or ether

- TE buffer, pH 8.0 (see Appendix 4, Section 2)
- 10× kinase buffer (0.5 M Tris-HCl, pH 7.6, 0.1 M MgCl$_2$, 50 mM DTT, 1 mM spermidine)
- [γ-^{32}P]ATP at 3000 Ci/mmol and 10 mCi/ml
- T4 polynucleotide kinase
- 0.5 M EDTA, pH 8.0

Method

The following protocol includes dephosphorylation (steps 1–2) with CIAP. For molecules with free 5'-OH groups (e.g. most chemically synthesized oligonucleotides) this is not necessary and the procedure should begin at step 3.

1   Dissolve the DNA in a small volume of 10 mM Tris-HCl, pH 8.0, then set up the following mix in a microfuge tube on ice: 10 pmol ends[a] DNA, 5 μl 10× CIAP buffer, sufficient water to bring the final volume to 50 μl, 0.05 units CIAP (total volume = 50 μl). Incubate at 37 °C for 30 min for duplex DNA or at 55 °C for 30 min for RNA. For DNA with blunt ends or recessed 5'-termini, better results may be obtained at the higher temperature, or by successive 15-min incubations at 37 °C and 55 °C.

2   Centrifuge for 2 s, then add an equal volume of buffer-saturated phenol. Extract twice with phenol and twice with chloroform–isoamyl alcohol or ether. Add a one-tenth volume of 5 M NaCl followed by two volumes of cold absolute ethanol for DNA or three volumes for RNA. Precipitate the nucleic acid at −20 °C overnight or at −80 °C for 30 min.

3   Resuspend the pellet in 10 μl TE buffer and set up the following mix in a microfuge tube on ice: 10 pmol ends[a] DNA, 5 μl 10× kinase buffer, sufficient water to bring the final volume to 50 μl, 200 μCi (65 pmol) [γ-^{32}P]ATP (3000 Ci/mmol), 10–20 units T4 polynucleotide kinase (total volume = 50 μl). Incubate at 37 °C for 30–45 min.

4   Terminate the reaction by addition of 2 μl 0.5 M EDTA, pH 8.0.

5   A small aliquot (1–2 μl) can be removed for determination of percentage incorporation (see *Protocols 1* and *2*). If incorporation is low, it is recommended that unincorporated nucleotides are removed by one of the methods described in *Protocols 3* and *4*. For oligonucleotides, unincorporated label is best removed by gel electrophoresis, HPLC, or TLC.

6   Denature the probe by heating to 95–100 °C for 5 min immediately before use.

7   As there is no net synthesis of DNA, the specific activity of the probe is:

$$\text{specific activity (d.p.m./μg)} = \frac{\text{total activity incorporated (d.p.m.)}}{\text{amount of substrate added (μg)}}$$

[a] 10 pmol ends = 3.3 μg of a 1 kb double-stranded molecule or 0.065 μg of a 20-mer oligonucleotide.

Protocol 9

5′-end labelling with T4 polynucleotide kinase: exchange reaction

Reagents

- 10× exchange buffer (0.5 M imidazole-HCl, pH 6.6, 100 mM MgCl$_2$, 50 mM DTT, 3 mM ADP, 1 mM spermidine)
- [γ-^{32}p]ATP at 3000 Ci/mmol and 10 mCi/ml
- T4 polynucleotide kinase
- 0.5 M EDTA, pH 8.0

Method

1 Set up the following mix in a microfuge tube: 10 pmol ends[a] DNA, 5 µl 10× exchange buffer, sufficient water to bring the final volume to 50 µl, 200 µCi (65 pmol) [γ-^{32}P]ATP (3000 Ci/mmol), 10 units T4 polynucleotide kinase (total volume = 50 µl). Incubate at 37 °C for 30–60 min.

2 Continue as in steps 4–6 for the forward reaction (*Protocol 8*).

3 As there is no net synthesis of DNA, the specific activity of the probe is calculated as in *Protocol 8*, step 7.

[a] 10 pmol ends = 3.3 µg of a 1 kb double-stranded molecule or 0.065 µg of a 20-mer oligonucleotide.

2.4.2 End-labelling with Klenow polymerase

Klenow polymerase is frequently used to fill in 3′-ends created by restriction enzyme cleavage (3). Its 5′→3′ polymerase activity uses the opposite, overhanging end as a template, adding nucleotides from the recessed 3′-end. The nucleotide that is chosen as the label depends on the sequence at the site, and it is often possible to introduce either a single or several labelled residues. By careful choice of label and restriction enzymes, it is possible to label only one species of site in a mixture of fragments cut with different enzymes. Again ^{32}P-labelled nucleotides are most frequently used, although ^{3}H, ^{35}S, ^{125}I, and biotin labels can be incorporated. Short reaction times are usually chosen, both for convenience and to minimize the possibility of removal of the label by the 3′→5′ exonuclease activity of the enzyme. This activity does, however, allow Klenow polymerase to be used to label blunt ends, by replacing the 3′-terminal nucleotide with a labelled equivalent. *Protocol 10* is suitable for a variety of situations but, as with all end-labelling procedures, the amount of label added should be matched with the available ends.

Protocol 10

End-repair with Klenow polymerase

Reagents

- 10× buffer (500 mM Tris-HCl, pH 7.5, 100 mM MgCl$_2$, 10 mM DTT). Most restriction enzyme buffers can also be used
- 5× nucleotide mix (containing each nucleotide at 100 µM except the one(s) to be used as the label)
- [α-^{32}P]dNTP(s) at 3000 Ci/mmol and 10 mCi/ml
- Klenow polymerase
- Chase solution (1 mM each of dATP, dCTP, dGTP, dTTP)
- 0.25 M EDTA, pH 8.0

Method

1 Set up the following reaction in a microfuge tube on ice: 1 µg^a DNA to be labelled, sufficient water to bring the final volume to 20 µl, 2 µl 10× buffer, 2 µl 5× nucleotide mix, 10 µCi (3 pmol) [α-^{32}P]dNTP(s) (3000 Ci/mmol), 2 units Klenow polymerase (total volume = 20 µl). Incubate at room temperature for 15 min.

2 If it is important that all end-labelled molecules are of the same length when using 3′-recessed DNA fragments, it is advisable to carry out a cold-chase step after labelling: add 2 µl of chase solution and incubate for a further 5 min at room temperature. If this is not necessary, proceed to step 3.

3 Terminate the reaction by addition of 2 µl 0.25 M EDTA, pH 8.0.

4 A small aliquot (1–2 µl) can be removed for determination of percentage incorporation (see *Protocols 1* and *2*). If incorporation is low, it is recommended that unincorporated nucleotides are removed by one of the methods described in *Protocols 3* and *4*.

5 Denature the probe by heating to 95–100 °C for 5 min immediately before use.

6 The specific activity of the probe is calculated using the formula in *Protocol 8*, step 7.

^{a} Equivalent to 1 pmol ends for a 3 kb fragment.

2.4.3 End-labelling with T4 DNA polymerase

The DNA polymerase from bacteriophage T4 has similar properties to Klenow polymerase, but its 3′→5′ exonuclease is about 250-fold more active. It is possible to control the relative levels of the polymerase and exonuclease activities by omitting or providing dNTPs. This allows the enzyme to be used for labelling recessed, blunt, and protruding 3′-ends and also for carrying out a replacement synthesis. In this procedure, the exonuclease is used to produce long recessed 3′-termini in the absence of dNTPs. These are subsequently filled in by the enzyme on the addition of nucleotides. Further details on the use of this enzyme for labelling are given in a previous volume of the *Practical Approach* series (13).

2.4.4 End-labelling with terminal deoxynucleotidyl transferase

Terminal deoxynucleotidyl transferase adds a series of nucleotides onto the 3'-end of single- or double-stranded DNA molecules (3). This activity does not require a template. Reaction conditions can, to some extent, be modified to determine the extent of addition of residues. For example, the concentration of labelled nucleotide can be chosen so that, on average, only one labelled residue is added per molecule of substrate. To produce probes with higher specific activities, progressively more residues can be added, but this may affect the specificity of the probe, particularly if it is short. The addition can be limited strictly to a single residue by the use of a nucleotide lacking a 3'-hydroxyl group, so that addition of further residues is blocked. Cordycepin 5'-[α-^{32}P]triphosphate (which is a 2'-hydroxy, 3'-deoxynucleotide) has been used as such a terminator, but more efficient reactions have been obtained with [α-^{32}P]2',3'-dideoxy-ATP (14). An appropriate procedure for TdT is given in *Protocol 11*.

Protocol 11

3'-end labelling with terminal deoxynucleotidyl transferase

Reagents

- 10× buffer (1.4 M sodium cacodylate, pH 7.2, 10 mM cobalt (II) chloride, 1 mM DTT). **Warning: cacodylate is a poisonous arsenic compound and must be handled with care. All contact with the skin must be avoided**

- [α-^{32}P]dideoxyATP (3000 or 5000 Ci/mmol)
- TdT
- 0.5 M EDTA, pH 8.0

Method

1 Dissolve purified DNA equivalent to 10 pmol of 3'-ends in 20 µl water.

2 Set up the following mix in the order given in a microfuge tube on ice: 10 pmol ends[a] DNA to be labelled, 5 µl 10× buffer, sufficient water to bring the final volume to 50 µl, 50 µCi (16 or 10 pmol) [α-^{32}P]ddATP (3000 or 5000 Ci/mmol), 10 units TdT (total volume = 50 µl). Mix gently by pipetting up and down once or twice, and cap the tube. Centrifuge for 2 s.

3 Incubate the reaction mix at 37°C for 60 min.

4 Terminate the reaction by the addition of 5 µl 0.5 M EDTA, pH 8.0.

5 A small aliquot (1–2 µl) can be removed for determination of percentage incorporation (see *Protocols 1* and *2*). If incorporation is low, it is recommended that unincorporated nucleotides are removed by one of the methods described in *Protocols 3* and *4*.

6 Denature the probe by heating to 95–100°C for 5 min immediately before use.

7 The specific activity of the probe is calculated using the formula in *Protocol 8*, step 7. With an *Alu*I digest of λ and [α-^{32}P]ddATP at 3000 Ci/mmol, we have achieved a specific activity of $> 8 \times 10^6$ d.p.m/µg.

[a] 10 pmol ends = 3.3 µg of a 1 kb double-stranded molecule or 0.065 µg of a 20-mer oligonucleotide.

Table 5 Autoradiographic and fluorographic properties of radioisotopes

	^{32}P	^{33}P, ^{35}S, ^{14}C	^{3}H	^{125}I
Sample preparation	Use a dry gel for good resolution, or a wet gel for convenience. Use a wet filter to allow reprobing. Sample can be wrapped	Dry the sample. Do not wrap. Make direct contact with the film	Dry the sample. Do not wrap. Make direct contact with the film	Use a dry sample for good resolution, or a wet one for convenience. Sample can be wrapped
Fluorography method	One or two intensifying screens	Impregnate with fluor	Impregnate with fluor	One or two intensifying screens
D.p.m. needed for good image after 24 h:[a]				
Direct autoradiography	3.5×10^2	4.0×10^3	$> 2.0 \times 10^6$	1×10^3
Preflashed fluorography	3.0×10^1	2.5×10^2	5.0×10^3	6.0×10^1
Enhancement by fluorography:[b]				
Dried gel	10	15	1000	16
Filter, TLC plate	14	10	100	17

[a] Results determined by loading sample into 5 mm polyacrylamide gel slots, followed by electrophoresis and drying. The units are d.p.m. of acid-precipitable material. This approximates to c.p.m. for all the isotopes except ^{3}H, for which 'counts' are approximately one-third less.

[b] Figures derived from the amount of radioactivity per cm^2 required to produce a detectable image in 24 h.

TdT can be used as an alternative to polynucleotide kinase for labelling oligonucleotides. With double-stranded templates, the most efficient labelling is obtained with 3'-protruding or blunt ends, although blunt ends with high GC contents tend to label less efficiently than those with more AT base pairs. Again, [3]H- and ^{35}S-labelled nucleotides can be used, but reaction rates are lower with ^{35}S-labelled analogues.

2.5 Detection of radioactively-labelled DNA and RNA

2.5.1 Autoradiography

It is possible to obtain an accurate representation of the spatial distribution of radiolabels in samples such as hybridized filters, polyacrylamide and agarose gels and TLC plates by the technique of autoradiography. The sample is placed in close contact with a medical X-ray film, which is coated with an emulsion containing silver halide crystals. The β-particles and/or γ-rays from the radioactive sample then convert silver ions to silver atoms within the crystal lattice, producing a 'latent image' of the sample. When the film is developed under standard photographic conditions, those crystals containing aggregates of silver atoms are reduced by the developer to metallic silver. This produces a blackening in the emulsion layer of the film, producing a visible image of the radioactive sample. Any remaining silver halide which has not been reduced is subsequently solubilized by the fixing agent.

Protocol 12 explains how this procedure is carried out in practice and *Table 5* gives details relevant to each of the common radioisotopes. Briefly, the filter or gel is simply placed in contact with a sheet of suitable X-ray film under 'safelight' illumination in a darkroom. Commercially available film cassettes are generally used as containers because they are light-proof and ensure good contact between the sample and the film. After the required period of exposure, the cassette is opened in the darkroom and the film developed. Exposure is generally at room temperature, although there is some evidence that storage at low temperature ($-70°C$) decreases background on long (> 1 week) exposures.

This basic procedure is often termed 'direct autoradiography' to distinguish it from fluorography (Section 2.5.2). Direct autoradiography can be employed with all radioisotopes commonly used in molecular biology and biochemistry (i.e. [3]H, ^{14}C, ^{35}S, ^{32}P, ^{33}P and ^{125}I). It offers the benefit of good resolution and, because the amount of conversion to silver is directly proportional to the number of radioactive emissions captured, the image can be readily quantified by densitometry to determine the relative amount of radioisotope present at any point on the sample. For techniques such as DNA sequencing, where resolution is of paramount importance, it is the procedure of choice. Using either ^{35}S or ^{32}P as the label, overnight exposure of a sequencing gel to X-ray film is sufficient to produce an image of acceptable intensity.

Protocol 12

Autoradiography and fluorography: practical considerations

1 Ensure that samples are correctly prepared. With ^{3}H, ^{35}S, ^{33}P and ^{14}C, samples should be dry and unwrapped so that they can be placed in direct contact with the X-ray film. ^{32}P- and ^{125}I-labelled samples can be either dry or wet although, if the latter, they should be covered in Saran wrap to prevent leakage of moisture onto the film. In general, dried gels are recommended, particularly for fluorography. However, hybridization filters may have to be kept wet to allow re-probing. Orientation of the film and the sample can be recorded by marking the sample with radioactive ink (standard ink containing a small amount of labelled nucleotide) or a phosphorescent pen, or by cutting both the film and the sample with scissors.

2 All operations should be carried out in a darkroom, under safelight conditions. These can be achieved by masking a standard 15 W bulb with a red filter such as a Kodak Wratten 6B or GBX2, Agfa R1 or CEA 4B.

3 For direct autoradiography, place both the sample and the film in contact in an appropriately-sized film cassette. Ensure that the cassette is dry and uncontaminated. If the radioactive material is on one surface of the sample only (e.g. on a hybridization filter), then this surface should be placed in contact with the film, particularly if ^{3}H, ^{14}C, ^{33}P or ^{35}S are being detected. Similarly, if single-sided film is being used, the sample should be placed against the side containing the photographic emulsion (usually the matt side). For fluorography, the film may be pre-flashed using a modified electronic camera flash (see *Protocol 13*). For ^{32}P and ^{125}I fluorography, it is also necessary to include intensifying screens (e.g. Amersham Pharmacia Hyperscreen). If one screen is used, this is placed behind the film on the opposite side from the sample and, if two are to be used, the second screen is placed behind the sample. Again, it is necessary to ensure that the correct (matt) side of the screen is used. Close the cassette carefully to ensure a light-tight seal. Leave a fluor-impregnated sample 30–60 s under safelights before exposure to the film to allow any light-induced fluorescence to decay.

4 Store the assembled cassette for the required length of time, either at room temperature in a dark cupboard for direct autoradiography, or at $-70\,^\circ$C for fluorography. It should be stored away from any source of radioactivity, particularly ^{32}P and ^{125}I, but even the soft β-emitters, such as ^{35}S and ^{14}C, can give rise to penetrating bremsstrahlung radiation, which can blacken films within cassettes. Cassettes should therefore not be stored in close proximity to each other.

5 Prior to film development, it is advisable to allow cassettes stored at $-70\,^\circ$C to warm to room temperature before opening to avoid condensation. Alternatively, film can be removed rapidly to developer in the darkroom and the remainder of the cassette contents allowed to warm before re-exposure. A slightly longer development period may then be needed.

6 For manual development, the X-ray film is placed successively in developer, water

Protocol 12 continued

(to stop the development), and then fixer. The fixed film is finally rinsed in running water for 5 min, before being left to air-dry. The precise times for each stage are dependent on the film type and the developer and fixer used (see the manufacturer's instructions), but generally times of 2–3 min in developer, 30 s in the stop bath and 2–5 min in fixer are appropriate. Films with a high silver density (e.g. Amersham Pharmacia Hyperfilm β-max) will require longer fix periods. Vertical tanks or horizontal trays can be used, but ensure that all solutions are regularly replenished. Poor developer will give an image of reduced intensity, while poor fixer will give a cloudy film. Once the film has been fixed, normal lighting can be used. A stop solution of 3–5% acetic acid can be used in place of water after the developer stage. Developers and fixers are available from a number of suppliers including Kodak, Ilford, Dupont and Agfa.

7 A variety of automatic X-ray film processors are available from suppliers such as Kodak and Amersham Pharmacia. These include some small, semi-portable machines. They are suitable for most screen-type films and some can be adjusted to cope with the longer fixing cycle required by high silver density films.

2.5.2 Fluorography

The major disadvantage of direct autoradiography is that, for many applications, it does not give the required degree of sensitivity, and unacceptably long exposure times are sometimes necessary. This is true, for example, when using filter hybridization with a ^{32}P-labelled probe to detect a unique sequence in mammalian DNA. For weak β-emitters such as ^{3}H, and to a lesser extent ^{14}C, ^{33}P or ^{35}S, many of the emissions fail to reach the film because they are absorbed within the sample. The thicker the sample, the more severe this effect. For a strong β-emitter such as ^{32}P or an X-ray-emitter such as ^{125}I, most of the emissions are too energetic to be captured within the film and pass right through it. Both of these problems can be overcome to some extent by converting the emitted energy to light by use of a fluor. This process is termed fluorography and practical details are described along with those for autoradiography in *Protocol 12* and *Table 5*.

The weaker emitters are brought into contact with the fluor by impregnation of the sample itself. The earliest methods for polyacrylamide gels used aqueous sodium salicylate (15) or the scintillator PPO with DMSO or glacial acetic acid as the solvent (16). Although these procedures give very good results, commercially-available aqueous systems such as Amplify (Amersham Pharmacia) and Enlightening (Dupont) are less hazardous, and more rapid and convenient. *Protocol 13* gives a generalized procedure for gel fluorography. *Table 5* shows that in dried gels 1.5 mm thick the greatest enhancement of sensitivity is seen for ^{3}H, which is the weakest β-emitter. Less signal enhancement is seen for thinner samples.

Protocol 13

General procedure for gel fluorography

Reagents

- Fixing solution
- Pre-soaking solution
- Scintillant cocktail

Method

When using a commercial scintillant cocktail, follow the manufacturer's instructions.

1 Soak the gel[a] in the fixing solution for the appropriate period of time. For sodium salicylate fluorography, fix in 9:2:9 methanol–acetic acid–water for 30 min. If using PPO in DMSO, fix in the same solution for 1 h. The PPO–acetic acid system does not require a fixing step.

2 Transfer the gel to the pre-soaking solution for the appropriate period. For sodium salicylate fluorography, pre-soak in water for 30 min. For PPO-DMSO, pre-soak in DMSO for 30 min, followed by transfer to fresh DMSO for a further 30 min. For PPO-acetic acid, pre-soak in glacial acetic acid for 5 min.

3 Transfer the gel to the scintillant cocktail for the appropriate period. The cocktail is either:

   (a) 1 M sodium salicylate, pH 5.0–7.0 (use 10 gel volumes and leave for 30 min)

   (b) 22.2% PPO in DMSO (use four gel volumes and leave for 3 h)

   (c) 20% PPO in glacial acetic acid (use four gel volumes and leave for 1.5 h)

4 If using PPO, rinse the gel for 1 h in 20 volumes of water. Gels treated with sodium salicylate do not need to be rinsed.

5 Dry the gel under vacuum: at 80 °C for 2 h for sodium salicylate; at room temperature for 1 h for PPO-DMSO; at 70 °C for 2 h for PPO-acetic acid.

6 Set up the film exposure (see *Protocol 12* and *Table 5*).

[a] The protocol is designed for 1 mm gel with a polyacrylamide content of less than 15%. Increase the times for thicker or more concentrated gels, and include 2% glycerol in the sodium salicylate scintillant cocktail to avoid cracking.

For ^{32}P β-particles and ^{125}I X-rays, fluorography is achieved by use of a screen containing a dense inorganic scintillator, usually calcium tungstate, to absorb and convert to light those emissions passing through the X-ray film. A photographic image is therefore effectively superimposed upon the autoradiographic image. Enhancements in sensitivity of around tenfold can be achieved with a single intensifying screen placed behind the X-ray film. A second screen placed behind the hybridization filter or gel can produce further enhancement of up to twofold. A variety of intensifying screens are available, but blue-light-emitting screens, which produce a good balance between sensitivity and clarity of image, are generally used. Examples are Dupont Cronex and Amersham Pharmacia Hyperscreen. It is also necessary to use an appropriate film, as will be discussed below. Using such a system, it is possible to reduce the exposure time for detection of a unique mammalian sequence with a ^{32}P-probe to between 4 h and overnight.

The light generated by fluorographic techniques produces an image by a rather different mechanism than with radioactive emissions, and this necessitates the use of different exposure conditions. The conversion of several silver ions to atoms is necessary to catalyse the subsequent reduction of a halide crystal. While this can be achieved by a single 'hit' with a β-particle or γ-ray, a photon of light will produce only a single conversion. Capture of several photons within a crystal is therefore necessary. While two or more conversions within a crystal are stable, a single conversion is not and reverts back to the ionic form with a half-life of milliseconds. A double hit within this period becomes unlikely at low incident light intensity. The half-life is increased significantly at low temperature and, in practice, exposure of the sample at $-70\,^{\circ}$C is necessary to optimize sensitivity when using fluorography.

While improving sensitivity, fluorography does have two significant disadvantages. First, some decrease in resolution occurs owing to dispersion of the primary emission and the secondary scintillation. This depends on the path length of the emission and is greatest for high-energy particles with intensifying screens, as emissions travel beyond the film and are then reflected back as light. The problem is most apparent when using two intensifying screens, but is nevertheless still acceptable for most filter hybridization applications. At the opposite extreme, very little loss of resolution occurs with ^{3}H.

The second disadvantage of fluorography derives from the instability of a single ion-to-atom transition, as discussed above. Because of this, the response of the film to low levels of light is disproportionately low and is only partially corrected by low-temperature exposure. It is therefore also necessary to presensitize the film by exposing it to a brief flash of light before exposure to the sample. Pre-flashing the film, so that its absorbance at 450 nm after development is increased by 0.1–0.2 units compared with a control, effectively introduces a stable pair of silver atoms per crystal. It will thus be possible for even single photons to lead to a stable latent image. *Protocol 14* gives a procedure for pre-flashing and for calibration of an electronic flash. It should be noted that pre-flashing above 0.2 absorbance units will disproportionately increase the response of the film to low levels of light and should be avoided. The effect of pre-flashing is greatest for longer exposures. If exposures are relatively short (< 24 h) and quantification is not necessary, then pre-flashing can be omitted. As this results in suppression of less intense signals, visual resolution can appear to be enhanced, although direct autoradiography remains the best means of achieving maximal resolution. It should also be noted that direct quantification is not possible with very intense images, as the response of X-ray film is linear only within a limited range.

Protocol 14

Procedure for pre-flashing film

Equipment

- X-ray film
- Electronic flash unit (e.g. Amersham Pharmacia Sensitize Pre-Flash)
- Densitometer or spectrophotometer
- Orange filter (e.g. Kodak Wratten number 21 or 22)
- Solutions or equipment for film development

Method

1 Film should be hypersensitized by exposure to a short duration ($\approx$ 1 ms) flash of light. Such a flash is provided by conventional electronic flash units. Mains-operated units are recommended in preference to battery-operated flashes, as a more reproducible performance is obtained over a period of time.

2 The intensity of the flash is attenuated by covering the aperture with an orange filter such as a Kodak Wratten number 21 or 22 and, if necessary, further layers of white paper or partly-exposed film. The conditions for achieving optimal pre-fogging are determined experimentally, as in step 3.

3 Place an appropriate screen-type film on a table below the modified flash unit. A series of separate exposures will be required and this is most readily achieved by masking the sheet of film with opaque material, leaving only a window exposed. Alternatively, a separate piece of film can be used for each exposure. Vary the exposure by adjusting the height of the flash unit or by adding or removing layers of paper from the flash unit aperture. All exposures and conditions should be carefully recorded.

4 Develop the film as usual and measure the absorbance changes relative to an un-exposed control. This can be achieved by densitometry, or by cutting pieces of film small enough to fit into cuvettes and measuring the absorbance in a spectrophoto-meter. The absorbance of the pre-flashed film should be increased by 0.15 units at 540 nm compared with a control.

5 Confirm the optimal exposure conditions by pre-flashing and developing a whole sheet of film, and check also for uniformity of exposure by measuring the absorb-ance of a number of sections across the surface of the film. The most uniform pre-exposures are obtained by maintaining a long distance (at least 50 cm) between the flash unit and the film.

6 Monitor the exposure conditions by checking, as in step 5, initially at weekly inter-vals and then monthly, to ensure that the performance of the flash unit is constant over a period of time.

2.5.3 Technical details for autoradiography and fluorography

The low energy β-particles from ^{3}H, ^{14}C, ^{33}P and ^{35}S have a short path length and are readily absorbed within a sample, so that it is necessary to dry both the gels and the filters to maximize the signal. A layer of Saran wrap between the sample and the film is sufficient to block all ^{3}H β-particles, as well as a significant proportion of ^{14}C, ^{33}P and ^{35}S emissions, which means that direct exposure is necessary. A dry sample is therefore needed to avoid artefactual blackening of the film by moisture. Hybridization filters should be exposed with the sample side next to the X-ray film to avoid absorption of emissions by the filter itself. For high-energy emitters (^{32}P and ^{125}I), these precautions are not necessary, although resolution can be improved by drying gels as the sample is made thinner. This also avoids problems with frozen wet gels, which take longer to thaw for re-exposure, are likely to leak and, in some cases, are physically unstable (especially agarose gels). However, it is often advantageous to expose hybridization filters wet and sealed in Saran wrap, as this facilitates the subsequent stripping and re-probing.

Direct autoradiography is most efficient with a 'direct' type of X-ray film such as Kodak BioMax MR or Amersham Pharmacia Hyperfilm β-max. The latter has, in addition, emulsion with a high silver content for reduced exposure times, and this is coated on only one side of the plastic base for enhanced resolution. Such films have a lowered sensitivity to blue light, making them unsuitable for use in fluorography. A number of blue-light-sensitive films are available for use with intensifying screens or with fluor-impregnated substrates; these also give acceptable results in direct autoradiography. Included among these are Kodak BioMax MS, Fuji RX, and Amersham Pharmacia Hyperfilm-MP. The plastic base is usually either blue (e.g. Fuji RX) or clear (e.g. Hyperfilm). Film lacking the surface anti-scratch layer (e.g. Amersham Pharmacia Hyperfilm-^{3}H) is also available for detecting the low-energy β-particles of ^{3}H, which would normally be absorbed by this layer. This type of film requires very careful handling. Light-sensitive paper is also available (e.g. Amersham Pharmacia Hyperpaper) which produces a dark image on a white paper background. This has been found to give good resolution and sensitivity for direct autoradiography, particularly with ^{35}S-labelled sequencing gels. The resulting picture can be readily reproduced, for example by photocopying.

Optimal exposure times are usually determined by a mixture of experience and trial and error. For common applications, standard exposure times can be used. For example, DNA sequencing gels with either ^{35}S or ^{32}P as the label are usually exposed overnight, while filter hybridizations of mammalian DNA probed with ^{32}P for a single-copy gene are often exposed for about 4 h and overnight. A quick scan with a β-monitor for ^{35}S and ^{32}P or an X-ray monitor for ^{125}I will give an appropriate estimate of the exposure time needed. As a rough guide, *Table 5* gives an indication of the amounts of radioactivity that will produce an image of acceptable intensity after overnight autoradiography and fluorography.

Faint images on an X-ray film can be enhanced approximately tenfold in intensity by treatment of the film with [^{35}S]thiourea, which binds ^{35}S covalently to silver grains to form silver [^{35}S]sulphide (17). This method is used with developed X-ray film and can be employed long after the initial sample has decayed. Image intensification of autoradiographs can also be achieved with Kodak Chromium Intensifier.

The following is a summary of the main features of direct autoradiography and fluorography:

- Direct autoradiography
  - maximum resolution for all radioisotopes
  - exposure at ambient temperature (or $-70°C$ for low backgrounds with long exposures)
  - quantitative results
- Fluorography
  - maximum sensitivity for all radioisotopes
  - expose at $-70°C$
  - pre-flash for quantitative results and/or long exposures
  - impregnate the sample with fluor for ^{3}H, ^{14}C, ^{33}P or ^{35}S
  - use intensifying screens for ^{32}P, ^{125}I

The approach adopted depends to a large extent on the relative importance of sensitivity and resolution, and also on whether there is a requirement for direct quantification of intensity. *Table 6* summarizes some of the problems that can be encountered in autoradiography and presents the most likely causes and solutions.

2.5.4 Detection of radiolabelled DNA and RNA by phosphor imaging

An alternative to conventional autoradiography and fluorography is provided by phosphor imaging (18). This technique requires specialized equipment consisting of a phosphor-imaging screen linked to the appropriate computer and hardware. The chemical coating on the surface of the phosphor screen captures the radioactive image within a filter or gel and sends it to the computer, which displays the image on a monitor and stores the file for future use. Phosphor imaging can be used with all of the radioisotopes commonly employed for labelling DNA and RNA, although some commercial systems have different screen types optimized for different isotopes. The screen can be reused on multiple occasions, but they must be handled carefully to prevent damage.

Phosphor imaging has several advantages over the conventional film-imaging techniques. It is more rapid, with images obtained in a few hours at most, and this greater sensitivity means that most signals can be detected directly without the need for fluorography. Phosphor imaging also gives a more linear response between the amount of signal and the density of the image, which makes it more suitable for quantitative analysis. Data capture on to disc provides a convenient means of recording images.

Phosphor imaging devices are available from several companies (e.g. Bio-Rad and Molecular Dynamics) and should be operated in accordance with the manufacturer's instructions.

Table 6 Guide to avoidance of artefacts in autoradiography and fluorography

Problem	Cause	Solution
Band resolution is worse than expected	Film and substrate were not in good contact	Use a good X-ray cassette
	Poor fluorography technique or wrong choice of fluor	If using PPO, observe all washing steps. If using salicylate, change to PPO or a commercial system, or reduce the soaking time
	Inherent characteristic of ^{32}P or ^{125}I	Use another radioisotope if possible
Film is partially or wholly fogged	Poor light-tightness in the darkroom, safelight too bright or too near or has the wrong filter	Remedy or repair
	High radiation background	Carry out exposure in a low background area. Segregate cassettes for ^{32}P and ^{125}I
	Old or dirty chemicals, or film stored badly	Observe good housekeeping practice; change solutions frequently
	Radioactive contamination of the cassette	Wipe cassettes out after use
	Old or damaged cassette	Replace
	Light emission from fluor in the substrate	Allow the substrate 20–60 s 'dark adaptation' before loading into the cassette
Film shows sharply defined artefacts	Rough handling of the film before exposure and/or developing	Handle the film carefully, do not subject it to pressure. Wear cotton gloves in hot weather
	Chemical contamination of the cassette or substrate, e.g. by water, acetic acid, etc.	Wipe cassettes out after each use. Ensure that gels and plates are dry. Thaw films exposed at −70°C thoroughly before developing
	Static electricity	Remove the film from the box slowly. Avoid wearing nylon overalls. Remove the Saran wrap from dried gels at least 1 min before exposing the film

3 Non-radioactive labelling and detection methods

As outlined in Section 1.3.2, there are two distinct types of methodology for non-radioactive labelling of nucleic acids. The first approach is indirect labelling, in which the primary label has to be located by a signal-generating detector complex after hybridization. Indirect labelling is often based on the high affinity between biotin, which can be used as the primary label, and streptavidin, which is usually conjugated to the detector complex. Other indirect methods depend on recognition of a primary label such as digoxygenin by an antibody which is incorporated in the detector complex. The signal generated by the detector complex can be colorimetric and hence recorded by standard photography, or chemiluminescent, in which case a film or phosphor imaging system is used.

The second approach is direct labelling, which uses a marker that generates its own signal. The commonest examples are fluorescent labels, which have been developed over the last decade for their specialized applications in automated DNA sequencing (Chapter 6, Section 6), fluorescent *in situ* hybridization (Chapter 5, Section 4.2) and, more recently, microarray technology. Fluorescent labels are detected by CCD imagery or by confocal microscopy. Their major attraction is the availability of labels with fluorescent emissions in different parts of the visible spectrum, allowing two or more labels to be detected individually from within a mixture. This has led to single-tube DNA sequencing, in which the four separate reactions carried out during conventional chain termination sequencing are combined in a single tube; the products of the reactions are discernible because they are labelled with different fluorochromes (Chapter 6, Section 6).

Many non-radioactive labelling and detection systems are available in kit form from commercial suppliers. The descriptions provided in this chapter relate to the most popular systems but are by no means comprehensive.

3.1 Biotin as a label for nucleic acids

Biotin is a small, water-soluble vitamin which can be chemically linked to purine or pyrimidine bases. In the form of a biotinylated nucleotide, it may be incorporated into nucleic acid probes by enzymatic methods including nick translation, random or specific primer labelling, end-labelling, and *in vitro* RNA synthesis (see Sections 2.3 and 2.4). It may also be incorporated by direct means, using intercalation (19) or photo-activatible groups (20).

The first nucleotide analogues containing biotin were based on dUTP and UTP and contained biotin attached to the 5-carbon of the pyrimidine ring through an allylamine linker arm. They function as substrates for a range of nucleic acid polymerases *in vitro*, albeit with reduced reaction rates, and allow the synthesis of stable probes which can be detected with avidin-linked detection systems (21). The length of the allylamine linker arm influences the ability of the biotin incorporated to interact with avidin, and a nomenclature based on linker arm length in carbon atoms has been introduced: bio-11-dUTP, bio-16-UTP, and so on. A different chemical procedure has been used to produce dATP and dCTP nucleotide analogues containing biotin. These compounds are synthesized by the attachment of the linker arm to the amino nitrogen at the 6-position of adenine and the 4-position of cytosine.

Another approach to the incorporation of biotin into nucleic acids has been the construction of photo-activatible analogues. Photobiotin (20) is typical of this approach. It is based on aryl azide, a compound which is stable in the dark but when photo-activated generates the highly reactive aryl nitrene which reacts with the bases of nucleic acid molecules. By constructing a compound containing biotin and aryl azide separated by a linker arm, single or double-stranded DNA can be labelled simply by mixing with photobiotin in the dark and exposing to short-wavelength visible light. It is reported that one biotin molecule can be incorporated per 100–400 nucleotides without interfering with the recognition of the complementary polynucleotide, and that the label can be detected with a variety of signal-generating systems.

A chemical approach to the biotinylation of nucleic acids has been described (22), involving transamination of unpaired cytosine residues with sodium bisulphite and ethylenediamine, followed by reaction of the primary amino groups formed with a succinimide ester of biotin. The effect of this modification on base pairing, and hence on reassociation kinetics, can be minimized by limiting the proportion of cytosine residues that are modified.

Biotin-labelled probes are localized either with antibodies or, more commonly, with avidin or streptavidin, through a variety of signal-generating systems. Avidin is a 58 kDa glycoprotein with four binding sites and an extremely high affinity for biotin: the association constant for the interaction is 10^{15}/M (i.e. some 10^6 times greater than the typical affinity of an antibody for its antigen). Streptavidin is an avidin-like protein isolated from the bacterium *Streptomyces avidinii*. It is widely used instead of avidin, its main advantage being a near-neutral isoelectric point, which leads to a reduction in non-specific binding to DNA, thereby minimizing background signals.

Many signal generation systems have been used with streptavidin, attached either covalently or via one or more of the biotin binding sites, to produce a range of detection procedures for biotinylated probes. Colloidal gold (23), fluorescein (24) and even ^{35}S-radiolabelled streptavidin (C. A. Read, unpublished data) have all been used for detection of biotin labels. For filter hybridization, streptavidin labelled with the enzymes horseradish peroxidase or alkaline phosphatase are most commonly used (25). The high substrate turnover of alkaline phosphatase makes this enzyme particularly suitable for such experiments. A substrate mix consisting of BCIP and NBT is converted by alkaline phosphatase to an insoluble blue product that is deposited on the filter over a period of up to 4 h; longer incubation times tend to increase the background coloration.

Another substrate system for use with alkaline phosphatase in filter hybridization experiments is an aryl phosphate-substituted 1,2-dioxetane (26). The enzymatic cleavage of this stable compound yields an unstable aryloxide form which provides a luminescent signal. By combining this with a fluorescent enhancer system, a sustained and essentially constant light output of many hours can be produced, enabling long integrations of signal on film or in electronic light-collection devices (27). It is claimed that less than 1000 alkaline phosphatase molecules can be detected in ideal conditions using this substrate, which makes it one of the most sensitive detection systems currently available.

3.1.1 Labelling a DNA molecule with biotin

Protocols 5 (nick translation) and 6 (random primer labelling) can be modified for biotin-labelling. A biotinylated nucleotide analogue (e.g. bio-11-dUTP, bio-7-dATP) may be used in place of the radioactive nucleotide (follow the manufacturer's instructions regarding concentration). In general, a labelling procedure that gives between one and five biotin groups per 100 bp of probe will perform adequately. In most filter hybridization experiments there is no need to separate unincorporated nucleotides from the labelled probe but, if desired, this is easily achieved by adding carrier DNA and precipitating with ethanol, using ammonium acetate instead of sodium acetate. Alternatively, the labelled probe can be separated on a Sephadex G-50 column, using dextran blue (which migrates with the probe) to colour the aliquots containing labelled DNA.

The incorporation of biotin label into the probe can be checked by the procedure set out in *Protocol 15*.

Protocol 15

Checking the efficiency of a biotin-labelling reaction

Equipment and reagents

- Vacuum oven
- 10× and 2× SSC (prepare as a 20× stock: 3.0 M NaCl, 0.3 M trisodium citrate)
- Nitrocellulose filter
- Blocking buffer (100 mM Tris-HCl, pH 7.5, 3 mM $MgCl_2$, 100 mM NaCl, 0.5% Tween-20)

Method

1 Prepare several dilutions in double-distilled water of the labelled DNA from 1 ng/ml to 10 pg/ml and denature by heating at 100 °C for 2–5 min.

2 Add an equal volume of 10× SSC to each dilution and apply 2 µl spots to a piece of nitrocellulose filter.

3 Wash the filter briefly in 2× SSC, air-dry and bake for 30 min at 80 °C in a vacuum oven.

4 Incubate the filter in blocking buffer for 30 min and follow the detection procedure outlined in steps 3–7 of *Protocol 16*. An active probe will give a signal for the 100 pg dot in approximately 15 min and for the 10 pg dot within 1 h.

3.1.2 Hybridization analysis with biotinylated probes

Hybridization conditions vary depending on the base composition of the probe, the extent of homology between the probe and the target sequence, and the size of the probe. It has been reported that probes that have been biotinylated by nick translation have a lower T_m than non-biotinylated probes (21). In practice, hybridization at 65 °C in an aqueous salt solution containing 10–500 ng of biotinylated probe per millilitre will be satisfactory for most experiments. Alternatively, formamide can be added to 45% and the hybridization carried out at 42 °C. The procedures used are exactly the same as for radiolabelled probes and are described in Chapter 5, Section 3. After hybridization, the bound probe is detected as described in *Protocol 16*.

Protocol 16

Detection of signals after hybridization analysis with a biotinylated probe

Reagents

- Blocking buffer (see *Protocol 15*)

- Streptavidin–alkaline phosphatase complex or conjugate

- Dilution buffer (100 mM Tris-HCl, pH 7.5, 3 mM $MgCl_2$, 100 mM NaCl, 0.05% Tween-20)

- Detection buffer (100 mM Tris-HCl, pH 9.5, 10 mM $MgCl_2$, 100 mM NaCl)

- Substrate solution. To prepare a 10 ml working strength solution: (a) weigh out 3.3 mg NBT (Sigma N6876) into a 1.5 ml microfuge tube and add 44 μl 70% dimethylformamide in distilled water; (b) weigh out 1.65 mg BCIP (Sigma B6777) into a 1.5 ml microfuge tube and add 33 μl 100% dimethylformamide; (iii) add both solutions to 10 ml detection buffer, mix gently and use immediately

Method

1   Transfer the filter to blocking buffer and shake at room temperature for 1 h. This step inhibits non-specific binding of the detection system.

2   Dilute the streptavidin–alkaline phosphatase complex or conjugate, according to the manufacturer's instructions, in dilution buffer. Prepare approximately 1 ml of this solution per 10 cm^2 of filter.

3   Remove the filter from the blocking buffer and drain excess liquid. Place the filter, DNA side up, in a fresh dish and immediately cover it with the diluted streptavidin–alkaline phosphatase. The filter must not be allowed to dry or intrusive background will result. Gently rock the tray for 10 min at room temperature, ensuring that the solution completely covers the filter at all times.

4   Wash the filter in excess (e.g. 200 ml) blocking buffer for 5 min at room temperature with vigorous shaking. Repeat twice.

5   Wash the filter in detection buffer for 5 min, shaking at room temperature. Repeat once. This step prepares the filter for incubation with the substrate. Meanwhile, the substrate solution containing NBT and BCIP can be prepared.

6   Remove the filter from the detection buffer and drain off excess liquid. Place the

Protocol 16 continued

filter, DNA side up, in a clean dish and add the substrate solution (1.5 ml per 10 cm² of filter). Put a lid on the dish and incubate the filter in the dark at room temperature without agitation until the signal is visible (up to 4 h incubation may be required).

7 Remove the membrane from the substrate solution and rinse it in double-distilled water. A copy of the result can be obtained by photocopying the filter.

3.2 Labelling with digoxygenin

Digoxygenin is a plant steroid obtained from *Digitalis pupurea*. It is used as an indirect label in much the same way as biotin, except that the detector complex is bound to the labelled molecules via an antidigoxygenin monoclonal antibody. Kits using this system are marketed by Roche Diagnostics. Enzymatic reactions are used to incorporate digoxygenin-11-dUTP into polynucleotides, with random priming preferred to nick translation because of greater efficiency. The detector complex contains either alkaline phosphatase (in which case colorimetric detection is carried out in a manner similar to that described in *Protocol 16*) or horseradish peroxidase. As an alternative to colorimetric detection, kits based on bioluminescence and chemiluminescence are available. All of these should be used in accordance with the manufacturer's instructions.

3.3 Fluorescent labelling

Indirect labelling with biotin and digoxygenin has comparable sensitivity to radiolabelling, largely because signal enhancement occurs during the detection procedure, with each label generating multiple signals through the enzyme's catalytic activity. Direct detection of fluorescent labels is less sensitive, although some signal enhancement is possible through the use of energy transfer dyes, which interact to produce an increased fluorescence (28), and special labels that incorporate multiple fluorescing moieties have also been developed (29). In general, however, fluorescent labelling is less suitable than indirect detection methods for applications other than those particular techniques, such as DNA sequencing, where the ability to mix different fluorescent labels in a single reaction offers a distinct advantage.

Fluorescent markers can be incorporated into DNA molecules by random priming but the bulky nature of the fluorescein, rhodamine and cyanine dyes means that fluorescently-tagged nucleotides are not good substrates for all DNA polymerases (30). End-labelling is possible, including the addition of a fluorescent dideoxyNTP to the 3'-end of a polynucleotide; this is the basis of the use of fluorescent labelling in DNA sequencing (Chapter 6, Section 6). Fluorescent nucleotides also act as efficient substrates in chemical synthesis of DNA. The resulting oligonucleotides are used as labelled primers in DNA sequencing, PCR and other procedures.

3.4 Direct enzyme labelling with horseradish peroxidase

This method results in horseradish peroxidase becoming directly crosslinked to a denatured double- or single-stranded DNA to form a stable enzyme-labelled probe (31). The principle of the labelling reaction is the ability of the negatively-charged phosphate backbone of a nucleic acid molecule to bind positively-charged proteins by electrostatic interactions. Horseradish peroxidase molecules cross-linked to polyethyleneimine (a small positively-charged polymer) will interact electrostatically with DNA, and the attachment can be converted into a covalent bond by mixing in a dilute solution of glutaraldehyde. The reaction is reliable and fast, with only 20 min needed to produce a probe containing, on average, one enzyme molecule every 30–50 bp. The probe can be used in over-night hybridizations at 42 °C with little loss in enzyme activity, and remains stable when stored at −20 °C for 6 months (32).

Buffers containing formamide as a hybrid-destabilizing reagent (see Chapter 5, Sections 3.2.3 and 3.3.6) have a detrimental effect on enzyme activity in this system. Buffers containing 6 M urea have a similar strand-denaturation effect (33) but without affecting peroxidase activity during the final detection, and hence urea is used in preference to formamide. Post-hybridization washing has to be performed at or below 42 °C, and urea is again used to create the necessary denaturing conditions, with a low salt concentration used to produce a high stringency wash.

Detection of the labelled molecules can be by colorimetric assay, but the commercial product marketed by Amersham Pharmacia achieves a higher signal through the use of an enhanced chemiluminescent substrate. In a complex reaction involving many intermediates and rate-limiting steps, oxidation of the diacylhydrazide luminol by a peroxide ion, released from hydrogen peroxide by peroxidase, can be enhanced by the addition of certain phenolic compounds to give a sustained output of light at approximately 420 nm (34). The light can be captured on blue-sensitive X-ray film in a standard cassette, or on Polaroid instant film in a camera luminometer. Low-light imaging devices such as CCD cameras can be used to create an image from which accurate signal quantification and spatial measurements can be made (35). Exposure times from seconds up to an hour are sufficient to visualize single-copy mammalian genes.

Acknowledgements

We thank all our colleagues who have contributed ideas, comments and data to this review, in particular Dennis Harris and John McCombe.

References

1. Rabin, M., Uhlenbeck, O. C., Steffensen, D. M., and Mangel, W. F. (1984). *J. Virol.*, **49**, 445.
2. Thein, S. L. and Wallace, R. B. (1986). In *Human genetic diseases: a practical approach* (ed. K. E. Davies), p. 33. IRL Press at Oxford University Press, Oxford.
3. Brown, T. A. (ed.) (1998). *Molecular biology labfax*, 2nd edn, Vol. 1. Academic Press, London.

4. Rigby, P. W. J., Dieckmann, M., Rhodes, C., and Berg, P. (1977). *J. Mol. Biol.*, **113**, 237.

5. Bertera, A. L., Cunningham, M. W., Evans, M. R., and Harris, D. W. (1990). In *Principles of gene manipulation* (ed. P. Greenaway), Vol. 1, p. 99. JAI Press.

6. Feinberg, A. P. and Vogelstein, B. (1983). *Anal. Biochem.*, **132**, 6.

7. Feinberg, A. P. and Vogelstein, B. (1984). *Anal. Biochem.*, **137**, 266.

8. Krieg, P. A. and Melton, D. A. (1984). *Nucl. Acids Res.*, **12**, 7057.

9. Zinn, K., di Maio, D., and Maniatis, T. (1984). *Cell*, **34**, 865.

10. Melton, D., Krieg, P. A., Rebagliati, M. R., Maniatis, T., Zinn, K., and Green, M. R. (1984). *Nucl. Acids Res.*, **12**, 7035.

11. Cox, K. H., DeLeon, D. V., Angerer, L. M., and Angerer, R. C. (1984). *Dev. Biol.*, **101**, 458.

12. Berkner, K. L. and Folk, W. R. (1977). *J. Biol. Chem.*, **252**, 3176.

13. Cunningham, M. W., Harris, D. W., and Mundy, C. R. (1990). In *Radioisotopes in biology: a practical approach* (ed. R. J. Slater), p. 137. IRL Press at Oxford University Press, Oxford.

14. Yousaf, S. I., Cartel, A. R., and Clarke, B. E. (1984). *Gene*, **27**, 309.

15. Chamberlain, J. P. (1979). *Anal. Biochem.*, **98**, 132.

16. Bonner, W. M. and Laskey, R. A. (1974). *Eur. J. Biochem.*, **46**, 83.

17. Askins, B. S. (1976). *Appl. Optics*, **15**, 2960.

18. Johnston, R. F., Pickett, S. C., and Barker, D. L. (1990). *Electrophoresis*, **11**, 355.

19. Sheldon, E. L., Kellogg, D. E., Watson, R., Levenson, C. H., and Erlich, H. A. (1986). *Proc. Natl Acad. Sci. USA*, **83**, 9085.

20. Forster, A. C., McInnes, J. L., Skingle, D. C., and Symons, R. H. (1985). *Nucl. Acids Res.*, **13**, 745.

21. Langer, P. R., Waldrop, A. A., and Ward, D. C. (1981). *Proc. Natl Acad. Sci. USA*, **78**, 6633.

22. Viscidi, R. P., Connelly, C. J., and Yolken, R. H. (1986). *J. Clin. Microbiol.*, **23**, 311.

23. Tomlinson, S., Lyga, A., Huguenel, E., and Dattagupta, N. (1988). *Anal. Biochem.*, **171**, 217.

24. Brigati, D. J., Myerson, D., Leary, J. J., Spalholz, B., Travis, S. Z., Fong, C.K. Y. *et al.* (1983). *Virology*, **126**, 32.

25. Leary, J. J., Brigati, D. J., and Ward, D. C. (1983). *Proc. Natl Acad. Sci. USA*, **80**, 4045.

26. Schaap, A. P., Sandison, M. D., and Handley, R. S. (1997). *Tetrahedron Lett.*, **28**, 1159.

27. Pollard-Knight, D., Simmonds, A. C., Schaap, A. P., and Brady, M. A. W. (1990). *Anal. Biochem.*, **185**, 353.

28. Ju, J., Ruan, C., Fuller, C. W., Glazer, A. N., and Mathies, R.A. (1995). *Proc. Natl Acad. Sci. USA*, **92**, 4347.

29. Berlin, Y. A., Korshun, V. A., and Boreskov, Y. G. (1991). *Nucl. Acids Symp. Ser.*, **25**, 85.

30. Voss, H., Nentwich, U., Duthie, S., Wiemann, S., Benes, V., Zimmermann, J., and Ansorge, W. (1997). *Biotechniques*, **23**, 312.

31. Renz, M. and Kurz, C. (1984). *Nucl. Acids Res.*, **12**, 3435.

32. Pollard-Knight, D., Read, C. A., Downes, M. J., Howard, L. A., Leadbetter, M. R., Pheby, S. A. *et al.* (1990). *Anal. Biochem.*, **185**, 84.

33. Dutton, F. L. and Chovnick, A. (1987). *Anal. Biochem.*, **164**, 227.

34. Thorpe, G. H. G., Kricka, L. J., Moseley, S. B., and Whitehead, T. P. (1985). *Clin. Chem.*, **31**, 1335.

35. Boniszewski, Z. A. M., Comley, J. S., Hughes, B., and Read, C. A. (1990). *Electrophoresis*, **11**, 432.

Chapter 5

Immobilization of nucleic acids and hybridization analysis

N. J. Dyson

The MGH Cancer Center, Harvard Medical School, 13th Street, Charlestown, Massachusetts 02129, USA

Additional material provided by T. A. Brown

1 Introduction

Nucleic acid hybridization is an important component of many molecular biology techniques. Filter hybridization methods exploit the specificity of molecular hybridization for the detection of rare sequences in a complex mixture. The applications of filter hybridization are widespread and are a fundamental part of any analysis of gene structure or gene expression.

In filter hybridization, denatured nucleic acids are immobilized on an inert support so that reannealing of sequences within the sample is prevented but the nucleic acids are still available for hybridization. A suitably-labelled probe is added and the filter is incubated under conditions that promote hybridization between the immobilized target and the nucleic acid probe. The filter is washed until only specifically hybridized probe remains and the label is detected. The intensity of the hybridization signal is an indication of the abundance of target sequences within the sample and, in several types of analysis, the position of the signal on the filter gives further qualitative information.

The first part of this chapter contains methods for the immobilization of nucleic acids on nylon and nitrocellulose filters. Most of the remainder of the chapter deals with the hybridization analysis of these filters. Parameters that affect both the hybrid stability and rates of hybridization are briefly described. Specific protocols for the basic techniques are given together with indications of how these can be adapted for related applications.

2 Immobilization of nucleic acids on filters

2.1 Types of filter

While several types of support can be used for immobilization of nucleic acids, nitrocellulose or nylon are the most popular matrices and are suitable for most applications. Filters may be purchased from commercial suppliers in many pre-cut sizes or in sheets and rolls which can be cut to size. Because blotting and

Table 1 Properties of materials used for immobilization of nucleic acids[a]

Property	Nitrocellulose	Supported nitrocellulose	Nylon	Positively charged nylon	Activated papers
Binding capability	ssDNA, RNA, protein	ssDNA, RNA, protein	ssDNA, dsDNA, RNA, protein	ssDNA, dsDNA, RNA, protein	ssDNA, RNA
Capacity (μg bound nucleic acid cm^{-2})	80–100	80–100	400–600	400–600	2–40
Tensile strength	Poor	Good	Good	Good	Good
Mode of nucleic acid attachment	Non-covalent	Non-covalent	Covalent	Covalent	Covalent
Lower size limit for efficient nucleic acid binding	500 nucleotides	500 nucleotides	50 nucleotides/bp	50 nucleotides/bp	5 nucleotides
Suitability for reprobing	Poor because of fragility	Poor because of loss of signal	Good	Good	Good
Typical commercial products[b]	Amersham Hybond-C Schleicher and Schuell BA83, BA85 Pall Biotrace	Amersham Hybond-C extra, super Stratagene Duralose-UV	Amersham Hybond-N Dupont GeneScreen ICN Biotrans Pall Biodyne Stratagene Duralon-UV	Amersham Hybond-N+ Bio-Rad ZetaProbe Dupont GeneScreen*Plus* Pall Biodyne B, Biodyne Plus ICN Biotrans+ Schleicher and Schuell Nytran	Schleicher and Schuell ABM and APT papers

[a] Modified from Brown, T. A., *Molecular Biology Labfax*, Vol. 2, 2nd edn, pp. 2–3, (1998), permission of the publisher, Academic Press, London.

[b] This list is representative of the matrices available and is not meant to be exhaustive since new products are regularly being marketed.

hybridization are such routinely-used techniques, most laboratories have old boxes of filters. Check the quality of these supplies; it is a good general rule not to use filters that are yellowed with age, or covered in dust, that carry any surface blemishes or marks, or filters that do not wet quickly and evenly. Dirt and grease interfere with the binding of nucleic acids to filters and cause background in hybridization. Filters should not come in contact with skin and should be handled with a pair of blunt forceps. Suppliers of various types of filters are given in *Table 1*.

2.2 Choice of filter

Many of the most commonly-used applications (that is, the preparation of Southern blots, Northern blots, dot-blots, colony and plaque replica filters) can be adequately performed on either nylon or nitrocellulose filters. Approximately the same maximum sensitivities can be achieved in a Southern blot under optimal conditions with nylon, nitrocellulose or charged nylon. However, it is important to note that a protocol that is designed to give optimal results with one type of filter may give poor results on another matrix, since the binding properties of the matrices are different. Some applications exploit features of nucleic acid binding that are restricted to specific types of filter. There are three particular situations in which a particular type of filter is preferred.

(a) When the target material contains small DNA or RNA fragments (< 500 nucleotides). These are not efficiently retained by nitrocellulose filters. Nylon membranes can be used for fragments down to 50 nucleotides long. For sizes below 50 nucleotides activated papers give best results.

(b) When the blot will be sequentially hybridized with several different probes (i.e. 'reprobed'). Nylon membranes are preferred because they are more robust than nitrocellulose filters, which tend to fall apart during multiple probing. Target nucleic acids can be covalently bound to nylon membranes, a feature that reduces the leaching of target from the filter and preserves signal intensity during repeated hybridizations.

(c) When the probe has a non-radioactive label. Some non-radioactive DNA and RNA detection systems give best results on nitrocellulose filters. Check the information provided by the supplier of the label to determine the most appropriate support matrix for a particular application.

When choosing the filter, it is worth remembering that nitrocellulose is usually baked at 80 °C (preferably in a vacuum oven) to immobilize the nucleic acids. Nylon membranes can be treated in a similar way but better results are obtained by crosslinking the nucleic acid to the filter by UV irradiation. The UV source is normally a transilluminator (wavelength 312 nm gives optimal results) but a dedicated UV box can be purchased (e.g. Stratalinker from Stratagene). Positively-charged nylon can be used without the need for baking or UV crosslinking.

2.3 Characteristics of filter matrices

2.3.1 Nitrocellulose

The first studies demonstrating DNA binding to a nitrocellulose filter were described by Nygaard and Hall (1). Nitrocellulose binds single-stranded DNA extremely efficiently in solutions of high ionic strength (2,3). Hence 20× SSC (3 M NaCl, 0.3 M trisodium citrate) and 20× SSPE (3 M NaCl, 0.2 M sodium phosphate, pH 7.4, 0.02 M EDTA) are commonly used solutions for DNA transfer to nitrocellulose. The binding capacity of nitrocellulose for single-stranded DNA is very high (80 μg/cm^2), provided that the length of the nucleic acid is greater than 500 nucleotides. Nitrocellulose filters with a pore size of 0.45 μm are normally used. Smaller-sized fragments are not bound efficiently by these filters and are poorly retained. Nitrocellulose with smaller pore sizes (0.1–0.2 μm) is occasionally used with small fragments.

Despite the widespread use of nitrocellulose, the mechanism by which DNA is retained by the filter is not clearly defined. The binding of nucleic acids to nitrocellulose is not covalent and the target is continually leached off the filter during hybridization and washing (4). Binding of nucleic acids is probably the result of hydrophobic interactions and target retention is enhanced by the exclusion of water.

A major problem of working with nitrocellulose is that it is fragile and prone to damage. This is particularly true if the filter is exposed to acid or alkali. For this reason, nitrocellulose is not recommended for procedures that involve harsh treatment of the filter such as alkali transfer, or stripping the filter with alkali or by boiling.

2.3.2 Supported nitrocellulose

Attempts have been made to improve the tensile strength of nitrocellulose filters by attaching the nitrocellulose matrix to a backing film. These 'supported nitrocellulose' filters have the same nucleic acid binding characteristics of ordinary nitrocellulose, but combined with a greater rigidity that reduces breakage of the filter during reprobing. Unfortunately, signal intensity drops off rapidly during reprobing due to loss of the bound target DNA, and the support material tends to increase the amount of non-specific background hybridization. Supported nitrocellulose filters are therefore only recommended for those applications where a nitrocellulose matrix must be used in an experiment where reprobing will be necessary.

2.3.3 Nylon

Nylon membranes, like nitrocellulose filters, have a high binding capacity for single-stranded nucleic acids (up to 600 μg/cm^2 is claimed) but, unlike nitrocellulose, nylon also efficiently retains double-stranded DNA. Nucleic acids will bind to nylon in solutions of low ionic strength so that high-salt transfer solutions are no longer necessary (but are still often used).

The tensile strength of nylon membranes is a major advantage in filter hybridizations. Nylon membranes are difficult to tear and will not break during use. The nylon matrix will withstand treatment with acids, alkalis or chaotropic agents. Consequently, nylon membranes are amenable to procedures that involve harsh treatment (alkali transfer; see ref. 5) or prolonged use (repeated stripping/reprobing).

As with nitrocellulose, nucleic acids can be bound to nylon by baking, which produces a non-covalent linkage. A covalent attachment of nucleic acids can be achieved by UV irradiation on a dry nylon membrane (6). This stable association reduces the losses of target that occur by leaching during prehybridization, hybridization, or during the procedures necessary to remove hybridized probe from a membrane. As a result, nylon membranes with covalently-linked DNA show good sensitivities even after multiple stripping and reprobings. The effectiveness of UV cross-linking depends on both the wavelength and the dosage of radiation and it is important that a UV source is calibrated for hybridization efficiency (see Section 3.15). Short-wave UV causes nicking of the DNA and may result in losses of DNA fragments. Prolonged UV exposure results in over-coupling of DNA and a reduction in hybridization efficiency.

Nylon membranes efficiently retain DNA fragments down to 100 nucleotides long and have also been used successfully with small oligonucleotides.

2.3.4 Charged nylon membranes

Both positively- and negatively-charged membranes are commercially available. Nylon modified to contain positively-charged amino groups is increasingly popular for immobilization of DNA. Positively-charged nylon shares most of the DNA-binding characteristics of uncharged nylon membranes but has the advantage that, under alkaline conditions, DNA becomes covalently bound to the matrix. With the use of 0.4 M NaOH as a transfer solution, denatured DNA becomes irreversibly coupled to the membrane during transfer and additional baking or irradiation steps are unnecessary (5).

It should be noted that alkali transfer is not suitable for RNA blots since RNA is hydrolysed by prolonged alkali treatment. However, charged nylon membranes work well with RNA blots prepared by more standard techniques (7).

2.3.5 Activated papers

Activated papers are used for the covalent immobilization of DNA and RNA (8,9). Two types are commonly used: DBM cellulose and DPT cellulose. The synthesis of DBM cellulose was described by Noyes and Stark (10). These filters have a lower binding capacity (2–40 μg/cm^2) than nylon or nitrocellulose and are most suitable for procedures in which covalent coupling of short oligonucleotides or small quantities of nucleic acid is required. DBM cellulose and DBM papers have been widely used for the isolation of specific RNAs. DBM and DPT can be synthesized in the laboratory but are more usually obtained from a commercial supplier (see *Table 1*) as ABM or APT paper. These papers are activated immediately before use to produce diazo groups on their surfaces that covalently bind to amino groups (11).

Table 2 Solutions for immobilization of nucleic acids and hybridization analysis

Solution	Composition	Notes
Stock solutions		
20× SSC[a]	3 M NaCl, 0.3 M trisodium citrate, pH 7.0	20× SSC that is used in hybridization solutions must be filtered before use
20×SSPE[a]	3.6 M NaCl, 0.2 M NaH$_2$PO$_4$, 0.02 M EDTA, pH 7.7	Can be used as an alternative to SSC in all applications, but especially in formamide hybridization solutions where its higher buffering capacity is an advantage
20× SET[a]	3 M NaCl, 0.4 M Tris-HCl, pH 7.8, 20 mM EDTA	Often used instead of SSC in hybridization solutions for oligonucleotide probes
100× Denhardt's[a]	2% bovine serum albumin, 2% Ficoll (mol. wt 400 000), 2% polyvinylpyrrolidone (mol. wt. 400 000)	–
Non-homologous DNA	10 mg/ml calf thymus, salmon sperm or herring sperm DNA, in water	Prepare as described in Appendix 4, *Protocol 3*
20% SDS	–	Dissolve SDS in water overnight with stirring. Store at room temperature
1 M Sodium phosphate buffer, pH 6.8	–	Mix 25.5 ml of 1 M NaH$_2$PO$_4$ and 2.45 ml of 1 M Na$_2$HPO$_4$. Filter-sterilize and store at room temperature
Deionized formamide	–	Prepare as described in Appendix 4, *Protocol 4*
Blotting solutions[a]		
Denaturing solution	1.5 M NaCl, 0.5 M NaOH	–
Neutralizing solution	1.0 or 1.5 M NaCl, 0.5 M Tris-HCl, pH 7.0	Use 1.0 M NaCl for dot blotting, and 1.5 M NaCl for Southern blotting and preparing plaque and colony filters

Hybridization solutions[a]

Aqueous hybridization solution	5× SSC, 5× Denhardt's, 1% SDS, 10% dextran sulphate (mol. wt 500 000), 0.3% tetrasodium pyrophosphate, 100 μg/ml non-homologous DNA	Mix together: 25 ml of 20 × SSC, 5 ml of 100× Denhardt's, 5 ml of 20% SDS, 10 g dextran sulphate, 0.3 g tetrasodium pyrophosphate, 1 ml of 10 mg/ml non-homologous DNA, plus 60 ml of water. Allow the dextran sulphate to dissolve and then adjust the volume to 100 ml with water. Do not add SDS directly to SSC because it will precipitate. See Section 3.5 for a description of the functions of the components of the solution
Formamide hybridization solution	5× SSPE, 5× Denhardt's, 50% formamide, 1% SDS, 50 mM sodium phosphate, pH 6.8, 10% dextran sulphate (mol. wt 500 000), 100 μg/ml non-homologous DNA	Mix together: 25 ml of 20× SSPE, 5 ml of 100× Denhardt's, 50 ml deionized formamide, 5 ml of 20% SDS, 2 ml of 1 M sodium phosphate buffer, pH 6.8, 10 g dextran sulphate, 1 ml of 10 mg/ml non-homologous DNA, plus 5 ml of water. Allow the dextran sulphate to dissolve and then adjust the volume to 100 ml with water. Do not add SDS directly to SSPE because it will precipitate. See Section 3.5 for a description of the functions of the components of the solution
Oligonucleotide hybridization solution	6× SET, 10 × Denhardt's, 0.1% SDS	Mix together: 30 ml of 20× SET, 10 ml of 100 × Denhardt's, 0.5 ml of 20% SDS plus 59.5 ml water. Do not add SDS directly to SET because it will precipitate
Wash solutions	2× SSC, 0.1% SDS	Mix 100 ml of 20× SSC + 5 ml 20% SDS and make up to 1 l with water
	0.2× SSC, 0.1% SDS	Mix 10 ml of 20× SSC + 5 ml 20% SDS and make up to 1 l with water
	0.1× SSC, 0.1% SDS	Mix 5 ml of 20× SSC + 5 ml 20% SDS and make up to 1 l with water

[a] Batches of these solutions can be prepared in advance, aliquoted and stored at −20°C. The solutions are stable for over 6 months.

2.4 Dot-blots

The simplest type of hybridization analysis is carried out using dot-blots (also called spot-blots or slot-blots) which are used to measure the abundance of target sequences in a sample. Multiple samples are individually spotted onto a filter and the quantity of a specific sequence in each sample is determined after hybridization with a labelled probe by comparing the signal produced with control samples containing known amounts of the target. Dot-blotting technique is simple and enables the simultaneous screening of many samples relatively rapidly. The analysis can be very sensitive; using radioactive probes of high specific activity (10^8–10^9 d.p.m./μg) as little as 1 pg of target can be detected in an overnight exposure of the hybridized filter (target is defined as sequences in the sample that are complementary to the probe).

A disadvantage of dot-blots compared with Southern blots (Section 2.6) is that there is less discrimination between correct hybridization and cross-hybridization. The signal seen in a dot-blot is the sum of all the hybridizing species within the sample, whereas in a Southern blot a band of strongly-hybridizing material can be picked out of a background smear. As a result, hybridization backgrounds are higher in dot-blots and it is important that the correct controls are included.

Samples can be dotted on to a filter by hand or applied using a multiple filtration device or manifold. A manifold enables a uniform application of many samples in a short period of time. If many dot-blots are to be attempted regularly then a manifold is a worthwhile investment. Nitrocellulose, supported nitrocellulose, nylon or positively-charged nylon can be used for dot-blot analysis. Activated papers may also be used but only if the amount of nucleic acid in the sample is small (< 2 μg) because these filters have a lower binding capacity.

2.4.1 DNA dot-blots

The quantity of DNA that is loaded on to a dot-blot will depend on the relative abundance of the target DNA within the sample. Using high-specific-activity ^{32}P-labelled probes (10^8–10^9 d.p.m./μg) the limit of detection in an overnight exposure is typically 1–5 pg target DNA. Approximately 10–100 pg target routinely gives easily identifiable signals. If more than 1 ng of target is present in the sample, then the signal will be detectable in shorter exposures of the film. If the specific activity of the probe is reduced then proportionally more target must be loaded to give a signal within the same time-scales.

When the sample is human genomic DNA and the probe will recognize a single-copy gene, a minimum of 10 μg DNA should be loaded per sample. It is estimated that a single-copy sequence of 1 kb represents approximately 3 pg of a 10 μg sample of human DNA. If the target DNA is present in more copies then the amount of DNA loaded can be decreased proportionately. High molecular weight DNA is not necessary for dot-blot analysis and it is preferable to reduce the size of genomic DNA by sonication or digestion with a restriction enzyme. These treatments make it easier to estimate accurately the concentration of the DNA and facilitate binding of DNA to the filter.

Methods for the preparation of DNA dot-blots on nylon membranes are shown in *Protocols 1* and *2*. In these, DNA is denatured by boiling the sample prior to application. This treatment also partially depurinates the DNA so that subsequent alkali treatment causes the phosphodiester bond to break at the site of depurination (12). This is particularly important for samples containing supercoiled DNA which renatures very rapidly and must be linearized or nicked to reduce reannealing. When samples are spotted onto the membrane with an automatic pipette (*Protocol 1*) it is important not to press on the membrane as this may cause background signals (samples can be dotted without touching the surface of the membrane at all). The samples should be applied to give a regular pattern. It is best to keep the dots small (less than 4 mm diameter) in order to concentrate the hybridization signal.

When all the samples have been applied, the membrane is treated with denaturing solution to ensure that all the samples are fully denatured on the membrane. The membrane is quickly neutralized and the DNA immobilized in its denatured state. This extra denaturation step usually improves the signal obtained in hybridization and is recommended for several reasons. First, some renaturation of DNA may occur during application of the sample to the membrane. This is particularly the case when the sample contains plasmid DNA. Second, when many samples are being analysed, their application may take considerable time and the amount of reassociation that occurs after boiling may vary between samples. Third, the alkali treatment assists in the nicking of supercoiled plasmid DNA.

Protocols 1 and *2* can be modified for use with nitrocellulose filters. Since high-salt solutions are normally used for binding of DNA to nitrocellulose, the filter must be wetted in water and equilibrated in 20× SSC. Samples are denatured by boiling and snap-cooled on ice, and an equal volume of 20× SSC is added to the sample before the sample is applied to the filter. The filter is then treated with denaturing (1.5 M NaCl, 0.5 M NaOH) and neutralizing solutions (1.5 M NaCl, 0.5 M Tris-HCl, pH 7.0), as described in *Protocol 1*. The use of high-salt solutions is not recommended with positively-charged nylon membranes. Instead, DNA can be dotted onto and covalently bound to positively-charged nylon membranes in 0.4 M NaOH.

Protocol 1

DNA dot-blots; manual application to a nylon membrane

Equipment and reagents

- UV transilluminator
- Nylon membrane
- Whatman 3MM paper
- Boiling water bath
- 20× SSC (see *Table 2*)
- Denaturing solution (see *Table 2*)
- Neutralizing solution (see *Table 2*)

Protocol 1 continued

Method

1   Cut a piece of nylon membrane to the appropriate size and lightly pencil a grid of squares with 6 mm sides.[a]

2   Add 20× SSC to the DNA sample to give a final concentration of 6× SSC. Denature the DNA by incubating the sample in a boiling water bath for 10 min. Snap-cool the DNA on ice.[b] Spin the tubes for 10 s in a microfuge to ensure that all the sample is at the bottom of the tube. Keep the samples on ice.

3   Wet the membrane with 6× SSC and lay the membrane on a piece of 3MM paper to remove any liquid from the surface. Lay the membrane on the lid of a plastic box so that only the edges of the membrane are in contact with the lid.

4   Apply each sample in a 2 μl volume to the membrane. If the sample is in a larger volume, then it must be applied in repeated 2-μl aliquots allowing the membrane to dry between the repeated applications.

5   When all the samples have been spotted, place the membrane on a stack of three sheets of 3MM paper that are saturated with denaturing solution and incubate for 10 min. The membrane will also become wet with denaturing solution (but should not become submerged). Transfer the membrane to a stack of 3MM paper soaked in neutralizing solution and incubate for a further 5 min.

6   Allow the membrane to dry. The DNA can now be immobilized either by baking at 80 °C for 2 h, or (more preferably) by UV crosslinking on a transilluminator.

[a] Membranes should be handled with gloved hands and a pair of blunt forceps. Grease from fingers interferes with DNA binding and causes high backgrounds on hybridization.

[b] The lids of some types of microfuge tube need to be pierced with a needle to prevent the tubes popping open in boiling water.

Protocol 2

DNA dot-blotting to a nylon membrane: use of a manifold

Equipment and reagents

- Dot-blotting manifold apparatus
- Water pump
- UV transilluminator
- Nylon membrane
- Whatman 3MM paper

- Boiling water bath
- 20× SSC (see *Table 2*)
- Denaturing solution (see *Table 2*)
- Neutralizing solution (see *Table 2*)

Method

1   Cut a piece of nylon membrane and a piece of 3MM paper to the size of the manifold and wet both pieces of paper in 6× SSC.

Protocol 2 continued

2   Lay the paper on the base of the manifold and place the membrane on top of it. Put the top section of the manifold in place.

3   Dilute the samples to a volume of 200–400 μl (depending on the depth of the wells in the manifold), with a final concentration of 6× SSC and denature the DNA by incubating the sample in a boiling water bath for 10 min. Snap-cool the DNA on ice. Spin the tubes for 10 s in a microfuge to ensure that all the sample is at the bottom of the tube. Keep the samples on ice.

4   Apply low suction to the manifold from a water pump. Adjust the pressure so that it takes approximately 5 min for 400 μl of 6× SSC to be drawn through the wells. Leave the pump on.

5   Add the samples to the wells.

6   Dismantle the apparatus. Treat the membrane with denaturing and neutralizing solutions and immobilize the DNA as described in *Protocol 1*, steps 5–6.

2.4.2 Controls for DNA dot-blots

It is essential to include both positive and negative control samples in dot-blot experiments. Negative control samples that lack the target sequence should be prepared by the same method as test samples when analysing cloned DNA; a sample of 10 μg non-homologous DNA (see *Table 2*) is a useful negative control. Positive control samples that contain known quantities of target sequences should also be used. Typically, this might be serial dilutions of unlabelled probe DNA in 6× SSC and carrier DNA. Note that if the samples are crude preparations then the quality of the DNA in the sample may be much poorer than the purified probe DNA, leading to incorrect quantification. Sample impurities can either inhibit DNA hybridization (leading to false-negative results) or non-specifically trap nucleic acids (leading to false-positive results).

2.4.3 RNA dot-blots

Dot blots can also be used for simple hybridization analysis of RNA samples. The principle of the method is the same as that used for DNA dot-blots, but because of the different properties of RNA the protocol is slightly different. Although RNA is single-stranded it is capable of forming stable secondary structures that may need to be broken to enable efficient hybridization with a nucleic acid probe. RNA can be denatured in several ways (see Section 2.7); the method in *Protocol 3* uses a mixture of formamide and formaldehyde. After denaturation, salt is added to the solution to facilitate the binding of RNA to the nylon membrane. The membrane is not treated before the RNA is immobilized by baking or UV crosslinking.

Protocol 3

RNA dot-blots: manual application to a nylon membrane

Equipment and reagents

- Formamide–formaldehyde solution (500 μl deionized formamide + 162 μl 37% formaldehyde + 100 μl 10× MOPS buffer. For preparation of deionized formamide, see Appendix 4, *Protocol 4*. 10× MOPS buffer is 0.2 M MOPS, pH 7.0, 0.5 M sodium acetate, 0.01 M EDTA)

- UV transilluminator
- Nylon membrane
- 20 × SSC (see *Table 2*)

Method

1 Cut a piece of nylon membrane to the appropriate size and lightly pencil a grid of squares with 6 mm sides.[a]

2 Place up to 20 μg of RNA in 2 μl of water in a microfuge tube and add 6 μl of formamide-formaldehyde solution.

3 Incubate the samples at 65 °C for 5 min. Snap-cool the tubes on ice and add 8 μl of 20× SSC.

4 Wet the membrane in 10× SSC and place it on the lid of a plastic box so that only the edges of the membrane are in contact with the lid. Apply the samples in 2-μl aliquots so that the diameter of the spot does not exceed 4 mm.[b]

5 Allow the membrane to air-dry then immobilize the RNA by baking or, preferably, by UV irradiation on a transilluminator. The membrane is ready for prehybridization or may be stored at room temperature in a desiccator.

[a] Since RNA applied in formaldehyde solutions is not efficiently retained by nitrocellulose, this method is not recommended for use with nitrocellulose filters. RNA can be denatured with glyoxal and applied to nitrocellulose as described by Thomas (13).

[b] Samples can also be applied using a manifold. After denaturation, dilute the samples to 200–400 μl with ice-cold 10× SSC and apply the samples as described in *Protocol 2*.

2.5 Preparation of plaque and colony replicas on nitrocellulose

A recombinant library contains a large number of clones, each of which contains a different sequence. The clones containing the sequence of interest usually represent only a small proportion of the population and it is necessary to screen large numbers of clones in order to purify and isolate the desired one. In the absence of any selectable biochemical function, either nucleic acid probes or antibodies can be used to identify a recombinant clone. Normally, the screen is performed on bacterial colonies that contain plasmids or cosmids, or on bacteriophage plaques. The library is spread out on agar plates and, when the colonies or plaques have grown to a suitable size, a replica of the plate is prepared on a nitrocellulose or nylon membrane. In nucleic acid screening, the DNA from the library is denatured, immobilized to the filter and hybridized with a labelled probe. Each hybridized filter contains DNA from many different clones. From the position of the hybridization signal it is possible to identify the position of the (potential) positive clone on the original plate.

The initial screen is usually carried out with a high density of clones. As a result, potential positives are often closely surrounded by undesired clones and further screening is necessary to purify a single clone. In secondary screening, recombinants are plated at low density so that each clone is easily identified and can be picked with less chance of contamination. Typically, several rounds of screening are performed until a single clone is identified.

Plaque and colony replica filters can be prepared on nylon or nitrocellulose using similar protocols. The filters can be purchased in the appropriate sizes and sterilized before use. The following sections contain methods for the preparation of replica filters on nitrocellulose for screening of bacteriophage, plasmid or cosmid libraries. Factors influencing the choice and construction of these libraries are described in Chapters 2 and 3.

2.5.1 Plaque screening

The procedure for the preparation of replicas of a bacteriophage λ library on nitrocellulose filters (*Protocol* 4) is based on the method described by Benton and Davis (14). In this protocol host bacteria are plated in a thin layer of agarose on the surface of normal agar plates. Phage are allowed to multiply and form plaques in the bacterial lawn. A replica of the plate is prepared on a nitrocellulose filter. Intact phage and unpackaged DNA are adsorbed by the filter which is then treated with alkali to lyse the phage particles and to denature the DNA. The single-stranded DNA is able to bind to the filter in high-salt solutions and the filter is neutralized to maintain its integrity. The filter is equilibrated with 6× SSC to remove excess salt and the DNA is immobilized by baking the filter under a vacuum. Filters prepared in this way are then hybridized with a labelled probe (see Section 3.10), and positive plaques picked from the original plate.

Phage are plated to give a plaque density approaching $200/cm^2$. Choosing the correct size of plaque is one of the most important parameters. Plaques must be sufficiently large to contain DNA, but if they are too big then the signal is difficult to purify later. Plaques will appear within 6–10 h after plating and should be clearly visible but not confluent when replicas are taken. It is best not to allow phage to grow overnight but to chill the plate at 4°C and continue growth the next day.

Before a replica is taken, the plate is incubated at 4°C for 30–60 min to help prevent the agarose from sticking to the filter. It is essential that the correct orientation of the filter on the plate is unambiguous. Some workers dip a needle in India ink to make the orientation marks clear. Two filters should be hybridized from each plate to help eliminate background spots. Up to five replicas can be taken per plate, but the signal on the last replica is often considerably reduced compared with the first. It is a good idea to increase gradually the length of time that the filter is in contact with the plate by roughly 1 min for each replica filter.

The denaturation and neutralization steps are carried out by placing the filter sequentially on saturated pads of 3MM paper. During this treatment no pools of liquid should appear on the surface of the filter, as this can lead to diffuse or smeary spots. Neutralization should bring the nitrocellulose to a neutral pH; if the pH remains high then the filter will become brittle during baking and may disintegrate later. The filter is equilibrated with $6\times$ SSC to remove excess salt before baking. The filter can be stored dry between two sheets of paper but is normally hybridized shortly after preparation. The agar plate is stored at $4\,^{\circ}C$.

A similar method to that described for bacteriophage λ can also be used for screening the male-specific bacteriophage M13. Since M13 produces a single-stranded DNA molecule denaturation and neutralization steps are unnecessary and can be omitted. Note that the probe DNA must contain sequences complementary to the DNA strand packaged by the phage.

Protocol 4

Preparation of plaque replica filters on nitrocellulose

Equipment and reagents

- Vacuum oven
- 22×22 cm nitrocellulose filters
- Whatman 3MM paper
- BBL top agarose with 10 mM $MgSO_4$. BBL top agarose is 180 mg BBL trypticase peptone, 50 mg NaCl, 60 mg agarose per 10 ml. Check the pH and adjust to 7.0–7.2 with NaOH. Sterilize by autoclaving at $121\,^{\circ}C$, 103.5 kPa (15 lbf/in^2), for 20 min. See Appendix 4 for preparation and use of $MgSO_4$ supplements

- L Dil buffer (10 mM Tris-HCl, pH 7.6, 10 mM $MgSO_4$, 1 mM EDTA)
- Plating cells (see Chapter 2, *Protocol 10*)
- BBL agar plates (18 g BBL trypticase peptone, 5 g NaCl, 6 g Bacto-agar per 1 l). Check the pH and adjust to 7.0–7.2 with NaOH. Sterilize by autoclaving at $121\,^{\circ}C$, 103.5 kPa (15 lbf/in^2), for 20 min
- Denaturing solution (see *Table 2*)
- Neutralizing solution (see *Table 2*)
- $20\times$ SSC (see *Table 2*)

Method[a]

1. Take an aliquot of phage containing approximately 100 000–150 000 p.f.u. in no more than 2 ml L Dil, add 2 ml plating cells, adsorb for 15 min at $37\,^{\circ}C$, and add 30 ml BBL top agarose with 10 mM $MgSO_4$ that has been pre-cooled to $42\,^{\circ}C$.

2. Plate on prewarmed 23×23 cm BBL plates.[b] Incubate the plates at $37\,^{\circ}C$ until plaques cover each plate but are not confluent. Move the plates to a cold room and incubate at $4\,^{\circ}C$ for at least 1 h before applying the filter.

3. Label the filter with a ballpoint pen and carefully place the filter on the agar surface. It is best to lay one edge of the filter on to the agar and, avoiding air bubbles, slowly lower the rest of the filter as it wets. Do not move the filter on the plate. If there are any problems, lift the filter off and start again.

4 Leave the filter on the plate for 1–5 min. Stab a needle through the filter at several asymmetric points to record the position and orientation of the filter on the plate.

5 With a pair of blunt forceps, remove the filter slowly from the plate and place it face up on a clean piece of filter paper. Up to five replicas can be taken from each plate.

6 Place the filter on a stack of three sheets of 3MM paper that are saturated with denaturing solution and incubate for 5 min. The 3MM paper should be wet enough to allow immediate saturation of the filters but not so wet that pools of solution appear on the filter. Transfer the filter to a similarly prepared pad of 3MM paper saturated in neutralizing solution. After 5 min, transfer the filter to a stack of 3MM paper saturated in 6× SSC for 2 min.

7 Place the filter on a clean and dry sheet of 3MM paper and allow it to air-dry. Bake for 2 h at 80 °C under vacuum and use immediately or store in a cool dry place.[c]

[a] See also Chapter 2, Protocol 14.

[b] Plates and filters with other dimensions can be used. For 82-mm diameter plates/filters, use 0.2 ml of plating cells, 5000 p.f.u. of phage and 3 ml top agarose; for 150-mm diameter plates/filters use 0.5 ml of plating cells, 20 000–30 000 p.f.u phage and 7 ml of top agarose.

[c] Nylon membranes can be baked or UV-irradiated to immobilize the DNA.

2.5.2 Screening bacterial colonies

The procedure for the screening of bacterial colonies (*Protocol 5*) is based on the methods of Grunstein and Hogness (15) and Hanahan and Meselson (16). Bacterial colonies are grown on agar plates and replica filters are prepared on nitrocellulose. These filters are placed on agar plates and the bacteria grown until the colonies are ready for screening. The bacteria are lysed and denatured DNA is immobilized on the filter, using the same procedure as for plaque filters. Replica filters for each plate are then hybridized with a labelled probe (Section 3.10).

Protocol 5

Preparation of replica filters of bacterial colonies on nitrocellulose [a]

Equipment and reagents

- Vacuum oven
- Nitrocellulose filters
- Whatman 3MM paper
- Denaturing solution (see *Table 2*)
- Agar plates containing the appropriate antibiotic(s) for selective growth of the *E. coli* host strain (see Appendix 4)
- Agar plates containing 50 μg/ml chloramphenicol. Prepare a 50 mg/ml stock solution of chloramphenicol, filter-sterilize and add the appropriate amount to autoclaved media cooled to 50 °C, immediately before use
- Neutralizing solution (see *Table 2*)
- 20× SSC (see *Table 2*)

Protocol 5 continued

Method

1 Prepare agar plates containing the appropriate antibiotic and plate out the plasmid or cosmid library.[b]

2 Incubate the plates at 37°C until the colonies are just visible.

3 Select the appropriate filter size.[c] Label the filter with a ballpoint pen and carefully place the filter on the agar surface. It is best to lay one edge of the filter on the agar and, avoiding air bubbles, slowly lower the rest of the filter as it wets.

4 Leave the filter on the plate for 1–5 min. Stab a needle through the filter at several asymmetric points to record the position and orientation of the filter on the plate.

5 With a pair of blunt forceps, slowly remove the filter from the plate and place it, colony side up, on a fresh agar plate containing the appropriate antibiotic. Avoid trapping air bubbles between the filter and the agar. Repeat the procedure to take further replica filters of the plate. Store the master plate at 4°C.

6 Incubate the nitrocellulose filters at 37°C until colonies are approximately 1 mm in diameter, but are clearly distinct. The growth will be quicker for the first replica filter than for later replicas.

7 After the colonies have grown, cosmid or plasmid copy number may be amplified by transferring the nitrocellulose filter to an agar plate containing 50 µg chloramphenicol and incubating for 4–10 h at 37°C. This step is useful for some vectors, such as pBR322, but is not necessary for vectors with a naturally high copy number (e.g. the pUC series).

8 Remove and treat the filter as described in *Protocol 4*, steps 5–7.

[a] See Chapter 2, *Protocol 15* for an example of the use of this method in screening a cosmid library.

[b] A 10 cm plate can hold up to 10 000 uniformly distributed colonies.

[c] It is important that the nitrocellulose discs are detergent-free and sterile. Nitrocellulose filters can be wrapped in aluminium foil and autoclaved.

2.6 Southern blotting

Southern blotting is the name given to a technique, originally described by Southern (2), for the transfer of DNA from a gel to a filter. The term 'Southern blotting' is commonly used to encompass DNA transfer from any type of gel to any type of filter matrix. DNA that is separated by electrophoresis and transferred in this way can be analysed by hybridization with a suitably-labelled nucleic acid probe. A resulting specific signal not only gives a measure of the quantity of a specific sequence within the sample but also gives information regarding the size of DNA fragments carrying the target sequence. Southern blotting methods are described below.

2.6.1 DNA transfer to nitrocellulose in high-salt solutions

A procedure for DNA transfer from agarose gels to nitrocellulose is given in *Protocol 6*. Briefly, the procedure is as follows. DNA samples are separated by electrophoresis in an agarose gel. The DNA is stained with ethidium bromide and the gel photographed. The DNA is depurinated and denatured within the gel prior to transfer. This is accomplished by treating the gel with acid and alkali solutions. Finally, the gel is equilibrated in a neutral pH, high-salt buffer. Transfer of the DNA to nitrocellulose is achieved by capillary diffusion of a high-salt solution through the gel and filter into a stack of absorbent towels. Provided that diffusion is uniform through the gel the spatial distribution of DNA fragments that is produced by electrophoresis is maintained on the nitrocellulose filters. After transfer, the filter is rinsed and baked to immobilize the DNA.

A detailed description of agarose gel electrophoresis is given in Volume I, Chapter 5. The percentage of agarose within the gel will depend on the size of DNA fragments to be detected and may vary from 0.5 to 2.5%. The type of running buffer used in the gel is not important provided that the DNA bands are well resolved. DNA transfer is likely to be reduced from gels that are excessively thick (> 7–8 mm) or contain a high percentage of agarose. Thin agarose gels (2–3 mm thick) or low percentage gels allow more rapid transfer but are more difficult to handle and require more care when setting up the blot.

The quantity of DNA that needs to be loaded on the gel will depend on the relative abundance of the target sequence within the sample. Using radioactive probes of high specific activity (10^8–10^9 d.p.m./μg), between 1 and 100 pg of target DNA is routinely detectable in an overnight exposure. Note that a 1 kb sequence present in single copy in human genomic DNA comprises 3 pg of a 10 μg sample. If a cloned piece of DNA is to be analysed then the quantity of target DNA that is immobilized may be in considerable excess over the amount of labelled probe. In circumstances such as this, where the target DNA is abundant (> 10 ng per band), probes with lower specific activity (10^6 d.p.m./μg) are sufficient.

Protocol 6

Southern blotting to nitrocellulose

Equipment and reagents

- Agarose gel containing 0.5 μg/ml ethidium bromide
- Shaking platform
- Blotting apparatus (see *Figure 1*)
- Small weight (≈ 500 g)
- Vacuum oven
- Nitrocellulose filter

- Whatman 3MM paper
- Paper towels
- 0.25 M HCl
- Denaturing solution (see *Table 2*)
- Neutralizing solution (see *Table 2*)
- 20× SSC (see *Table 2*)

Method

1 Preparing the gel: electrophorese the DNA samples in an agarose gel containing 0.5 μg/ml ethidium bromide and photograph the DNA bands.

2 Depurination: rinse the gel with distilled water and place it in a glass dish with 0.25 M HCl. Use between five and 10 times the gel volume for this and subsequent washing steps. Leave the gel for 30 min at room temperature on a shaking platform set at low speed.

3 Denaturation: remove the HCl and rinse the gel with distilled water. Add the appropriate volume (see step 2) of denaturing solution and incubate at room temperature with shaking for 20 min. Pour off the solution and replace with fresh denaturing solution. Incubate for a further 20 min, shaking gently.

4 Neutralization: remove the denaturing solution and rinse the gel with distilled water. Add the appropriate volume of neutralizing solution and incubate at room temperature with shaking for 20 min. Remove the solution and replace with fresh neutralizing solution. Incubate for a further 20 min, shaking gently.

5 Preparing the blotting apparatus: a typical blotting apparatus is shown in *Figure 1*. Fill the reservoir with at least 500 ml of 20× SSC and set up a flat platform. Wet three sheets of 3MM paper in 20× SSC and use to cover the platform. The ends of the 3MM paper should dip into the 20× SSC reservoir and serve as a wick. Smooth the 3MM paper over the platform with a gloved hand by rolling a clean pipette over it. This removes any air bubbles that may be trapped. Measure the gel and cut a piece of nitrocellulose to the exact size, using a pair of sharp scissors or a scalpel blade. Wet the filter by laying it on the surface of distilled water in a glass container. The filter should wet quickly and evenly. Shake the container to submerge the filter. Pour off the water and soak the filter in 20× SSC. Cut five sheets of 3MM paper to the same size as the filter.

6 Assembling the blot: place the neutralized gel on the saturated 3MM paper on the level platform and remove any air bubbles that may be trapped between the paper and the gel by rolling a clean pipette over the surface of the gel or by gently pressing the gel with a gloved hand. Take the wetted filter from the 20× SSC container and place it on the top of the gel. Remove any bubbles that may be trapped between the filter and the gel. Place Saran wrap or parafilm around the edges of the gel so that no part of the nitrocellulose filter can directly contact the 3MM paper wick. Place the five sheets of dry 3MM paper on top of the filter and stack paper towels to a thickness of 4 cm on top of the 3MM paper. Put a glass plate on top of the paper towels and place a small weight (≈ 500 g) on top of the plate. Leave the blot overnight.

7 Disassembling the blot: remove the weight, paper towels, and 3MM paper; the gel will have compressed leaving a thin gel/filter sandwich. With a pencil, mark the position of the wells on the filter and cut one corner to allow orientation of the gel. Carefully peel off the filter and gently submerge it in a solution of 6 × SSC in a glass

container. Leave for 3 min, then lay the filter on a clean piece of 3MM paper to dry. The gel should be restained with ethidium bromide to check that most of the DNA has been transferred.

8   Immobilizing the DNA: after allowing the filter to air-dry, place it between two sheets of 3MM paper and bake at 80 °C in a vacuum oven for 2 h. The filter is now ready for prehybridization, but may also be stored in a dark, dry place.

Gels are stained with ethidium bromide and photographed before transfer to record the position of marker DNA bands of known size. It is a good idea to lay a ruler alongside the gel so that the distance migrated by each band from the well is clear on the photograph. By comparing the position of these bands with the position of bands subsequently located by hybridization, the size of DNA fragments that contain the target sequence can be estimated. Ethidium bromide staining will not significantly reduce the hybridization efficiency of DNA, but has been reported to reduce transfer efficiencies of RNA (17).

2.6.1.1 Depurination

Depurination of DNA occurs after exposure to hydrochloric acid. Acid treatment of the gel is often used to reduce the size of DNA since small DNA fragments are more easily transferred (9). This treatment is essential for DNA fragments that are greater than 5 kb, which would otherwise diffuse poorly through the gel. Some workers prefer to use UV irradiation to shear DNA as an alternative to depurination (18). Irradiation is a quicker alternative but may be less reproducible as the strengths and wavelengths of transilluminators vary considerably. The effectiveness of depurination (or irradiation) can be checked by transfer of end-labelled DNA markers. Poor transfer of large DNA fragments indicates that the depurination treatment is insufficient. However, if the depurination is too long then the DNA may be broken into very small fragments that are poorly retained by the filter and only the larger end-labelled fragments will be detectable on the filter.

It is often helpful to watch the colour of the xylene cyanol and bromophenol blue marker dyes during the depurination and denaturation steps. When the gel is fully equilibrated in the acid solution the dyes change from their normal light and dark blue colours to light green and bright yellow, respectively. The optimum length of time for depurination is to wait until the dyes change colour and incubate for a further 10 min. Although the length of this time will depend on the composition and thickness of the gel, the step usually totals 25–30 min. When the gel is treated with alkali solution, the dyes return to their normal blue colours.

2.6.1.2 Denaturation

Denaturation is necessary to produce single-stranded DNA that will be able to hybridize with the complementary DNA probe. Incompletely denatured DNA will provide a poor target for hybridization and, in order to ensure that denaturation is complete, the denaturation step is usually a little longer than the depurination step. There is no upper limit for the length of the denaturation step, but prolonged times will lead to diffusion of the DNA within the gel, producing fuzzy bands. Gloves need to be worn while handling the gel; a water-driven aspirator is a convenient way to remove the various solutions from the gel.

2.6.1.3 Neutralization

Neutralization of the gel is necessary because DNA is not retained by nitrocellulose at pH values greater than 9.0. It is a good idea to change the neutralizing solution on the gel at least once. The pH of the gel can be checked at the end of the treatment by pressing a pH stick against the dry surface of the gel. If the pH is not less than 9.0, then the neutralization step should be continued. As with the denaturation step, there is no upper limit to the length of the treatment but excessive times cause diffuse bands.

During the denaturation and neutralization steps, the gel is equilibrated with 1.5 M NaCl. High ionic conditions are necessary for efficient retention of DNA by nitrocellulose.

2.6.1.4 DNA transfer

An assembled blot allows transfer buffer (20× SSC) to be drawn from the reservoir through the gel and filter into the stack of absorbent paper. As the buffer flows, DNA is transferred by capillary diffusion from the gel and is retained by the nitrocellulose filter. The rate of DNA transfer varies considerably depending on the percentage and thickness of the gel and the size of the DNA fragments. Most workers find it convenient to allow DNA transfer to occur overnight. However, when low percentage gels are used and the DNA fragments are relatively small, more than 90% of the DNA may be transferred within the first 2 h (19).

Figure 1 illustrates a typical blotting set-up. It is important to ensure that the reservoir tray is clean and free of detergents. Rapid and effective DNA transfer is helped by the use of highly-absorbent paper towels. The purpose of the weight on top of the stack is to keep close contact between the various layers of the blot. Under an excessively heavy weight, DNA becomes trapped in a crushed gel and is not efficiently transferred. When setting up the blot, it is important to remove any air bubbles between the gel and the filter that would otherwise distort DNA transfer. Surrounding the gel with parafilm or Saran wrap prevents transfer buffer from short-circuiting between the wick and the absorbent stack.

After transfer, the position of the gel slots can be marked with a blunt pencil either from the depressions from the wells on the nitrocellulose filter or by turning the gel and filter over and marking through the gel slots. The filter is then rinsed briefly in 6× SSC to remove excess salt. If the salt is not removed, the filter becomes very brittle on baking. The filter should still be a clear white at this stage. Patches of yellow on the filter that have appeared during the blot indi-

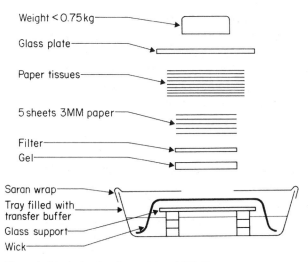

Weight < 0.75 kg
Glass plate
Paper tissues
5 sheets 3MM paper
Filter
Gel
Saran wrap
Tray filled with transfer buffer
Glass support
Wick

Figure 1 A typical set-up for nucleic acid blotting.

cate that the gel was incompletely neutralized before transfer. The gel should be restained with ethidium bromide to confirm that there is no DNA left in the gel. Large quantities of DNA stained with ethidium bromide sometimes produce a faint pink colour on the filter.

2.6.1.5 Baking and storage

Nitrocellulose filters are air-dried, placed between 3MM paper and baked at 80 °C to immobilize the DNA, which becomes more firmly bound by the exclusion of water. Nitrocellulose is inflammable and has the potential to explode if ignited under suitable conditions. The filters should therefore be baked under vacuum. Prolonged baking does not improve hybridization performance but causes the filters to become yellow and brittle. Baked filters can be stored for several months in a dry, dark place. For longer storage, filters are best kept in a desiccator.

2.6.2 Alkali blotting to positively-charged nylon membranes

Nylon membranes have several advantages over nitrocellulose for the preparation of blots. One method that exploits the properties of nylon membranes for DNA transfer is the alkali-blotting procedure originally devised by Reed and Mann (5). In alkali blotting, the DNA is transferred, preferably to a positively-charged nylon membrane, using 0.4 M NaOH as the transfer solution. Alkali blotting is not suitable for DNA transfer to nitrocellulose or for RNA transfers. Nitrocellulose will not retain DNA at a pH greater than 9.0 and becomes very fragile after prolonged alkali treatment. RNA is hydrolysed by NaOH and cannot withstand the transfer conditions. However, DNA can be blotted to uncharged nylon membranes in alkali conditions, but only if the alkali concentrations are reduced to 0.25 M and if 1.5 M NaCl is added to the transfer solvent (N. Dyson and A. McKie, unpublished observations).

A method for alkali transfer is given in *Protocol 7*. The major differences from the method used for transfer to nitrocellulose are that no neutralization step is necessary and 0.4 M NaOH is used as the transfer solution in place of 20× SSC. Because nylon membranes are hydrophilic, they do not need to be pre-wetted before use and a dry membrane can be placed directly on to the gel. Depurination of the DNA by acid treatment of the gel is still necessary for efficient transfer of large fragments, although DNA transfer in alkali is generally quicker and poor transfer of large fragments is less of a problem. The majority of the DNA is transferred within the first 2 h of blotting, although the blot may be left overnight if this is more convenient.

Protocol 7

Alkali blotting to a positively-charged nylon membrane

Equipment and reagents

- Agarose gel containing 0.5 μg/ml ethidium bromide
- Shaking platform
- Blotting apparatus (see *Figure 1*)
- Small weight (≈ 500 g)
- Positively-charged nylon membrane
- Whatman 3MM paper
- Paper towels
- 0.25 M HCl
- 0.4 M NaOH
- 20× SSC (see *Table 2*)

Method

1 Electrophorese the DNA samples in an agarose gel containing 0.5 μg/ml ethidium bromide and photograph the gel on a transilluminator.

2 Depurinate the DNA by treating the gel with 0.25 M HCl as described in Section 2.6.1.1 and *Protocol 6*, step 2.

3 Rinse the gel with water and denature the DNA by soaking the gel in five volumes of 0.4 M NaOH for 20 min.

4 Set up the transfer apparatus as shown in *Figure 1* and described in *Protocol 6*, steps 5–6, using 0.4 M NaOH as the transfer solution and a positively-charged nylon membrane.[a] Allow the transfer to take place for 2 h to overnight.

5 Disassemble the blot. With a pencil, mark the position of the well slots. Carefully peel off the membrane and gently submerge it in a solution of 2× SSC in a glass container. Lay the membrane on a clean piece of 3MM paper. The gel should be restained with ethidium bromide to check that most of the DNA has been transferred. The membrane may be prehybridized or dried and stored between sheets of 3MM paper in a cool place. For long-term storage the membrane is best kept under vacuum in a desiccator.

[a] Positively-charged nylon membranes will wet directly on contact with the gel and do not need to be pretreated.

Alkali blotting has several advantages over blotting in $20\times$ SSC. DNA is maintained in a denatured state during transfer, and is therefore fully denatured on the membrane. In high-salt transfer to nitrocellulose, DNA reassociation is possible during neutralization of the gel and may also occur during transfer. Although DNA transfer is slightly quicker in alkali solutions, since diffusion of the high-salt solution is slow, blots are usually left overnight as the prolonged exposure to alkali does no harm (20). Perhaps most important of all, DNA transferred to positively-charged nylon membranes becomes covalently immobilized during the transfer process (5). The covalent fixation of DNA is a major advantage since membranes prepared in this way can be reused many times but do not lose sensitivity after repeated stripping and reprobing. Since the DNA is bound during transfer, no further treatment such as baking or UV crosslinking is necessary before hybridization. This is an advantage over uncharged nylon membranes which give best results when the DNA is covalently attached by UV crosslinking (21). It should be noted that UV treatment of blots on positively-charged nylon causes a reduction in hybridization signal (5).

2.6.3 Alternative blotting methods

In conventional blotting techniques, the nucleic acids are transferred by capillary diffusion of an appropriate solvent. Alternative transfer methods have also been described and will be briefly mentioned here.

Electroblotting, in which charged macromolecules migrate in an electric field, is the transfer method of choice for protein (western) blotting (22). Electroblotting of DNA to DBM paper has also been described (23, 24). Electroblotting of DNA to nitrocellulose is possible (25) but difficult because of the high-salt conditions needed for nucleic acid binding to the filter. Using nylon membranes and low-salt solutions, electrophoretic transfer of nucleic acids can be completed in 2–3 h from a suitably pretreated gel.

Vacuum blotting was first described as a method for protein transfer from polyacrylamide gels (26), but has become popular as an approach for nucleic acid transfer from agarose gels and several types of apparatus for vacuum blotting are now available. In vacuum blotting, the transfer buffer (water for nylon; $20\times$ SSC for nitrocellulose) is stored in a reservoir above the gel and drawn through it by a vacuum pressure equivalent to 30–70 cm of water. The strength of the vacuum is important as too low a pressure results in little transfer, whereas too high a pressure causes the gel to collapse and traps the nucleic acids. Optimal transfer time is 15–30 min and produces blots with sensitivity equivalent to standard methods. Depurination and denaturation of the gel is still necessary but the length of these steps can be reduced to 15–20 min by drawing acid and alkali solutions through the gel.

There have also been attempts to improve the standard arrangement for capillary transfer. The most successful of these is a downward blotting procedure, in which the gel and filter are placed at the top of the paper stack (20). Buffer is drawn from a reservoir placed at one side of the stack, and passes downwards through the filters, rather than upwards, as in the standard apparatus.

The advantage is that the gel is not crushed and transfer times are consequently shorter, with 1 h being sufficient for a 4-mm gel of 1% agarose (27).

2.7 Northern blotting

Northern blotting is the term that refers to the transfer of RNA from a gel to a filter and is analogous to Southern blotting (8,28). An RNA sample is subjected to gel electrophoresis and transferred to a filter so that the separation achieved on the gel is maintained on the solid support. Signals obtained after subsequent hybridization with a suitable probe can be compared with control samples to provide information about the abundance (signal intensity) or size (distance of migration) of RNA transcripts.

RNA is prone to ribonuclease digestion not only during extraction and purification of the sample but also during blotting and hybridization. Probably as a result of this, the sensitivity of detection achieved on northern blots is generally more variable between experiments and between investigators than with Southern blots. For these reasons, a positive-control sample that is known to be intact is usually loaded on the gel. Wherever possible, solutions that are used for RNA should be prepared with sterile deionized water and treated with DEPC to inactivate ribonucleases (see Volume I, Chapter 4). It is good practice to reserve a new gel box for electrophoresis of RNA, since large quantities of ribonuclease are often used in (and contaminate) DNA mini-preparations.

Since RNA forms extensive secondary structures, electrophoresis of RNA is carried out under denaturing conditions. Several methods for denaturation are commonly used: employing glyoxal, formaldehyde, methylmercuric hydroxide or dimethyl sulphoxide. The preparation and electrophoresis of RNA samples is described in Volume I, Chapter 5. A method for RNA transfer from formaldehyde gels to nylon membranes is given in *Protocol 8*. This method is robust and routinely give blots with good hybridization performance.

Protocol 8

Northern blotting to nylon membranes

Equipment and reagents

- RNA samples in a denaturing formaldehyde gel (0.7–2% agarose, 2–6 mm thick; see Volume I, Chapter 5)
- UV transilluminator
- Blotting apparatus (see *Figure 1*)
- Small weight (≈ 500 g)
- Nylon membrane
- Whatman 3MM paper
- Paper towels
- 20× SSC (see *Table 2*)
- 5 μg/ml ethidium bromide in 0.5 M ammonium acetate

Method

1 Electrophorese the RNA samples in a denaturing formaldehyde gel.

2   Cut off the marker lanes for staining with ethidium bromide.[a] Set up the transfer apparatus as shown in *Figure 1* and described in *Protocol 6*, steps 5–6, using 20 × SSC as the transfer solvent.[b]

3   Allow the transfer to proceed overnight. Dismantle the blot and mark the position of the wells on the membrane with a blunt pencil. Stain the gel with 5 μg/ml ethidium bromide in 0.5 M ammonium acetate to check that most of the RNA has transferred.[c]

4   Air-dry the membrane and wrap in Saran wrap. Immobilize the RNA by UV cross-linking on a transilluminator.[d] The membrane can be prehybridized immediately or stored dry in a cool place. For long-term storage the membrane is best kept under vacuum in a desiccator.

[a] Rinse the formaldehyde from the gel to prevent high background staining with ethidium bromide by soaking twice for 20 min in 0.5 M ammonium acetate.

[b] Fomaldehyde gels are more brittle than standard agarose gels.

[c] Transfer of RNA longer than 5 kb is improved by treating the gel for 30–45 min in 50 mM NaOH, 1.5 M NaCl and neutralizing for 20 min in 0.5 M Tris-HCl, pH 7.4, 1.5 M NaCl before transfer.

[d] Baking the membrane for 2 h at 80 °C after UV irradiation gives a slight increase in signal intensity.

RNA up to approximately 5 kb long can be transferred from a formaldehyde gel to a nylon membrane without the need for treatment of the gel. The efficiency of transfer can be checked by staining the gel with ethidium bromide after blotting. The gel should not be stained with ethidium bromide before blotting as this reduces RNA transfer (17); unlabelled markers can be cut from the gel and processed separately. Alternatively, radiolabelled markers can be used and detected directly on the filter. The transfer of larger transcripts is improved by a controlled partial hydrolysis of the RNA by treating the gel in 0.05 M sodium hydroxide. RNA can also be blotted to nitrocellulose from formaldehyde gels but since nitrocellulose does not effectively retain RNA in concentrated formaldehyde solutions, it is necessary to soak the gel in transfer buffer before setting up the blot. Northern blots are not treated after transfer before UV cross-linking and/or baking, as this may result in losses of RNA from the filter. Excess salt can be removed by rinsing the membrane in 6× SSC prior to hybridization.

3 Hybridization analysis of immobilized nucleic acids

3.1 Hybridization theory

Double-stranded DNA can be denatured into two single strands of complementary sequence by heating or by treatment with alkali or other helix-destabilizing agents, such as formamide. When incubated under the appropriate conditions, the complementary strands will reassociate (or renature) to reform a duplex structure. The term *hybridization* refers to this formation of sequence-specific base-paired duplexes. Hybridization occurs not only between single-stranded DNA molecules but also between RNA molecules of complementary sequence. Similarly, hybridization can be used to produce DNA–RNA hybrids.

3.2 Factors affecting the T_m

The thermal stability of nucleic acid hybrids is expressed as the T_m, or melting temperature. Parameters affecting the T_m of hybrids have been carefully studied for hybrids in solution but, in general, there have been fewer systematic studies of the behaviour of immobilized nucleic acids, although the factors involved are thought to be similar.

The melting temperature of perfectly base-paired DNA duplexes in solution depends on ionic strength, base composition and the presence of specific denaturing agents. The effects of all these parameters are combined (19) to produce the following equation:

$$T_m = 81.5 + 16.6(\log M) + 0.41(\%G + C) - 0.61(\%form) - \frac{500}{L}$$

where T_m is the melting temperature, M is the molarity of monovalent cations, ($\%G + C$) is the percentage of guanosine and cytosine nucleotides in the DNA; (*%form*) refers to the percentage of formamide in the solution and L is the length of the base-paired duplex.

3.2.1 Ionic strength

Salt concentrations have an important effect on the T_m of a duplex. The T_m increases by 16.6°C for each tenfold increase in monovalent cations within the range of 0.01–0.40 M NaCl (29). Above this concentration, an increase is seen up to 1.5 M Na$^+$ but the effect is greatly reduced. Note that in 5× SSC (1 M Na$^+$ concentration) the above equation simplifies to:

$$T_m = 81.5 + 0.41(\%G + C) - \frac{500}{L}$$

Divalent cations also affect hybrid stability and divalent cation impurities have been observed to have stabilizing effects on duplexes washed in low ionic strength solutions.

3.2.2 Base composition

In sodium chloride solutions, AT base-pairs are less stable than GC base-pairs. The adjustment that is made in the T_m calculation is valid for (% G + C) values from 30 to 75% (30). The effects of base composition are not seen if hybridization is carried out in 2.4 M tetramethylammonium chloride (see Section 3.12) (31).

3.2.3 Helix-destabilizing agents

Formamide has a destabilizing effect on nucleic acid duplexes and its effect on the T_m of DNA duplexes has been extensively studied (29,32). Formamide has been shown to reduce the T_m of poly(dA:dT) by 0.75°C per 1% formamide, and the T_m of poly(dG:dC) by 0.5°C per 1% formamide (33). It is estimated that an 'average' figure for random DNA sequences is ≈ 0.6°C per 1% formamide (19).

The effects of formamide on the T_m of RNA-RNA or RNA–DNA duplexes are less clear and may vary considerably between different sequences. At 80% formamide, DNA–DNA hybrids have been found to be less stable than DNA–RNA hybrids containing the same sequence (33,34). As a result of this difference in stability, hybridizations can be carried out in 80% formamide to preferentially select RNA–DNA hybrids over DNA–DNA hybrids (35).

As an alternative to formamide, 6 M urea has also been used in hybridization reactions (36). Although the effects of urea are less well known, inclusion of 6 M urea causes a reduction of approximately 30°C in the T_m and is approximately equivalent to the inclusion of 50% formamide (36).

3.2.4 Mismatching

A crude approximation is that the T_m of a DNA duplex is reduced by 1°C for each 1% mismatch (1 mismatch in 100 bp) (37). Various studies have shown that the effect of a 1% mismatch may vary between 0.5 °C and 1.4 °C, depending on the base composition of the sequence and the distribution of mismatches within the duplex (30,38). Clusters of mismatches, or those in GC-rich sequences, have less effect on the T_m than those that are distributed at regular intervals through the sequence.

3.2.5 Length of the duplex

Differences in the lengths of duplexes that are > 500 bp have little effect on thermal stability. For short duplexes, the adjustment $500/L$ is inaccurate. An approximate T_m can be calculated for duplexes of less than 50 bp from the equation (39):

$$T_m = 4(\text{number of GC base pairs}) + 2(\text{number of AT base pairs})$$

3.3 Factors influencing the rate of hybridization

Hybridization of two nucleic acid sequences begins with the formation of a short duplex region between correctly base-paired complementary sequences (nucleation). This event is followed by a zippering up of the two strands by the progressive addition of base-paired nucleotides to either end of the duplex. The zippering up of the two strands occurs rapidly and, under most experimental conditions, it is the rate of nucleation that determines the rate of hybridization. The nucleation rate is dependent on the random collision of complementary sequences. The quantity of duplex formed by hybridization is determined primarily by the concentration of complementary nucleic acid sequences, the length of time available for hybridization and the rate of hybridization.

Rates of hybridization have been investigated extensively in solution hybridization reactions (30,40,41), but are far less characterized in filter hybridization. In many cases, the effects on hybridization of DNA immobilization are unclear, although experimentally determined values show that the rate of hybridization on a filter may be reduced tenfold (42). Parameters that affect the stability of nucleic acid duplexes generally also influence the rate of hybridization reactions, as described below.

3.3.1 Temperature

The maximum rate of hybridization occurs at a temperature 20–25 °C below the T_m of a DNA–DNA duplex (30,37). At low temperatures, the rate of hybridization is slow but cross-hybridization between related but non-identical sequences is more frequent as mismatched duplexes become stable. At temperatures approaching the T_m of the duplex, the effective rate of hybridization is slow because correctly-paired duplexes tend to dissociate. Temperature has similar effects on hybridization rates for DNA–RNA hybrids, except that the maximal rate of hybridization is obtained at 10–15 °C below the T_m (43).

3.3.2 Salt conditions

The ionic strength of the hybridization solution profoundly affects the rate of hybridization (44). DNA hybridizes at very slow rates at low salt concentrations and a maximum rate of hybridization is seen with Na^+ concentrations of approximately 1.5 M (44). The relationship between Na^+ concentration and hybridization rate is not linear. Below 0.1 M Na^+ a twofold increase in Na^+ concentration can cause a five- to ten-fold increase in the rate of hybridization. In contrast, the difference in rates of hybridization between 1 M Na^+ and 15 M Na^+ is small. In a typical filter hybridization, a hybridization solution might contain 5× SSC, corresponding to 1 M Na^+.

3.3.3 Mismatches

As described before, a duplexed nucleic acid that contains mismatched nucleotides has a reduced stability compared with one comprising perfectly-paired sequences (37). In addition, mismatched sequences hybridize at a slower rate. It is estimated that the optimal rate of hybridization (at 25 °C below the T_m) is reduced by a factor of two for each 10% mismatch (37).

3.3.4 Viscosity

Increasing the viscosity of hybridization solutions is reported to reduce the rate of solution hybridization, presumably by lowering the rate of collision between complementary sequences (44,45). However, the effect of viscosity on filter hybridization rates has not been extensively studied and the inclusion of some inert polymers (see Section 3.3.5) that increase viscosity has also been shown to increase hybridization rate.

3.3.5 Addition of inert polymers (dextran sulphate, PEG)

The inclusion of the inert polymer dextran sulphate increases the rate of hybridization of both solution and filter hybridization reactions (46,47). This increase is thought to be due to an effective increase in nucleic acid concentration caused by exclusion of nucleic acids from the volume occupied by the polymer. The addition of 10% dextran sulphate increases the rate of solution hybridization by a factor of ten. In filter hybridization experiments, the increase depends on the type of probe used. Using single-stranded probes of either RNA or DNA, a three- or ten-fold increase can be seen; with double-stranded probes the yield of hybrid may increase by up to 100-fold (47). This increase in signal is thought to result from the formation of extensive networks containing concatenates of many probe molecules (9). These networks still contain single-stranded regions of DNA that are able to hybridize to the target and produce a greater hybridization signal than would be expected from a single probe molecule. This increase in signal is very useful for qualitative experiments where only a positive signal is required. For quantitative experiments, the amount of signal produced through networking is misleading and may lead to an overestimation of differences in levels.

The addition of 6–8% polyethylene glycol (PEG 6000) can be used as an alternative to dextran sulphate (36,48). PEG has an advantage over dextran sulphate in that it produces less viscous solutions at low temperatures. PEG is claimed to give a slightly better acceleration of hybridization rates than dextran sulphate when single-stranded probes are used (48).

3.3.6 Formamide

Formamide is often added to hybridization solutions to lower the T_m of the duplexes and enable hybridization at lower temperatures. At lower incubation temperatures, the losses of target nucleic acid from the filter during hybridization (that occur with a non-covalently immobilized target) are reduced and the stability of RNA probes is increased.

The inclusion of 30–50% formamide has been shown to have little or no effect on the rate of hybridization on a filter (49). Apparently, the addition of 20% formamide reduces the rate of filter hybridization by approximately one-third (50). In solution hybridization, 80% formamide has been shown to reduce the rate of hybridization threefold (for DNA–DNA duplexes) or 12-fold (for DNA–RNA duplexes) (51).

3.3.7 Probe complexity

Nucleic acids which have high complexity (no significant sequence repetition) hybridize in solution more slowly than nucleic acids that contain repetitive sequences (low complexity) (44,52). Presumably, this observation reflects differences in the rates of nucleation and similar effects may also occur in filter hybridization. The difference in hybridization rate between the two extremes of complexity may be as great as 400-fold (42) but in most practical situations the complexity of the probe is not a parameter that is easily altered.

3.3.8 Base composition

The base composition has only a minor effect on the rate of hybridization (44,53) and for most experiments this effect can be ignored.

3.3.9 pH

The rate of hybridization is constant between pH 5 and pH 9 (44). Hybridization reactions are normally carried out at neutral or slightly acid pH.

3.3.10 Length of the probe

The rate of hybridization of DNA in solution is directly proportional to the length of the nucleic acid (44). If the DNA strands are of different lengths, the rate of DNA–DNA hybridization in solution is dependent on the length of the shorter fragment (54). In filter hybridization, the rate of hybridization is directly proportional to the length of the probe only when the amount of DNA target on the filter is in excess and the rate of hybridization is limited by diffusion. In circumstances where the filter contains only a trace amount of target and the probe is in excess, the rate of hybridization (in the nucleation-limited reaction) is independent of the length of the probe (42,43). However, it should be noted that if the length of the probe is extended with sequences that are unable to hybridize, the rate of hybridization is significantly reduced. This is seen for single-stranded probes but is even more apparent for double-stranded probes, which also reassociate more rapidly in solution during hybridization.

3.4 Filter hybridization

Target nucleic acids that are immobilized on a membrane filter are able to hybridize with a labelled nucleic acid probe. Unhybridized probe is washed away and specifically hybridized probe is detected. For efficient hybridization to take place both the probe and the target must be single-stranded.

Filter-hybridization involves three basic steps: prehybridization, hybridization and washing. The purpose of the prehybridization step is to block any sites on the filter that might bind the labelled probe and produce background signal. Typically, the prehybridization solution contains protein, charged polymers and random-sheared carrier DNA that act as blocking agents. Prehybridization can take from 15 min to 4 h.

In the hybridization step, the filter is incubated in a solution that contains the labelled probe. The hybridization solution contains the blocking agents used in the prehybridization step and the incubation is at optimal conditions for the desired duplex formation. Hybridization can be carried out for 1–48 h but is typically carried out overnight.

After hybridization, the filter is washed to remove the probe. The stringency of the washing can be altered so that only probe molecules that are completely base-paired with the immobilized target are left on the filter (washing to high stringency), or so that probe molecules that have hybridized to related sequences, forming duplexes that contain mismatches, are also left on the filter (washing to low stringency). Washing the filter may take 30–60 min.

3.4.1 Concentration of the probe

The initial rate of filter hybridization depends on the concentration of target on the filter and of probe in solution. Increasing the concentration of the probe will increase the initial rate of hybridization and will drive the reaction to completion more quickly. When double-stranded probes are used, the reassociation of the probe in solution is favoured by high concentrations of probe and occurs rapidly. When probing RNA with a double-stranded DNA probe, the rate of reassociation of probe can be slowed by adding formamide to the hybridization solution.

Unfortunately, probe concentrations cannot be increased without limit. In addition to the impracticability of producing vast quantities of labelled probe, non-specific irreversible background is produced by long incubations at high probe concentrations (e.g. 100 ng/ml probe in an overnight hybridization). Typical probe concentrations for overnight hybridization are 1–10 ng/ml when hybridization solutions contain dextran sulphate, or 1–100 ng/ml in the absence of dextran sulphate.

3.4.2 Hybridization and time

The objective in a filter hybridization reaction is to maximize the amount of duplex that is formed between target and probe. Superficially, it might appear that long hybridization incubations would be expected to give best results. However, this is not the case because as the hybridization reaction proceeds other factors come into play:

- double-stranded probe reassociates in solution
- the discrimination of the probe between correct and closely-related sequences often changes during hybridization
- non-specific background forms on the filter.

When a double-stranded probe is used to probe nucleic acid on a filter, two competing reactions occur: reassociation of the probe in solution and hybridization of the probe to the target on the filter. It is assumed that the target is immobilized on the filter and is unable to reassociate. As has been described above, the rate constant for hybridization in solution is much greater than the rate constant for hybridization to target on a filter, and solution hybridization may be considered to occur much more rapidly. Compounding this, it has been estimated that as much as 30% of an input double-stranded DNA probe may be reassociated or unavailable for hybridization to the target (42). Although hybridizations are often carried out overnight, the majority of a double-stranded probe will have reassociated after 6–8 h of incubation.

The discrimination between hybridization to perfectly-matched sequences and hybridization to closely-related sequences may change with incubation time

(37). In a filter hybridization, where a single-stranded probe is in excess over the target on the filter, the homologous reaction between correctly-paired sequences proceeds quickly and reaches completion. The heterologous reaction of the probe with related sequences is a slower reaction but will eventually catch up with the homologous reaction if the incubation is sufficiently long. A net result is that the discrimination of the hybridization (correct sequence versus related sequence) is best early in the incubation and becomes poorer as the reaction proceeds.

In hybridizations where the filter-bound target is in excess over the probe in solution, the discrimination of the hybridization does not change during the reaction because all the sequences on the filter are in continuous competition for the small quantity of probe.

In hybridization where double-stranded probe is in excess over filter-bound target, the discrimination of the hybridization declines in a manner similar to that described for a single-stranded probe in excess, but the decrease is smaller. This difference is due to the reassociation of double-stranded probe in solution and puts the target in excess over probe at later time-points.

Non-specific background caused by irreversible sticking of labelled probe to the filter increases in a time-dependent manner. Under normal reaction conditions (provided that the filter is correctly blocked) non-specific background is not usually a problem, but hybridizing for more than 24 h may cause the signal to become obscured by the background.

3.4.3 Nucleic acid probes

Any form of nucleic acid that can be labelled can be used as a hybridization probe. In most circumstances, low-complexity probes are prepared from cloned sequences. In order to minimize background signal, labelled probe should be separated from unincorporated or free label (see Chapter 4, *Protocols 3* and 4).

Probes used in filter hybridization can be labelled with radioactive or non-radioactive markers (see Chapter 4). Hybridization with non-radioactive probes can often be driven with high probe concentration without excessive backgrounds. Detection of non-radioactive signals is often achieved in 10 min to 4 h, considerably less than the days to weeks that may be necessary for autoradiographic detection of a radiolabelled probe.

Nucleic acid probes need to be greater than 15 nucleotides and less than 1.5 kb long. Probes shorter than 15 nucleotides have been used for hybridization analysis but are not recommended for use with genomic DNA because of the probability that several complementary sequences will be found (see Chapter 7, Section 2.2.2). Nucleic acid probes greater than 1.5 kb have been correlated with high backgrounds, and it is advisable to keep the probe size less than this. DNA probes can be prepared from longer sequences than 1.5 kb by methods such as nick translation (Chapter 4, Section 2.3.1) or random primer labelling (Chapter 4, Section 2.3.2), both of which produce smaller DNA fragments.

^{32}P-labelled probes of specific activity greater than 1×10^8 d.p.m./μg are needed to detect a single-copy sequence in an overnight exposure of the filter to film. DNA probes with high specific activity can be produced by nick-translation (typical specific activity 1–5×10^8 d.p.m./μg) or random primer labelling (typical specific activity 1–5×10^9 d.p.m./μg). RNA probes with high specific activities (1–2×10^9 d.p.m./μg) can be produced by RNA polymerase transcription systems (Chapter 4, Section 2.3.3).

The specific activity of probes produced by end-labelling techniques (Chapter 4, Section 2.4) depends on the length of the DNA. DNA probes prepared by end-labelling techniques are used only for high-sensitivity detection with oligonucleotide probes for which specific activities higher than 10^9 d.p.m./μg are routine.

3.5 Hybridization of a Southern blot on a nitrocellulose filter

A basic procedure for filter hybridization is given in *Protocol 9*. This protocol describes hybridization of a DNA dot-blot or Southern blot on a nitrocellulose filter with a ^{32}P-labelled DNA probe.

Protocol 9

Basic method for filter hybridization: hybridization of a Southern blot or a DNA dot-blot with a radiolabelled, double-stranded DNA probe

Equipment and reagents

- Hybridization oven
- Boiling water bath
- Nitrocellulose filter carrying a Southern blot or dot-blot

- Aqueous hybridization solution (see *Table 2*)
- 20$\times$ SSC (see *Table 2*)
- Radiolabelled, double-stranded probe DNA
- Wash solutions (see *Table 2*)

Method

1 Warm the aqueous hybridization solution to 68 °C.

2 Wet the filter in 5$\times$ SSC and place the damp filter in one of the cylinders of the hybridization oven.[a]

3 Add at least 1 ml of warmed aqueous hybridization solution per 10 cm^2 of filter.[b] Carry out the prehybridization by rotating the cylinder in the hybridization oven for 3 h at 68 °C. The filter should be able to move freely in the cylinder and the solution should sweep gently over the surface of the filter.

4 Denature the double-stranded probe by incubating in a boiling water bath and snap-cooling on ice.[c] Add the denatured probe to the prewarmed hybridization solution. There should be at least 1 ml of hybridization solution per 10 cm^2 filters, and final probe concentration should be 10 ng/ml (for a probe of specific activity 10^8 d.p.m./μg) or 2 ng/ml (for a probe with specific activity 10^9 d.p.m./μg). Mix well.[d]

5 Remove the prehybridization solution from the cylinder and replace with the hybridization solution plus labelled probe.[e]

Protocol 9 continued

6 Rotate the cylinder overnight at 68 °C.

7 Remove the hybridization solution and dispose in the appropriate way. Add 5–10 ml of low-stringency wash solution (2× SSC, 0.1% SDS) per 10 cm^2 of filter. Rotate the cylinder for 5 min at room temperature.

8 Repeat step 7 twice with fresh aliquots of low-stringency wash solution.[d]

9 Increase the stringency of the wash by washing the filter twice in 0.2× SSC, 0.1% SDS at room temperature for 5 min.

10 The stringency of the washes can be increased further by increasing the washing temperature and/or lowering the salt concentration. Washing twice for 15 min at 42°C in 0.2× SSC, 0.1% SDS is a moderately stringent wash. Washing twice for 15 min at 68°C in 0.2 × SSC, 0.1% SDS or 0.1× SSC, 0.1% SDS is a high-stringency wash. It is not necessary to wash for greater than 30 min for each level of stringency. Washing solutions should be prewarmed to the appropriate temperature before use.

11 Finally rinse the filter in 2× SSC at room temperature and lay the filter on blotting paper to remove excess liquid. Do not allow the filter to dry if it is to be rewashed or stripped for reprobing.

12 Autoradiograph the filter (see Section 3.16).

[a] If a hybridization oven is not available then it is possible, though inadvisable, to carry out the hybridization in a heat-sealed plastic bag. This procedure is inadvisable because of the difficulty in repeatedly sealing and opening the bag without contaminating the work area with radioactive material. If a bag is used then it should be double-sealed along all sides approximately 1 cm from the edge of the filter. Use the same volumes of hybridization solution as described in the main protocol. Ensure that all air bubbles are removed. A simple way to do this is to seal the fourth side of the bag 2 cm from the edge of the filter, then snip off a corner and displace the bubbles from the bag through the small hole by running a ruler along the bag or running the bag against the edge of a bench, then reseal the bag. The filter should be able to move freely in the bag and the solution should sweep gently over the surface of the filter (this will require that the bag is placed on a rocking platform in an incubator). The filter should be removed from the bag and placed in a glass dish in order to carry out the washes.

[b] More than one filter can be hybridized at one time. Up to 10 filters can be hybridized in a single cylinder or bag, but ensure that air bubbles are not trapped between the filters and that the filters do not stick together during the incubations.

[c] DNA probes can also be denatured by treatment with alkali. Add one-tenth the original volume of 3 M NaOH. After 5 min at room temperature add one-fifth the original volume of 1 M Tris-HCl, pH 7.0 and one-tenth the original volume of 3 M HCl.

[d] Radioactive probes must be handled in accordance with local safety procedures. Hybridization and wash solutions must be disposed of in the appropriate way. Note that all wash solutions will be radioactive.

[e] Do not allow the filter to dry out at any time during the prehybridization, hybridization and wash steps.

Essentially the same solution is used for prehybridization and hybridization, the labelled probe being the only new component added for hybridization. Hybridization solution (see *Table 2*) can be prepared in advance and stored frozen in aliquots for at least 6 months. This solution is a combination of components that are chosen either to minimize the sticking of labelled probe to the filter or to promote hybridization. It includes 5× SSC to give ionic conditions that produce high rates of hybridization. Dextran sulphate accelerates the rate of hybridization by excluding DNA from part of the hybridization solution and raising the effective probe concentration. The addition of SDS to washing solutions was first described by Botchan *et al.* (55) and its inclusion in hybridization solutions considerably reduces background signal. Denhardt's solution (56) is a blocking solution containing Ficoll, polyvinylpyrrolidone (which like nucleic acids is a polyanion polymer) and bovine serum albumin to saturate the DNA-binding sites that remain on the filter and block binding of the probe to the filter.

Phosphate buffer or sodium pyrophosphate is also included in hybridization solutions to maintain a neutral pH during hybridization. Pyrophosphate may also help to reduce background with probes that contain unincorporated label.

Prehybridization of nitrocellulose filters is carried out for 3 h. This step can be reduced, but the chances of getting bad backgrounds increase. When the prehybridization is completed it is best to remove the prehybridization solution and replace it with an equal volume of prewarmed solution that contains the DNA probe, rather than simply adding the denatured probe into the prehybridization solution.

DNA probes prepared by nick-translation, with a specific activity of 1×10^8 d.p.m./µg, are added to give a final concentration of 2–10 ng/ml. Probes with a specific activity of 1×10^9 d.p.m. µg^{-1} are usually added to give a lower final concentration at 1–2 ng/ml. These concentrations are good for high-sensitivity detection of 1–5 pg target DNA. If the target sequences are more abundant, lower specific activity probes or lower concentrations of probe can be used. Higher probe concentrations can also be used with probes of 1×10^8 d.p.m./µg prepared by nick-translation. Up to 100 ng/ml of probe can be added to hybridization solutions containing dextran sulphate, but the hybridization time must be reduced to 1 h to prevent excessive background signal.

Hybridization solutions can be saved and reused until the probe has become degraded or has decayed to an unacceptably low specific activity. To reuse the solution, the probe must be denatured, which can be accomplished by heating the hybridization solution in a boiling water bath for 10 min.

After hybridization the filters are washed in solutions containing SSC and SDS. The SDS assists in the removal of unbound probe. Initially, the filters are washed in low-stringency conditions (2× SSC, room temperature), then the stringency of the washes is gradually increased by decreasing the salt concentration and increasing the temperature. The progress of the washing can be followed by monitoring the filter with a Geiger counter between washing steps or by counting aliquots of the wash solutions in a scintillation counter. When washing is complete there should be little or no background signal on the filter.

The hybridization signal can often be located on the filter with a Geiger counter but this is not essential as autoradiography has much greater sensitivity. It is important that the filter is not allowed to dry out during hybridization or during washing as this results in background signals. Probe that is in contact with the filter as it dries becomes immobilized and is not removed by washing.

3.6 Hybridization of nylon membranes

Nylon membranes can be hybridized using essentially the same method as that described for nitrocellulose filters (*Protocol 9*) with only a few modifications. The time required for prehybridization of nylon membranes is less than that for nitrocellulose filters and the prehybridization step can be reduced to 15–30 min. Nylon membranes can be wetted directly in the prehybridization solution.

3.7 Alternative or additional blocking agents

For hybridization with RNA probes, 100 μg/ml of yeast tRNA is often used as a blocking agent in place of denatured DNA. The preparation of sheared DNA solutions can be tedious and some workers choose to use tRNA in all hybridizations. Heparin (57) or dried milk powder (58) are less commonly used but are effective alternatives to Denhardt's solution in prehybridization solutions.

Hybridization backgrounds with some probes can be reduced by the addition of 10 μg/ml of poly(A) or poly(C) homopolymers. Poly(A) is usually added to prehybridization and hybridization solutions when either the probe or the target contains AT-rich sequences. Examples are with poly(A)$^+$ RNA immobilized on the filter or hybridizations with probes prepared from poly(A)$^+$ RNA or poly(A)$^+$-derived cDNA. Similarly, poly(C) is added to 10 μg/ml if the probe or target contains sequences generated by oligo(dG) and oligo(dC) homopolymer tailing.

3.8 Hybridization with RNA probes

In vivo labelled mRNA, tRNA or rRNA sequences can be prepared by standard techniques and used as hybridization probes. Alternatively, purified RNA can be labelled with T4 polynucleotide kinase (Chapter 4, Section 2.4.1). However, the most common type of RNA probe is ^{32}P-labelled RNA of high specific activity, synthesized by *in vitro* transcription of cloned sequences in RNA polymerase transcription systems (Chapter 4, Section 2.3.3). RNA probes with specific activities greater than 1×10^9 d.p.m./μg are routinely synthesized and added to hybridization solutions at 1–5 ng/ml. RNA probes, being single-stranded, do not need to be denatured before use and hybridize to target sequences in the absence of any competing reassociation of the probe in solution. Hybridizations with RNA probes are typically carried out overnight at 42 °C in a hybridization solution containing 50% formamide (see *Table 2*).

Background with RNA probes can be considerably reduced by treating the filter with ribonuclease during the washing procedure, to remove unpaired loops, single-stranded tails and unhybridized probe. Filters are washed for 30 min at room temperature with $2\times$ SSC containing $25\,\mu g/ml$ RNase A and $10\,units/ml$ RNase Tl. The RNase wash can be included between the low-stringency and high-stringency washes, or on a filter that has been fully washed and exposed to film. Some RNA probes have been found to give poor results in hybridizations at $42\,°C$ but give improved signal and reduced background when hybridized at higher temperatures. In a few cases, hybridization at $65\,°C$ in 50% formamide is optimal. If the length of an RNA probe greatly exceeds $1.5\,kb$ the probe may be more prone to background problems. The size of RNA probes can be reduced by controlled, partial alkaline hydrolysis (59).

3.9 Hybridization to RNA dot-blots and northern blots

RNA is more rapidly degraded than DNA, particularly in solutions that are at high temperatures and at neutral or alkaline pH. Consequently, RNA blots are usually hybridized at $37–42\,°C$ in slightly acidic solutions that also contain 50% formamide. The composition and preparation of the formamide prehybridization and hybridization solution is shown in *Table 2* (modified from refs. 56 and 60). Ideally, water that is used in the preparation of RNA hybridization solutions should be DEPC-treated to remove ribonucleases but, in practice, deionized sterile water usually gives satisfactory results.

The quantity of target RNA sequences that can be detected by hybridization is similar to the values seen for DNA detection. As little as $1\,pg$ of target RNA can be detected in an overnight exposure after hybridization with a probe labelled to $10^8–10^9$ d.p.m./μg, but an overnight sensitivity of between 5 and $25\,pg$ is probably more usual. Note that an mRNA sequence that represents 0.5% of total mRNA, equivalent to about 0.002% of total cellular RNA, constitutes $200\,pg$ of a $10\,\mu g$ sample of total RNA.

3.10 Hybridization to colony or plaque replica filters

Plaque and colony replica filters are hybridized using the method described in *Protocol 9* with only a few modifications. Colony filters are prehybridized in a large volume (1 ml prehybridization solution per $2–5\,cm^2$ of filter) and agitated by shaking, before prehybridization, to wash bacterial debris from the filter. Many workers also increase the blocking agents in the prehybridization and hybridization solutions to a final concentration of $10\times$ Denhardt's solution. Emphasis is placed on minimizing background signal, particularly spotty or blotchy backgrounds. At least two replica filters are hybridized for each plate and only spots that appear in the same position on each filter are potential positives. Because many filters are often hybridized together, it is important to ensure that no bubbles are trapped between the filters. Bubbles will cause localized drying of the filter and lead to clear patches and spots.

Dextran sulphate is often omitted from hybridization solutions when hybridizing plaque or colony replica filters as the target sequences are usually relatively abundant. Some batches of dextran sulphate cause spotty backgrounds and this effect is more likely if the probe concentration inadvertently exceeds $15\,ng/ml$.

3.11 Hybridization with oligonucleotide probes

Oligonucleotides are hybridized at a probe concentration of 0.125 ng/ml for each unique sequence. Different oligonucleotide probes may be pooled, provided that the total probe concentration does not exceed 20 ng/ml. Oligonucleotide probes are normally hybridized in solutions that do not contain dextran sulphate because the molar concentration of the probe is relatively high and the improvement in hybridization rate with dextran sulphate is quite small (19). The oligonucleotide hybridization solution shown in *Table 2* contains SET rather than SSC because this buffer tends to give better results with oligonucleotide probes, although the reason for the improvement is not clear.

Short DNA duplexes have a reduced melting temperature and the T_m of oligonucleotide probes (up to 50 nucleotides) is approximately predicted by the equation (39):

$$T_m = 4(\text{number of GC base pairs}) + 2(\text{number of AT base pairs})$$

although the actual T_m should be calculated experimentally. The hybridization and washing conditions used with oligonucleotide probes are adjusted to make allowance for the reduced T_m. Oligonucleotides are hybridized at a temperature between 5 and 10°C below the T_m for 14–48 h (61). Filters are then washed four times for 10 min in 6× SET, 0.1% SDS at the hybridization temperature; this washing is often sufficient. The filters are then examined with a Geiger counter and if they still show considerable activity above background, the wash temperature should be increased by 2–3°C and washing repeated. Alternatively, the filters can be blotted to remove excess solution and exposed while still damp and wrapped in Saran wrap. The washing temperature can gradually be increased until the T_m is reached. The final wash should be at the T_m for 2 min only. Suggested hybridization and washing conditions are shown in *Table 3*.

Hybridization with mixed oligonucleotide probes is often used when the nucleic acid sequence of the probe is interpreted from an amino acid sequence. Degenerate probes should be chosen from sequences that are rich in tryptophan or methionine (each of which are coded by a single nucleotide triplet) or amino acids which have two codons (asparagine, aspartic acid, cysteine, glutamic acid, glutamine, histidine, phenylalanine and tyrosine). Hybridization of degenerate probes is carried out at 5°C below the T_m of the least stable (most AT-rich) sequence in the mixture. The T_m of a hybrid less than 20 bp in length decreases by approximately 5°C for each mismatched base-pair (39,62).

Table 3 Hybridization temperatures for oligonucleotide probes

Oligonucleotide length (nucleotides)	Suggested hybridization temperature (°C)	Suggested final wash temperature in 6× SET (°C)
14	27	34
17	36	41
20	42	48
23	48	55

3.12 Hybridization in tetramethylammonium chloride

An alternative method for hybridization of oligonucleotide probes has been described in which a solution containing 3 M TMAC is used. In TMAC solutions, the T_m of the oligonucleotides is independent of the base composition and depends solely on the length of the sequence (31,63,64). Hybridization in TMAC is most useful when degenerate oligonucleotide probes are used, since all the sequences will have the same T_m, or with probes with high GC content.

3.13 Quantification of the hybridization signal

In many experiments, approximate quantification of results by eye, comparing the intensity and spread of signals, is sufficient. Provided that the hybridization is carried out with the concentration of probe in excess of the target on the filter, the intensity of the signal is a good measure of the quantity of target present. More accurate measurements of the signal from ^{32}P-labelled probes can be obtained in two ways.

(a) The β-emissions can be detected directly. This can be carried out with radio-activity detectors that are commercially-available for use on filters. A cheaper alternative is to cut the filters into small pieces and to count the signal in a liquid scintillation counter. A signal that is clear in an overnight exposure to film can be quantified in this way. Faint bands give only 10 c.p.m. above background and each sample may need to be counted for 5 or 10 min.

(b) The film can be scanned with a densitometer. It is necessary to use films that are not overexposed and it is a good idea to scan several different exposures of each filter. The area under the curve is used as a measure of the signal and can be determined by integration or by cutting the trace and weighing the paper.

Signal measurements are plotted on a graph against the quantity of target available for hybridization. The relationship should be linear but a plateau may occur at the highest loadings. Accurate quantification of samples is only possible on the part of the graph that is linear. Loss of linearity occurs if the hybridization probe is no longer in excess over target sequences or if densitometric traces are taken of overexposed films. Extensive networking of double-stranded probes can also cause loss of linearity and appears as a disproportionate increase in signal with increased quantities of target. Networking can be reduced by using smaller double-stranded probes, by leaving dextran sulphate out of the hybridization solution or by using single-stranded probes.

3.14 Removal of probe and rehybridization of filters

Once a successful result is obtained, the probe can be stripped from the filter and the blot rehybridized with a different probe. Ten rehybridizations of DNA blots on nylon membranes are not unusual. Nevertheless, the procedures required for probe removal are necessarily harsh and unless the target is covalently bound, target nucleic acids will also be lost from the filter, resulting in lower signals on reprobing. Target nucleic acids can be covalently bound (by UV crosslinking or by using positively-charged nylon membranes) to nylon membranes. This, coupled with their physical resilience, makes nylon membranes the membranes of choice if rehybridization of the filter is likely to be necessary. However, with careful handling, nitrocellulose filters can also be rehybridized up to five or six times.

For successful removal of probes, membranes must never be allowed to dry out after hybridization or washing as this causes the probe to become bound. Probe can be stripped using any of the three treatments listed in *Protocol 10*. The treatments are listed in order of increasing harshness so, if possible, the first method should be used. Alkali treatment is not suitable for use with RNA blots as the RNA becomes hydrolysed, but is particularly effective with RNA probes. The filter should be autoradiographed to check that the probe has been removed. The filter can be stored dry or immediately prehybridized and hybridized with a new probe.

Protocol 10

Protocols for removal of the probe from filters

A. Gentle treatment
Reagent
- 0.005 M Tris-HCl, pH 8.0, 0.002 M EDTA, 0.1× Denhardts solution

Method
1   Wash the filter for 1–2 h at 65 °C in the reagent above.
2   Autoradiograph the film to check that the probe has been removed.

B. Moderate treatment (not suitable for RNA blots)
Reagents
- 0.4 M NaOH
- 0.1 × SSC (see *Table 2*)
- 0.1% SDS
- 0.2 M Tris-HCl, pH 7.0

Method
1   Incubate the filter at 45 °C for 30 min in 0.4 M NaOH.
2   Rinse the filter twice at room temperature for 10 min in 0.1× SSC.
3   Rinse the filter twice at room temperature for 10 min in 0.1% SDS.
4   Rinse the filter twice at room temperature for 10 min in 0.2 M Tris-HCl, pH 7.0.
5   Autoradiograph the film to check that the probe has been removed.

C. Harsh treatment
Reagent
- 0.1% SDS

Method

1 Boil a solution of 0.1% SDS. Pour on to the filter and allow to cool to room temperature.

2 Autoradiograph the film to check that the probe has been removed.

3.15 Calibration of UV transilluminators

Ultraviolet irradiation is an effective means of covalently-binding nucleic acids to nylon membranes (6). The power and wavelength of UV radiation varies between transilluminators and it is essential to calibrate each one. Both suboptimal crosslinking (too little irradiation) and degradation of nucleic acids (too much irradiation) cause poor sensitivity of detection.

Prepare five or six identical strips of control DNA or RNA blots that contain a range of 1–100 pg of target material in carrier DNA. Dot-blots of serially diluted DNA or Southern blots containing lanes of 100 pg and 10 pg of a DNA digested with *Hind*III restriction endonuclease are suitable. Wrap each of the filters in Saran wrap and irradiate each strip, DNA side down, on the transilluminator for a different length of time. Time points of 30 s, 45 s, 1, 2, 5, and 10 min are recommended. Hybridize the strips together with a nucleic acid probe (1×10^8 d.p.m./μg) as described in *Protocol 9*. Following autoradiography, the optimal exposure time will be indicated by the filter with the strongest signal. Minor improvements in signal intensity can be obtained by baking the nylon membranes for 2 h at 80 °C after irradiation.

3.16 Autoradiography

Autoradiography is used for detecting radiolabelled probe on filters after hybridization (see Chapter 4, Section 2.5). Filters are wrapped in Saran wrap and exposed to X-ray film in a light-proof cassette. If the filter is to be rewashed or stripped and rehybridized, then the filter must be kept damp. The length of the exposure time needed depends on the signal intensity. A signal of 100 c.p.m. or more is easily detected in an overnight exposure at room temperature. Weaker signals (e.g. 5–10 c.p.m. above background) can be detected after overnight exposure at −70 °C with an intensifying screen. For very weak signals longer exposure times can be used (2 weeks or more) and detection sensitivity can be improved by using two intensifying screens and by preflashing the film.

3.17 Determination of T_m for filter hybridization

T_m can be measured using the simple procedure shown in *Protocol 11*, in which the stability of hybridized probe in the hybridization solution is monitored over a temperature range.

Protocol 11

Determination of T_m

Equipment and reagents

- Hybridization oven
- Scintillation counter and vials
- Dot-blot carrying 1–10 ng of plasmid DNA
- Appropriate ^{32}P-labelled probe
- Hybridization buffer (see *Table 2*)

Method

1 Prepare a dot blot[a] containing sufficient target nucleic acid to give an easily detectable signal (e.g. 1–10 ng of plasmid DNA).

2 Hybridize with a ^{32}P-labelled probe.

3 Wash the filter three times for 10 min each in hybridization buffer at the hybridization temperature.

4 Count the dot blot by Cerenkov counting (see Chapter 4, Section 2.1.2). Ensure that the filter does not dry out during counting as this will irreversibly bind the probe to the filter.

5 Raise the temperature of the water bath by 2–4 °C and prewarm a small volume of hybridization solution.

6 Wash the filter in hybridization solution at the new temperature for 15 min. Count the filter as before and repeat steps 4–6 until all the signal is washed from the filter.

7 Plot a graph of the percentage of hybrid melted against the washing temperature. The T_m is the temperature at which half the counts have been removed.

[a] Dots can be washed in different solutions to estimate the T_m in various wash conditions. The T_m of the hybrids in different solutions can be estimated by the equation:

$$(Tm)\mu2 = 18.5 \log\frac{\mu2}{\mu1}$$

where $\mu1$ and $\mu2$ are the ionic strengths of the two solutions.

When plasmid DNA is used to probe plasmid DNA, the hybrids formed melt over a narrow temperature range because only perfectly base-paired duplexes are formed. However, when hybridization also occurs between partly-related sequences the temperature profile obtained is broader (65). As a result, the melting profile of a probe hybridized to genomic DNA may be broad if the hybridization is carried out at low stringency ($T_m - 25 °C$) but narrow if the probe is hybridized at high stringency ($T_m - 8 °C$). The melting profile is also broad when the probe contains sequences of very different lengths—a feature that is more pronounced with short probes.

3.18 Troubleshooting guide

Problems can occur during blotting and hybridization for a wide variety of reasons. A troubleshooting guide is given in *Table 4*. This guide is not exhaustive, but lists some of the most common problems in Southern and northern blotting. The most likely causes are given, together with suggested remedies (see also ref. 66).

4 Advanced hybridization techniques

If we move beyond the essential techniques in molecular biology we encounter a number of versions of hybridization analysis that involve nucleic acids that are immobilized by means other than attachment to a filter. Three of these advanced applications are briefly considered here.

4.1 *In situ* hybridization to detect RNA transcripts in tissue sections

RNA molecules immobilized within tissue sections can be hybridized with DNA or RNA probes in order to determine which cells and tissues express a particular gene. Expression is indicated by a hybridization signal showing that a cell or tissue contains RNA transcripts of the target gene. This is *in situ* hybridization, one of several techniques (immunocytochemistry and *in situ* PCR are others) that can be used to study spatial patterns of gene expression in tissue sections. In many respects the procedures used are similar to those employed with filter-immobilized nucleic acids, with the obvious difference that the tissue is prepared initially by embedding, fixing and sectioning (67). Hybridization is carried out with sections attached to glass slides. Detection of radioactive probes can be by film autoradiography, but this provides only low spatial discrimination. To assign hybridization signals to individual cells a second type of autoradiography, using an emulsion, is needed. The emulsion is spread over the tissue section and left for the required time before being scraped off. Hybridization is indicated by silver grains that result from exposure of the emulsion; these adhere to the tissue section and can be viewed by dark-field microscopy.

4.2 *In situ* hybridization of chromosomes

This is a second version of *in situ* hybridization in which the target material is isolated chromosomes rather than tissue sections. The objective is to use a hybridization probe to determine the position within the chromosomes of one or more genes. A chromosome spread is made on a glass slide and the DNA denatured, usually by treatment with formamide. Radioactive probes can be used but these give relatively poor spatial resolution and indicate only the approximate position of the target gene. Higher resolution mapping is possible when the probe is labelled with a fluorescent label (see Chapter 4, Section 3.3); this technique is called fluorescent *in situ* hybridization or FISH (68). The results are monitored by confocal microscopy. By using fluorescent labels with different emission energies, and hence differently coloured fluorescences, the relative positions of two or more genes can be discerned in a single experiment (69).

FISH was initially developed as a low-resolution technique for use with stand-

Table 4 Troubleshooting guide for Southern and northern blotting and hybridization analysis

Problem	Suggested remedy
Poor signal	
Poor transfer of DNA or RNA	Load extra lanes of control samples on your gel. DNA transfer can be checked by re-staining the gel with ethidium bromide. If large DNA fragments are transferred poorly, check the depurination step (0.25 M HCl)
Poor retention of DNA on the filter, especially on reprobing	If using an uncharged nylon membrane, check that the UV transilluminator is correctly calibrated (see Section 3.15). Northern blots should not be treated with low-salt buffer before fixation. Note that very small DNA and RNA fragments may not be efficiently retained by the filter (<50 nucleotides with nylon membranes, <500 nucleotides with nitrocellulose filters)
Low probe activity	Best results are obtained with probes of specific activity > 10^8 d.p.m./µg. Check the incorporation in the labelling reaction
Low probe concentration	Ensure accurate measurement of template DNA for use in labelling reactions. Check the recovery of probe if purification is performed to remove unincorporated nucleotides
Incomplete denaturation of probes	Follow denaturation protocols exactly
Filter washed at too high a stringency	Estimate the T_m of the probe (see Section 3.17) and if necessary perform the final wash at a lower temperature or higher salt concentration
Spotty background over the whole filter	
Unincorporated label not removed from the probe DNA	Unincorporated nucleotides can give high backgrounds. Remove by one of the methods given in Chapter 4, *Protocols 3 and 4*
Particulate matter in the hybridization	Ensure the components are in solution, or pass the hybridization buffer through a 0.45 µm filter buffer
Residual agarose adhering to the filter after blotting	Rinse the filter carefully in 6× SSC after blotting
Baking or UV crosslinking the filter when it is still soaked in high salt	Rinse the filter carefully in 6× SSC after blotting
Patchy or uniform background over the whole filter	
Part of the filter dried out during the hybridization and/or washing steps	Never allow the filter to dry out during hybridization or washing. Increase the volumes of the hybridization and/or wash solutions
Filters have stuck together	Ideally filters should be hybridized individually

Cause	Remedy
Probe concentration is too high	Ensure that the amount of template DNA is accurately measured
Insufficient blocking reagents in the hybridization	See the hybridization protocols for recommended hybridization buffers
Probe not denatured	Non-denatured double-stranded probes often give high background
Extra bands in one or more lanes	
Stringency of the wash is too low	Increase the stringency by increasing the temperature and/or decreasing the salt concentration. Purify unique sequences
Probe contains highly repeated sequences	Increase the stringency by increasing the temperature and/or decreasing the salt concentration
Partial or incomplete restriction of the target DNA	Check restrictions before blotting
Smeared background within one or more lanes	
Genomic contamination of the probe	Ensure the probe is sufficiently purified
DNA or RNA degradation	Check the sample by running on a gel and staining with ethidium bromide
Insufficient blocking DNA in the prehybridization and/or hybridization buffers	Check the protocols for the correct blocking DNA concentration to use
Stringency of the washes is too low	Increase the stringency by increasing the temperature and/or decreasing the salt concentration
Cannot remove the probe prior to reprobing	
Filter has dried out after hybridization	DNA becomes bound by exclusion of water. The filter must stay moist during hybridization. Since ^{35}S-probed filters must be dried before autoradiography (see Chapter 4, Section 2.5), use of ^{35}S is not recommended if reprobing is desired
Loss of sensitivity after reprobing (normally, single-copy genes should be detectable after loading 5–10 μg of DNA on the gel, with 16–24 h autoradiography after the third reprobing)	
A problem with probe labelling, hybridization and/or washing in the reprobing step	See possible causes and remedies given above
Poor retention of DNA on the filter during the initial hybridization, probe removal and/or reprobing	Ensure that the UV irradiation of the filter to immobilize the DNA is optimal (see Section 3.15)

ard metaphase chromosome preparations. The demands of genome mapping studies, which require discrimination between DNA sequences that are close together on a chromosome, stimulated the development of higher-resolution FISH techniques, which centred on novel ways of preparing the chromosomes prior to hybridization (68). Mechanically stretched chromosomes can be generated by including a centrifugation step in the preparation protocol. These chromosomes are up to 20 times the normal metaphase length and provide a fivefold increase in resolution for gene mapping. However, the most important advance in the basic FISH technique uses purified DNA rather than chromosomes as the target material. The DNA is stretched over the surface of a coverslip and then hybridized with the probe. This is called fibre-FISH and enables markers that are less than 10 kb apart to be distinguished, compared with a lower limit of 1 Mb for metaphase chromosomes and 200–300 kb for mechanically stretched preparations.

4.3 Microarray and chip technologies

The most important recent development in hybridization analysis is probably the development of technologies for probing high-density arrays of immobilized DNA molecules and oligonucleotides. Two types of array are currently in use:

(a) A microarray is essentially a large dot-blot, but with the DNA samples immobilized on glass slides rather than filters (70). Densities of 80×80 dots per 18×18 mm slide are routinely used. Microarrays are primarily used in transcriptome studies, with cDNA probes used to identify genes that are expressed in particular tissues or in response to environmental changes, or to determine the expression patterns of genes whose functions are unassigned (71). Fluorescent probes are often used with the hybridization pattern across the array monitored by confocal microscopy.

(b) A DNA chip is a thin wafer of silicon on which an array of oligonucleotides has been synthesized (72). A density of up to one million oligonucleotides per cm^2 is possible, with hybridization between an oligonucleotide and the probe detected electronically. DNA chips are used for the same purposes as microarrays and can also be used to type DNA sequence variations such as single nucleotide polymorphisms.

References

1. Nygaard, A. P. and Hall, B. D. (1963). *Biochem. Biophys. Res. Commun.*, **23**, 98.
2. Southern, E. M. (1975). *J. Mol. Biol.*, **90**, 503.
3. Nagamine, Y., Sentenac, A., and Fromageot, P. (1980). *Nucl. Acids Res.*, **8**, 2453.
4. Haas, M., Vogt, M., and Dulbecco, R. (1972). *Proc. Natl Acad. Sci. USA*, **69**, 2169.
5. Reed, K. C. and Mann, D. A. (1985). *Nucl. Acids Res.*, **13**, 7207.
6. Church, G. M. and Gilbert, W. (1984). *Proc. Natl. Acad. Sci. USA*, **81**, 1991.
7. Li, J. K., Parker, B., and Kowalic, T. (1987). *Anal. Biochem.*, **163**, 216.
8. Alwine, J. C., Kemp, D. J., and Stark, G. R. (1977). *Proc. Natl. Acad. Sci. USA*, **74**, 5350.
9. Wahl, G. M., Stern, M., and Stark, G. R. (1979). *Proc. Natl. Acad. Sci. USA*, **76**, 3683.

10. Noyes, B. E. and Stark, G. R. (1975) *Cell*, **5**, 301.
11. Seed, B. (1982). *Nucl. Acids Res.*, **10**, 1799.
12. Parnes, J. R., Velan, B., Felsenfeld, A., Ramanathan, L., Ferrini, U, Apelia, E., and Seidman, J. G. (1981). *Proc. Natl. Acad. Sci. USA*, **78**, 2253.
13. Thomas, P. S. (1980). *Proc. Natl. Acad. Sci. USA*, **77**, 5201.
14. Benton, W. D. and Davis, R. W. (1977). *Science*, **196**, 180.
15. Grunstein, M. and Hogness, D. (1975). *Proc. Natl. Acad. Sci. USA*, **72**, 3961.
16. Hanahan, D. and Meselson, M. (1980). *Gene*, **10**, 63.
17. Thomas, P. S. (1980). *Proc. Natl. Acad. Sci. USA*, **77**, 5201.
18. Lis, J. T., Prestidge, L., and Hogness, D. S. (1978). *Cell*, **14**, 901.
19. Meinkoth, J. and Wahl, G. (1984). *Anal. Biochem.*, **138**, 267.
20. Chomczynski, P. (1992). *Anal. Biochem.*, **201**, 134.
21. Khandjian, E. W. (1987). *Biotechnology*, **5**, 165.
22. Towbin, H., Stachelin, T., and Gordon, J. (1979). *Proc. Natl. Acad. Sci. USA*, **76**, 4350.
23. Bittner, M., Kupferer, P., and Morris, C. F. (1980). *Anal. Biochem.*, **102**, 459.
24. Stellway, E. J. and Dahlberg, A. E. (1980). *Nucl. Acids Res.*, **8**, 299.
25. Smith, M. R., Devine, C. S., Cohn, S. M., and Lieberman, M. W. (1984). *Anal. Biochem.*, **137**, 120.
26. Peferoen, M., Huybrechts, R., and De Loof, A. (1982). *FEBS Lett.*, **145**, 369.
27. Brown, T. A. (1993). In *Current protocols in molecular biology* (ed. F. M. Ausubel, R. Brent, R. E. Kingston, D. D. Moore, J. G. Seidman, J. A. Smith, and K. Struhl). p. 2.9.7. John Wiley, New York.
28. Alwine, J. C., Kemp, D. J., Parker, B. A., Reiser, J., Renart, J., Stark, G. R., and Wahl, G. M. (1979). In *Methods in enzymology* (ed. R. Wu), Vol. 68, p. 220. Academic Press, New York.
29. Schildkraut, C. and Lifson, S. (1965). *Biopolymers*, **3**, 195.
30. Marmur, J. G. and Doty, P. (1961). *J. Mol. Biol.*, **3**, 584.
31. Melchior, W. B. and von Hippel, P. H. (1973). *Proc. Natl. Acad. Sci. USA*, **70**, 298.
32. McConaughy, B. L., Laird, C. L., and McCarthy, B. J. (1969). *Biochemistry*, **8**, 3289.
33. Casey, J. and Davidson, N. (1977). *Nucl. Acids Res.*, **4**, 1539.
34. Kafatos, F. C., Jones, C. W., and Efstratiadis, A. (1979). *Nucl. Acids Res.*, **7**, 1541.
35. Schmeckpepper, B. J. and Smith, K. D. (1972). *Biochemistry*, **11**, 1319.
36. Renz, M. and Kurz, C. (1984). *Nucl. Acids Res.*, **12**, 3435.
37. Bonner, T. T., Brenner, D. J., Neufield, B. R., and Britten, R. J. (1973). *J. Mol. Biol.*, **81**, 123.
38. Hyman, R. W., Brunovskis, I., and Summers, W. C. (1973). *J. Mol. Biol.*, **77**, 189.
39. Wallace, R. B., Shaffer, J., Murphy, R. F., Bonner, J., Hirose, T., and Itakura, K. (1979). *Nucl. Acids Res.*, **6**, 3543.
40. Kennel, D. E. (1971). *Progr. Nucl. Acid Res. Mol. Biol.*, **11**, 259.
41. Wetmur, J. G. (1976). *Ann. Rev. Biophys. Bioeng.*, **5**, 337.
42. Flavell, R. A., Birfelder, E. J., Sanders, J. P., and Borat, P. (1974). *Eur. J. Biochem.*, **47**, 535.
43. Birnsteil, M. L., Sells, B. H., and Purdom, I. F. (1972). *J. Mol. Biol.*, **63**, 21.
44. Wetmer, J. G. and Davidson, N. (1986). *J. Mol. Biol.*, **31**, 349.
45. Chang, C. T., Hain, T. C., Hutton, J. R., and Wetmur, J. G. (1974). *Biopolymers*, **13**, 1847.
46. Wetmer, J. G. (1971). *Biopolymers*, **14**, 2517.
47. Wahl, G. M., Stern, M., and Stark, G. R. (1979). *Proc. Natl. Acad. Sci. USA*, **76**, 3683.
48. Amasino, R. M. (1986). *Anal. Biochem.*, **152**, 304.
49. Hutton, J. R. (1977). *Nucl. Acids Res.*, **4**, 3537.
50. Howley, P. M., Israel, M. F., Law, M.-F., and Martin, M. A. (1979). *J. Biol. Chem.*, **254**, 4876.
51. Casey, J. and Davidson, N. (1977). *Nucl. Acids Res.*, **4**, 1539.

52. Britten, R. J. and Kohne, D. E. (1968). *Science*, **161**, 529.

53. Hutton, J. R. and Wetmur, J. G. (1973). *J. Mol. Biol.*, **77**, 495.

54. Chamberlin, M. E., Galau, G. A., Britten, R. J., and Davidson, E. H. (1978). *Nucl. Acids Res.*, **5**, 2073.

55. Botchan, M., Topp, W., and Sambrook, J. (1976). *Cell*, **9**, 269.

56. Denhardt, D. (1966). *Biochem. Biophys. Res. Commun.*, **23**, 641.

57. Singh, L. and Jones, K. W. (1984). *Nucl. Acids Res.*, **12**, 5627.

58. Johnson, D. A., Gautsch, J. W., Sportsman, J. R., and Elder, J. H. (1984). *Gen. Anal. Tech.*, **1**, 3.

59. Cos, K. H., DeLeon, D. V., Angerer, L. M., and Angerer, R. C. (1984). *Dev. Biol.*, **101**, 485.

60. Gillespie, D. and Spiegelman, S. (1965). *J. Mol. Biol.*, **12**, 829.

61. Suggs, S. V., Wallace, R. B., Hirose, T., Kiwashima, E. H., and Itakura, K. (1981). *Proc. Natl. Acad. Sci. USA*, **78**, 6613.

62. Wallace, R. B., Johnson, M. J., Hirose, T., Miyake, T., Kiwashima, E. H., and Itakura, K. (1981). *Nucl. Acids Res.*, **9**, 879.

63. Wood, W. I., Gitschier, J., Lasky, L. A., and Lawn, R. M. (1985). *Proc. Natl. Acad. Sci. USA*, **82**, 1585.

64. Jacobs, K. A., Rudersdorf, R., Neill, S. D., Dougherty, J. P., Brown, E. L., and Fritsch, E. F. (1988). *Nucl. Acids Res.*, **16**, 4637.

65. Sim, G. K., Kaftos, F. C., Jones, C. W., Koehler, M. D., Efstratiadis, A., and Maniatis, T. (1979). *Cell*, **18**, 1303.

66. Brown, T.A. (ed.) (1998). *Molecular biology labfax*, 2nd edn, Vol. 2, p. 15. Academic Press, London.

67. Seidman, J. G. (1998). In *Current protocols in molecular biology* (ed. F. M. Ausubel, R. Brent, R. E. Kingston, D. D. Moore, J. G. Seidman, J. A. Smith, and K. Struhl), p. 14.0.1. John Wiley, New York.

68. Heiskanen, M., Peltonen, L., and Palotie, A. (1996). *Trends Genet.*, **12**, 379.

69. Lichter, P. (1997). *Trends Genet.*, **13**, 475.

70. van Hal, N. L. W., Vorst, O., van Houwelingen, A. M. M. L., Kok, E. J., Peijnenburg, A., Aharoni, A. *et al.* (2000). *J. Biotechnol.*, **78**, 271.

71. DeRisi, J. L., Iyer, V. R., and Brown, P. O. (1997). *Science*, **278**, 680.

72. Ramsay, G. (1998). *Nature Biotechnol.*, **16**, 40.

Chapter 6
DNA sequencing

T. A. Brown

Department of Biomolecular Sciences, University of Manchester Institute of
Science and Technology, Manchester M60 1QD, UK

1 Introduction

The rapid generation of lengthy, contiguous DNA sequences is one of the corner-
stones of molecular biology. Not every molecular biologist wishes to sequence an
entire genome, but most will, at some stage in their careers, sequence a cloned
fragment from a genomic or cDNA library or use sequencing to check the
construction of a recombinant DNA molecule. This chapter describes the
methodology for carrying out these tasks.

Virtually all DNA sequencing is performed using the chain-termination
method first devised by Sanger et al. (1) in the 1970s. This technique involves a
DNA synthesis reaction in which a polymerase enzyme copies a DNA template
into new strands of DNA, each of which is terminated with a modified nucleotide
that blocks further strand extension. The modified nucleotides are 2',3'-dideoxy-
nucleoside triphosphates (ddNTPs) which cause chain termination because they
lack the 2'-hydroxyl that normally participates in phosphodiester bond synthesis.
Therefore, when the reaction is carried out in the presence of a ddNTP, chain
termination occurs at positions opposite complementary nucleotides in the template DNA (e.g. opposite thymidines if ddA is used). The strand synthesis reaction
is carried out four times in parallel, once with each of ddA, ddC, ddG and ddT,
and the products are examined to determine the lengths of the chain-terminated
molecules that have been synthesized. This can be achieved by electrophoresis
in thin, denaturing polyacrylamide gels in which molecules differing in length
by just a single nucleotide can be separated into a series of bands. The products
of the four reactions are run in four adjacent lanes and the sequence read from
the bottom of the gel upwards. The band that has moved the furthest is located
first: this represents the strand that has been terminated by incorporation of the
dideoxynucleotide at the first position in the template. The track in which this
band occurs is noted because this indicates the identity of the first nucleotide in
the sequence. The next band up corresponds to a DNA molecule that is one
nucleotide longer and hence indicates the identity of the second nucleotide in
the sequence. The process is continued until the individual bands become too
close together to be distinguished.

Manual DNA sequencing, in which the strand synthesis reactions are set up by hand, and the polyacrylamide gel is prepared, loaded and run by hand, and the sequence is read from an autoradiograph by a combination of hand and eye, is a cheap, accurate and rapid way of generating a DNA sequence. Most of this chapter describes how to carry out manual sequencing. Automated DNA sequencing, in which the strand synthesis reactions are carried out by hand before loading into a machine which separates the chain-terminated molecules and automatically reads the sequence, is expensive (both to purchase the machine and to perform individual sequencing experiments) but is an equally accurate and rapid way of generating a DNA sequence. Section 6 gives an introduction to automated sequencing. Thermal cycle sequencing, which is a type of chain-termination sequencing that uses techniques from PCR to generate the sequence, is covered along with other aspects of PCR in Chapter 7. For details of the second technique for carrying out DNA sequencing, the chemical degradation method of Maxam and Gilbert (2), which is now only used for specialist applications, the reader should consult a more advanced molecular biology manual (3).

2 Preparation of template DNA

The early methodology for chain-termination sequencing was dependent on the availability of single-stranded versions of cloned genes to act as the templates for the DNA sequencing reactions. The M13mp series of vectors (see Volume I, Chapter 9, Section 5) were primarily developed for this purpose and are still widely used for the generation of template DNA. Preparation of single-stranded DNA from M13 phage particles (Section 2.1) is relatively easy and the resulting DNA is very pure if the procedure has been carried out correctly. However, high purity is no longer such a strict requirement, since the special types of DNA polymerase that are now used in DNA sequencing are less sensitive to contamination than Klenow DNA polymerase, the enzyme on which the early methodology was based. This means that mini-preparations of plasmid DNA (Section 2.2) can now be used as the template for chain-termination sequencing. PCR products can also be sequenced, as described in Chapter 7.

2.1 Preparation of single-stranded DNA from M13 clones

The great advantage of using M13 as a cloning vector for sequencing is that phage particles containing single-stranded DNA are released from infected cells without lysis (see Volume I, Chapter 9, Section 5). The cells can be removed from the culture by a low-speed centrifugation, leaving a phage suspension that contains 10^{11}–10^{12} particles/ml and which is free from cellular debris. Preparation of single-stranded DNA from the suspension simply involves precipitation of the phage particles with PEG, removal of the protein coats by phenol extraction and concentration of the DNA by ethanol precipitation. A 1.5 ml culture grown in a glass tube and extracted as described in *Protocol 1* will yield a few micrograms of single-stranded DNA, sufficient for two or three sequencing experiments.

The quality of the sequence that is obtained depends very much on the quality of the DNA preparation and this is frequently an area where the beginner finds difficulty, although practice is often all that is needed. The notes in *Protocol 1* describe the most frequent problems and how to avoid them. It is easy to process 10–20 infected cultures in parallel so that template DNA from a number of different clones can be obtained at the same time. If a larger number of clones is being dealt with, then the extraction procedure can be scaled down and carried out in microtitre trays (*Protocol 2*). Each preparation now provides enough DNA for one or two sequencing experiments.

Protocol 1

Preparation of single-stranded DNA from M13 clones

Equipment and reagents

- Shaking incubator
- Toothpicks or Pasteur pipettes
- DYT medium (16 g Bacto-tryptone, 10 g Bacto-yeast extract, 5 g NaCl per 1 l). Check the pH and adjust to 7.0–7.2 with NaOH. Sterilize by autoclaving at 121 °C, 103.5 kPa (15 lbf/in^2), for 20 min
- 20% PEG 6000, 2.5 M NaCl

- TE buffer, pH 8.0 (see Appendix 4, Section 2)
- Buffer-saturated phenol (see Appendix 4, *Protocol 6*)
- 3 M sodium acetate, pH 5.5, prepared by adding glacial acetic acid to 3 M sodium acetate until this pH is obtained
- Chilled absolute ethanol

Method

1. Inoculate about 5 ml of DYT medium with a colony of a suitable host strain of *E. coli*. Incubate overnight at 37 °C without shaking.[a]

2. Briefly agitate the overnight culture to resuspend the bacteria. Add 15 µl aliquots to 1.5 ml aliquots of fresh DYT medium, in 10-ml sterile glass tubes with loose metal caps.[b]

3. Transfer a single plaque into each tube, with a toothpick or Pasteur pipette.[c]

4. Shake the tubes at 300 r.p.m. for 5–7 h. Good aeration is necessary.[d]

5. Transfer as much as possible of each culture to a 1.5 ml microfuge tube, and centrifuge for 5 min at maximum speed.

6. Remove 1 ml of each supernatant to a fresh tube, taking great care not to dislodge any of the bacterial pellet.[e] At this stage the supernatants may be stored for a day or so at 4 °C. If this is done, it is advisable to add 5 µl of chloroform to each to prevent bacterial growth and to centrifuge again before proceeding with step 7.

7. To each supernatant, add 200 µl of 20% PEG 6000, 2.5 M NaCl. Mix well and leave at room temperature for 30 min.

8. Centrifuge for 5 min. This should produce a pinhead-sized phage pellet.

9. Pour off the supernatants, turn each tube upside down to drain, and then carefully wipe the inside of each tube with an absorbent tissue.[f]

159

Protocol 1 continued

10 Resuspend each pellet in 100 μl TE buffer, pH 8.0, and add 100 μl buffer-saturated phenol. Vortex for 15 s and leave at room temperature for 5 min.

11 Vortex again, and centrifuge at maximum speed for 5 min. This should produce a sharp interface between the upper aqueous layer and the lower phenol layer. Transfer the aqueous layer to a fresh tube, add 10 μl of 3 M sodium acetate, pH 5.5, and 0.25 ml of chilled ethanol.[g]

12 Leave at −20 °C overnight or at −80 °C for 30 min to precipitate the DNA.

13 Collect the DNA by centrifugation at maximum speed at 4 °C for 10 min and pour off the ethanol.

14 Wash the pellet by adding another 1 ml of chilled ethanol. Centrifuge at maximum speed at 4 °C for 10 min and pour off the ethanol.

15 Allow the pellet to air-dry and then dissolve in 20 μl TE buffer, pH 8.0. Store at −20 °C, and avoid repeated freeze-thawing.[h]

[a] *Escherichia coli* XL1-Blue is an excellent host strain but any of those allowing Lac selection are suitable (see Appendix 3, Section 1.2). Note that the host strain should be stored on a proline selection plate, otherwise loss of the F′ plasmid from some cells may occur, resulting in low phage titres (see Appendix 3, Section 2.4).

[b] A rich medium such as DYT is recommended in order to maximize culture growth and phage yield.

[c] The M13 plaques should be used as soon as possible after plating out, preferably within 24 h, as longer storage on agar, even at 4 °C, results in deletion of all or part of the cloned DNA.

[d] The growth conditions for the infected culture are critical. Temperature control is important for good phage yields, so avoid reopening the incubator during culture growth as this may lead to a short 'cool-shock'. Angle the tubes to increase aeration. Do not incubate for longer than 7 h as this will result in the cells reaching the death phase and they will start to lyse, contaminating the supernatant with cell debris.

[e] One millilitre of supernatant is plenty. Do not be tempted to take more as this will inevitably lead to some cellular material being transferred.

[f] Careful removal of PEG is important as carryover will inhibit the action of the polymerase used in the DNA-sequencing reactions. Do not touch the phage pellet with the absorbent tissue. Aspiration with a pipette connected to a water pump is a possible alternative but this must be done gently to avoid dislodging the pellet.

[g] Avoid disturbing the interface between the organic and aqueous layers and do not transfer any phenol. Some protocols include a chloroform extraction at this stage, to ensure all phenol is removed from the aqueous layer. A chloroform step introduces its own contamination problems and is not recommended.

[h] Single-stranded DNA is not as stable as double-stranded DNA: ideally the template DNA should be used within 2 days of preparation. Cloned DNA breakdown also occurs in plaques and phage suspensions. If it is necessary to store material for future sequencing it is best to prepare a frozen or freeze-dried culture of transfected bacteria (see Volume I, Chapter 2, Section 4) and to repeat the template preparation when more DNA is needed.

Protocol 2

Preparation of single-stranded DNA in microtitre trays

Equipment and reagents

- Microtitre tray
- Shaking incubator
- Centrifuge with adapters for microtitre trays
- Toothpicks or Pasteur pipettes
- DYT medium (16 g Bacto-tryptone, 10 g Bacto-yeast extract, 5 g NaCl per 1 l). Check the pH and adjust to 7.0–7.2 with NaOH. Sterilize by autoclaving at 121 °C, 103.5 kPa (15 lbf/in^2), for 20 min

- Multipipette
- 20% PEG 6000, 2.5 M NaCl
- TE buffer, pH 8.0 (see Appendix 4, Section 2)
- Buffer-saturated phenol (see Appendix 4, *Protocol 6*)
- 3 M sodium acetate, pH 5.5, prepared by adding glacial acetic acid to 3 M sodium acetate until this pH is obtained
- Chilled absolute ethanol

Method

1. Prepare an overnight culture as described in *Protocol 1*, step 1. Adjust the OD$_{600}$ to 0.2 with DYT medium and place 200 μl in each well of a microtitre tray.

2. Transfer a single plaque into each well, with a toothpick or Pasteur pipette. Incubate with rapid shaking (300 r.p.m.) at 37 °C for 4–5 h.

3. Spin the microtitre tray in a plate centrifuge for 3 min at maximum speed.

4. Transfer 150 μl of each supernatant to a new microtitre plate using a multipipette.

5. Add 50 μl 20% PEG 6000, 2.5 M NaCl. Leave at room temperature for 30 min with shaking.

6. Spin in a plate centrifuge for 10 min at maximum speed.

7. Remove the supernatant with a multipipette.

8. Resuspend each pellet in 75 μl TE buffer, pH 8.0, add an equal volume of buffer-saturated phenol and shake for 10 min.

9. Spin in the plate centrifuge for 10 min at maximum speed, and remove the phenol layer.

10. Add 5 μl 3 M sodium acetate, pH 5.5, and 130 μl ethanol to each well. Leave at −20 °C overnight or at −80 °C for 30 min to precipitate the DNA.

11. Collect the DNA by centrifugation at maximum speed for 10 min and remove the ethanol.

12. Wash each pellet by adding another 200 μl of chilled ethanol. Centrifuge at maximum speed for 10 min and remove the ethanol.

13. Allow the pellets to air-dry and then dissolve each one in 10 μl TE buffer, pH 8.0. Store at −20 °C, and avoid repeated freeze-thawing.

It is sometimes useful to obtain a rough estimate of the sizes of the inserts in M13 DNA preparations and to know their orientations. Both can be determined by agarose gel electrophoresis of an aliquot of phage supernatant to which SDS has been added to strip off the protein coats. In *Protocol 3*, the size of the single-stranded DNA is estimated by comparison with preparations whose insert sizes are known. In *Protocol 4*, the orientation of the insert is checked by probing with a second single-stranded DNA whose orientation is known. Hybridization, which shows that the target and probe inserts have *opposite* orientations, is revealed by a change in mobility when the gel is run.

Protocol 3

Size analysis of M13 DNA preparations

Equipment and reagents

- Agarose gel and equipment for gel electrophoresis
- Phage supernatant (see *Protocol 1*, step 6)
- 2% SDS
- 6 × agarose gel loading buffer (see Volume I, Chapter 5, *Table 3*)

Method

1 Mix 20 μl phage supernatant with 1 μl SDS and 4 μl of a 6 × agarose gel loading buffer.

2 Run in an agarose gel with single-stranded phage DNA preparations with known inserts sizes as markers.

3 Stain with ethidium bromide and view by UV transillumination.

Protocol 4

Determining insert orientation in M13 template DNA

Equipment and reagents

- Agarose gel and equipment for gel electrophoresis
- Phage supernatants (see *Protocol 1*, step 6), including one whose insert orientation is known
- 2% SDS
- 20× SSC (see Chapter 5, *Table 2*)
- Light mineral oil
- 6 × agarose gel loading buffer (see Volume I, Chapter 5, *Table 3*)

Method

1 Add 20 μl of phage supernatant to 20 μl of supernatant whose insert orientation is known.

2 Add 1 μl 2% SDS and 4 μl of 20× SSC, and overlay with a small volume of light

mineral oil to prevent evaporation. Incubate at 65 °C for 1 h. It is useful to prepare a control with two phage suspensions whose inserts are known to be opposite orientations.

3   Remove the sample, being careful not to transfer any mineral oil, add 4 μl of 6× agarose gel loading buffer, and electrophorese in an agarose gel. Stain with ethidium bromide and view by UV transillumination. Note that the hybridization rarely goes to completion, so complementarity between the inserts is indicated by the presence of three bands, one from each of the separate DNAs, and one from hybridized molecules.

2.2 Preparation of template DNA from double-stranded plasmid clones

Most routine cloning experiments are performed with plasmid vectors and, frequently, the DNA to be sequenced is initially available in this form. Rather than re-cloning in an M13 vector, it is possible to use double-stranded plasmid DNA as the template for DNA sequencing, providing that the preparation method scrupulously removes contaminating cellular RNA. This is important because short RNA fragments can prime DNA synthesis, leading to spurious products during the sequencing reactions and additional bands on the polyacrylamide gel from which, in this case, the sequence cannot be read. *Protocol 5* (modified from ref. 4) is based on the alkaline lysis method for mini-preparation of plasmid DNA (5), with RNA removed by precipitation with lithium chloride. An alternative method is to treat the mini-prep with RNase A, followed by phenol extraction, but this can leave short oligoribonucleotides that are excellent DNA synthesis primers.

Immediately before use in sequencing, double-stranded plasmid DNA must be denatured to give single-stranded regions to act as templates for strand synthesis. This involves treatment with alkali as described in *Protocol 5* (modified from ref. 4). It is not necessary to linearize the plasmids before denaturation.

Regardless of how good the DNA preparative method is, sequencing with double-stranded DNA will be unsuccessful if the plasmids contain nicks. These will act as priming sites for DNA synthesis and, as with RNA contaminants, lead to the generation of spurious products during the sequencing reactions. The extent to which a plasmid DNA preparation is nicked depends not so much on the extraction method as on the characteristics of the host strain in which the plasmids are cloned. The host strain should always be *endA1* because the endonuclease coded by the active version of this gene is the main cause of plasmid nicking. *E. coli* XL1-Blue (see Appendix 3, Section 1.2) is a good choice if the vector is Lac selectable.

It is worth remembering that many of the commercial plasmid vectors that are frequently used in cloning experiments are phagemids, meaning that they carry filamentous phage origins of replication and can direct the synthesis of single-stranded DNA, secreted as extracellular phage particles (Volume I, Chapter 9, Sections 3.3 and 3.8). With one of these vectors, it is quicker and easier to prepare single-stranded rather than double-stranded DNA. Follow *Protocol 12* of Volume I, Chapter 8 to generate phage particles and then prepare single-stranded DNA as described in *Protocol 1* in this chapter. The resulting template DNA will give a clearer and longer DNA sequence than the double-stranded version of the phagemid. This presumes that the vector directs high-level synthesis of phage particles—some do not. To check this, assay the titre of the phage suspension by the method given in *Protocol 10* of Volume I, Chapter 2 (use DYT medium rather than SM buffer to prepare the phage dilutions). If the titre is $> 10^{11}$ then consider using the phagemid technology to prepare sequencing templates rather than making double-stranded DNA.

Protocol 5

Preparation of double-stranded DNA for sequencing

Equipment and reagents

- Vacuum desiccator
- Vortex mixer
- LB medium (10 g Bacto-tryptone, 5 g Bacto-yeast extract, 10 g NaCl per l). Check the pH and adjust to 7.0–7.2 with NaOH. Sterilize by autoclaving at 121 °C, 103.5 kPa (15 lbf/in²), for 20 min. See Appendix 4 for preparation and addition of antibiotics
- Glucose buffer (25 mM Tris-HCl, pH 8.0, 10 mM EDTA, 50 mM glucose)
- Lysis buffer (0.2 M NaOH, 1.0% SDS)
- Isopropanol

- 3 M sodium acetate, pH 4.8, prepared by adding glacial acetic acid to 3 M sodium acetate until this pH is obtained
- TE buffer, pH 8.0 (see Appendix 4, Section 2)
- 4 M LiCl
- Buffer-saturated phenol (see Appendix 4, *Protocol 6*)
- 24:1 chloroform–isoamyl alcohol (see Appendix 4, *Protocol 6*)
- Chilled 70% ethanol

Method

1 Prepare a 5-ml overnight culture of the host bacterium containing the recombinant plasmid, in LB medium containing the appropriate antibiotics.

2 Transfer 50 µl of the overnight culture to 5 ml of LB with antibiotics, and incubate with shaking until the OD_{600} reaches 1.0.[a]

3 Transfer as much as possible of the culture to a 1.5 ml microfuge tube, and pellet the cells by centrifugation for 5 min at maximum speed.

4 Remove the supernatant and resuspend the pellet in 150 µl glucose buffer. Leave at room temperature for 5 min.

5 Add 500 µl lysis buffer. Gently swirl or invert the tube to mix and leave at room temperature for 5 min.

6 Add 225 μl 3 M sodium acetate, pH 4.8. Gently swirl or invert the tube to mix. Place on ice for 45 min.

7 Spin in the microfuge at maximum speed for 5 min. Remove the upper 500 μl of the supernatant.[b] To this supernatant, add 500 μl isopropanol. Mix and leave at room temperature for 10 min.

8 Collect the DNA pellet by centrifugation at maximum speed for 5 min. Pour off the supernatant and dry the pellet in a vacuum desiccator.

9 Resuspend the pellet in 125 μl TE buffer, pH 8.0. Add 375 μl 4 M LiCl. Place on ice for 30 min.

10 Remove the precipitated RNA by centrifugation at maximum speed for 5 min at 4°C. Pour off and retain the supernatant.

11 Add 500 μl buffer-saturated phenol to the supernatant. Vortex for 15 s, and leave at room temperature for 5 min.

12 Vortex again, and centrifuge at maximum speed for 5 min. This should produce a sharp interface between the upper aqueous layer and the lower phenol layer. Transfer the aqueous layer to a fresh tube, and add 500 μl of 24:1 chloroform–isoamyl alcohol. Vortex for 15 s, and leave at room temperature for 5 min.

13 Vortex again, and centrifuge at maximum speed for 5 min. Transfer the aqueous layer to a fresh tube, and add two volumes of isopropanol. Mix and leave at room temperature for 30 min.

14 Collect the DNA pellet by centrifugation at maximum speed for 5 min. Discard the isopropanol and wash the pellet by adding 1 ml of chilled 70% ethanol. Centrifuge at maximum speed for 10 min at 4°C and pour off the ethanol.

15 Allow the pellet to air-dry and then dissolve in 50 μl TE pH 8.0.[c] Store at −20°C, and avoid repeated freeze-thawing.[d]

[a] This is mid-log phase. If the culture reaches the end of the exponential phase then the plasmid DNA will be extensively nicked and unsuitable as a sequencing template.

[b] The interface between the supernatant (containing supercoiled plasmid DNA) and 'pellet' (containing chromosomal DNA and cell debris) is indistinct. To avoid transferring debris, do not take more than 500 μl and remove this sample from the upper part of the supernatant.

[c] The yield of plasmid DNA can be checked by running 5 μl in an agarose gel. The ideal result is a single, clearly-visible band of supercoiled plasmid DNA. If two bands are seen, then some open-circular DNA is present, meaning that the preparation contains nicked molecules. If the upper, open-circular band has a similar or greater intensity than the lower, supercoiled band then the DNA is unsuitable for DNA sequencing, unless a short sequence (< 250 bp) is sufficient. There should be no background smear of chromosomal DNA and no fuzzy band of RNA near the bottom of the gel.

[d] Double-stranded DNA can be stored at −20°C for several months with little effect on its quality as a sequencing template.

Protocol 6

Denaturation of double-stranded DNA[a]

Equipment and reagents

- Vacuum desiccator
- 3 M sodium acetate, pH 6.0, prepared by adding glacial acetic acid to 3 M sodium acetate until this pH is obtained
- 2 M NaOH, 2 mM EDTA
- Chilled absolute and 70% ethanol
- TE buffer, pH 8.0 (see Appendix 4, Section 2)

Method

1 To 20 μl of double-stranded template DNA, add 2 μl of 2 M NaOH, 2 mM EDTA. Mix and incubate at 37 °C for 5 min.

2 Add 7 μl of chilled water and place on ice.

3 Add 7 μl 3 M sodium acetate, pH 6.0. Mix and check that the resulting pH is 7.0 by spotting 1 μl on to pH paper. Add more sodium acetate if necessary.[b]

4 Add 75 μl of chilled absolute ethanol and place at −80 °C for 30 min.

5 Collect the DNA by centrifugation at maximum speed for 10 min at 4 °C. Pour off the ethanol.

6 Wash the pellet by adding 400 μl of chilled 70% ethanol. Centrifuge at maximum speed at 4 °C for 10 min and pour off the ethanol.

7 Dry the pellet in a vacuum desiccator and resuspend in 7 μl TE buffer, pH 8.0.

[a] Carry out this procedure immediately before the double-stranded DNA is used as a sequencing template.

[b] Seven microlitres of sodium acetate solution is usually sufficient but it is advisable to check that the sample has reached neutral pH, as described. If additional sodium acetate is needed then prepare a fresh batch of this solution before using it again.

3 The sequencing reactions

A number of commercial kits are available for DNA sequencing. These provide the necessary reagents in premixed form and minimize the possibility of sequence failure due to operator error. For routine work where the length and quality of the resulting sequence are not of paramount importance, a kit is the best option. However, it is still important to understand the roles of the various components of the sequencing reactions and, hence, to appreciate how to modify the reactions in order to achieve a longer sequence or a clearer definition of the sequence in a critical area of the template DNA

3.1 Components of a chain-termination sequencing reaction

3.1.1 The primer

The primer is an oligonucleotide of 15–25 bases that acts as the starting point for strand synthesis and forms the 5′-end of each chain-terminated molecule. Its annealing position on the template DNA is critical because this defines the region of the template that is sequenced. Often a 'universal' primer is used—one which anneals to the vector DNA at a position adjacent to the site into which the inserted DNA has been ligated. A universal primer can therefore be used to obtain sequence from any fragment that has been inserted into a particular vector. Their versatility is even greater than this because a single cloning region is frequently shared between a series of vectors. For example, in most of the vectors that use Lac selection the cloning sites are contained within the *lacZ'* gene (see Volume I, Chapter 9), so a universal primer designed to anneal to a region of *lacZ'* will work with all vectors of this type.

A single experiment will, at best, generate 500 bp of sequence. However, many DNA inserts are longer than this and so will not be entirely sequenced in a single experiment. One way to extend the sequence information from an insert is to carry out two experiments, one with a 'forward' primer and one with a 'reverse' primer. These are pairs of universal primers that anneal at opposite sides of a cloning site and enable the inserted DNA to be sequenced from both ends when a double-stranded template is used. Alternatively, a special primer can be synthesized to anneal within the insert DNA, using sequence information from a previous experiment to dictate the appropriate sequence for this primer. This strategy, along with others for obtaining a contiguous sequence of a lengthy DNA molecule, is discussed in Section 5.

3.1.2 The DNA polymerase

Chain-termination sequencing was initially carried out with Klenow polymerase, a modified version of DNA polymerase I of *E. coli* that lacks the $5' \rightarrow 3'$ exonuclease activity of the natural enzyme. The absence of this exonuclease is important because the activity could remove nucleotides from the 5'-end of a chain-terminated molecule—the end defined by the primer—and hence destroy the relationship between chain length and the terminating ddNTP that is essential for reading the sequence. Klenow polymerase was not perfect for DNA sequencing because of two limitations. First, it has relatively low processivity (i.e. the length of polynucleotide that it synthesizes before detaching from the template is relatively short) and a slow elongation rate, which combine to limit the length of sequence that can be obtained and to produce shadow bands at all positions in the polyacrylamide gel. Shadow bands are caused by chain termination occurring in the absence of ddNTP incorporation and their presence makes its more difficult to read the sequence. Second, Klenow polymerase also displays a variable discrimination between ddNTPs and normal nucleotides, its ability to use a ddNTP as a substrate depending to a certain extent on the sequence surrounding the nucleotide being copied. This variability causes differences in band intensity and a variety of artefacts (e.g. the band for the first C in a run of Cs is always light). These artefacts are well-characterized but their presence can results in errors in reading the sequence, especially in automated systems.

Most sequencing is now carried out with Sequenase, a modified version of the

DNA polymerase I enzyme of T7 bacteriophage. The initial version of this enzyme, Sequenase 1.0, was produced by chemical modification (6) but it has now been superseded by the genetically engineered Sequenase 2.0 (7). The natural T7 enzyme lacks the $5' \rightarrow 3'$ exonuclease and the purpose of the modifications is to delete the complementary $3' \rightarrow 5'$ exonuclease—the proofreading function—which is so active in the T7 polymerase that chain termination occurs ineffectively because most ddNTPs are removed from the 3'-terminus as soon as they are attached. Sequenase is highly processive and has a rapid elongation rate, so shadow bands are reduced to a minimum. It incorporates ddNTPs in a more uniform fashion than Klenow polymerase so there is less variation in band intensities.

3.1.3 The nucleotides

In manual sequencing the chain-terminated molecules are labelled with ^{32}P, ^{33}P or ^{35}S by including a radioactive nucleotide in the sequencing reactions. Which nucleotide is labelled is determined by the GC content of the template DNA. If the GC content is $> 50\%$ then maximum labelling is achieved with radioactive dCTP, and if the GC content is $< 50\%$ radiolabelled dATP is used. A short exposure time is possible with ^{32}P but the high activity of this label means that the chain-terminated molecules are unstable, so the electrophoresis has to be carried out immediately after the sequencing reactions, and the bands on the resulting autoradiograph tend to be fuzzy. Labelling with ^{33}P or ^{35}S necessitates a longer exposure but gives a clearer sequence.

Occasionally, a modified nucleotide, other than the ddNTPs, is included in the sequencing reaction. The objective is to reduce formation of hairpin loops, as these change the electrophoretic mobilities of the chain terminated molecules and result in 'compressions' on the autoradiograph, these being groups of bands that run closely together, making it difficult to read the sequence. dITP and 7-deaza-dGTP are modified versions of dGTP that do not form stable base pairs with dCTP and so reduce the occurrence of hairpin loops when substituted for dGTP in the sequencing reactions.

3.2 Sequencing reactions with Sequenase

Protocols 7 and *8* comprise a standard procedure for carrying out chain-termination sequencing with Sequenase.

3.2.1 The annealing step

In *Protocol 7*, the primer and template are annealed by heating to 65 °C and then cooling slowly to room temperature. Some protocols attempt to save time by transferring the mix directly to 37 °C, but experience shows that more efficient primer annealing, and hence a stronger sequence, occurs using the procedure given here.

The primer–template ratio is important, especially in double-stranded sequencing where the primer and complementary strand of the template have to compete for annealing sites. The relevant factors are:

(a) The single-stranded annealing mix should contain 0.5 pmoles of primer and a 1:1 molar ratio of primer to template. This corresponds to 1.25 µg of an M13 clone with a 500 bp insert.

(b) The double-stranded annealing mix should contain 1.0 pmol of primer and a 2:1 molar ratio of primer to template. This corresponds to 1.0 µg of a pUC18 clone with a 500 bp insert.

(c) The volumes of template DNA given in *Protocol 7*, step 1 should contain approximately 1.25 µg of single-stranded DNA or 1.0 µg of double-stranded DNA, (presuming that the template preparations described in *Protocols 1, 2* and *5* were carried out efficiently). M13mp vectors are large (7.2 kb), and inserts of different sizes have relatively little effect on the total molecular mass of the molecule. Therefore, the recipe shown in *Protocol 7* is appropriate for any M13 clone. However, plasmid vectors are relatively small (pUC18 is 2.7 kb) and insert size has a significant effect on molecular mass. If the insert is > 2 kb then change the volumes shown in *Protocol 7*, step 1(b) to 3 µl primer + 6 µl template.

Although the primer–template ratio is important, in practice it is difficult to estimate the actual amount of DNA in a template preparation, and usually it is assumed that the preparation is satisfactory, this opinion only being revised if the bands in the resulting sequence are faint (too little template DNA) or too intense (too much template DNA). *Protocols 1, 2* and *5* give sufficient template DNA for at least two sequences, so, if necessary, a repeat experiment can be carried out with a different amount of template.

After annealing, the template–primer mixture can be stored at $-20\,°C$ for several months with little appreciable decline in quality, but usually the procedure moves directly on to *Protocol 8*, in which the sequencing reactions are carried out.

3.2.2 Strand extension and chain termination

The sequencing reactions involve a pulse-chase, the template initially being copied in the presence of limiting amounts of dNTPs, including the label, with the intention that the first segments of DNA to be synthesized are highly labelled, so that all bands—even those for the shortest chain-terminated molecules—are clearly visible on the autoradiograph. In the second stage of the reaction, the initiated chains are extended to completion with non-limiting amounts of dNTPs and inclusion of the appropriate ddNTP in each reaction. This is called the 'labelling-termination' strategy and it allows the operator some flexibility in determining which part of the template molecule is studied. It is quite possible with Sequenase to synthesize strands up to 3000 bp long, but the sequence of only 500 bp of this region will be interpretable on the autoradiograph because the shorter molecules will have run off the bottom of the gel and the longer molecules bunch together so their bands cannot be distinguished. The position of this 'window' within the 3000 bp of potential sequence can be controlled by taking account of the following factors:

(a) Chain termination does not occur during the labelling step, so increasing the dNTP concentration in the labelling mix will move the lower limit of the window away from the primer. The labelling mix described in *Protocol 8* contains 1.5 μM of each unlabelled nucleotide and should yield a sequence covering the 500 bp closest to the primer. By increasing the concentration of each nucleotide to 7.5 μM a sequence from approximately 150–650 bp from the primer can be read.

(b) The time allowed for the labelling reaction can also have an impact on the position of the window. In practice, extending the 5-min, room temperature incubation recommended in *Protocol 8* has little effect because the concentration of the labelled nucleotide becomes limiting before this incubation is completed. However, reducing the incubation to 1 min, or carrying it out on ice, will ensure that the sequence closest to the primer can be read, although the bands on the autoradiograph might be quite weak. An alternative and usually more successful way to achieve the same result is to change to a primer that anneals at a position that is further upstream on the template (see ref. 8).

(c) The ratio of dNTPs and ddNTPs in the termination mixes determines the amount of strand synthesis that occurs after the labelling step. If the ddNTP concentration is high, then shorter chains are produced; if it is low then the chains are, on average, longer. Unfortunately, with manual sequencing the theoretical possibility of accessing more sequence by decreasing the ddNTP concentration cannot be realized because limitations with the resolving power of gel electrophoresis prevent more than 500 bp of sequence being read. All that happens is that the bands that can be resolved are less intense because some of the label has been sequestered in longer molecules whose bands are clumped at the top of gel.

Protocol 7

Primer annealing

Equipment and reagents

- 65°C water bath
- 10× Sequenase buffer (400 mM Tris-HCl, pH 7.4, 20 mM MgCl$_2$, 500 mM NaCl, 50 mM DTT). This can be stored at −20°C for several months

- Universal sequencing primer (0.5 pmol/μl; New England Biolabs, Sigma, Stratagene, Promega, etc.)
- Single- or double-stranded template DNA (from *Protocols 1, 2* or *5*)

Method

1 Prepare an annealing mix in a microfuge tube comprising either:

   (a) 1 μl 10× Sequenase buffer, 1 μl universal sequencing primer (0.5 pmol/μl) and 8 μl single-stranded template DNA, or

---

Protocol 7 continued

   (b) 1 μl 10× Sequenase buffer, 2 μl universal sequencing primer (0.5 pmol/μl) and 7 μl double-stranded template DNA.

2 Place in the 65 °C water bath and incubate for 10 min.

3 Switch off the water bath and allow to cool overnight to room temperature.[a]

4 Proceed to *Protocol 8* or the annealed primer–template mixture can be stored at −20 °C for several months.

[a] Alternatively, transfer water at 65 °C into a glass dish, place the annealing mix in this dish, and allow to cool to room temperature over a period of at least 30 min.

Protocol 8

Sequencing reactions

Equipment and reagents

- A-termination mix (80 μM dATP, 80 μM dCTP, 80 μM dGTP, 80 μM dTTP, 8 μM ddATP, 50 mM NaCl, prepared in 10 mM Tris-HCl, pH 7.4, 0.1 mM EDTA). Prepare termination mixes from 20 mM stock solutions. Both the stock solutions and the termination mixes can be stored at −20 °C for several months

- C-termination mix (80 μM dATP, 80 μM dCTP, 80 μM dGTP, 80 μM dTTP, 8 μM ddCTP, 50 mM NaCl, prepared in 10 mM Tris-HCl, pH 7.4, 0.1 mM EDTA)

- G-termination mix (80 μM dATP, 80 μM dCTP, 80 μM dGTP, 80 μM dTTP, 8 μM ddGTP, 50 mM NaCl, prepared in 10 mM Tris-HCl, pH 7.4, 0.1 mM EDTA)

- T-termination mix (80 μM dATP, 80 μM dCTP, 80 μM dGTP, 80 μM dTTP, 8 μM ddTTP, 50 mM NaCl, prepared in 10 mM Tris-HCl, pH 7.4, 0.1 mM EDTA)

- Sequenase 2.0 (Amersham Pharmacia), diluted to 2 units/μl in Sequenase dilution buffer (10 mM Tris-HCl, pH 7.4, 5 mM DTT, 0.5 mg/ml bovine serum albumin) immediately before use and kept on ice

- Annealed primer–template mixture (from *Protocol 7*)

- Labelling mix (1.5 μM dCTP, 1.5 μM dGTP, 1.5 μM dTTP, prepared in 10 mM Tris-HCl, pH 7.4, 0.1 mM EDTA). Can be stored at −20 °C for several months[a]

- [α-^{35}S]dATP (10 mCi/ml, 500 or 1200 Ci/mmol)[b]

- Formamide dye mix (0.1% bromophenol blue, 0.1% xylene cyanol, 10 mM EDTA, pH 8.0, prepared in deionized formamide—see Appendix 4, *Protocol 4*)

Method

1 To a microfuge tube labelled 'A', add 2.5 μl of A termination mix. Similarly, prepare 'C', 'G' and 'T' tubes.

2 To the annealed primer–template mixture, add 2 μl labelling mix, 1.0 μl [α-^{35}S]dATP and 2 μl Sequenase 2.0. Leave at room temperature for 5 min. This is the labelling step.[c]

Protocol 8 continued

3 Add 3.5 μl of the mix prepared in step 2 to each of the four tubes A, C, G and T prepared in step 1. Incubate at 37 °C for 10 min. This is the termination step.[d]

4 Add 4 μl formamide dye mix to each tube. This stops the enzyme reaction and adds the components needed for loading the samples on to the polyacrylamide gel.

5 Proceed immediately to gel loading (*Protocol 10*) or store at −20 °C for up to 1 week.

[a] If labelled dCTP is being used, replace this mix with 1.5 μM dATP, 1.5 μM dGTP, 1.5 μM dTTP, prepared in 10 mM Tris-HCl, pH 7.4, 0.1 mM EDTA.

[b] See Section 3.1.3 for the possible use of [α-^{35}S]dCTP as the label. ^{35}S- rather than ^{32}P labelling is recommended for sequencing with Sequenase.

[c] Sequenase gives a more uniform band intensity if labelling is carried out in the presence of Mn^{2+} (9). To exploit this feature, add 1 μl of 0.15 M sodium isocitrate, 0.1 M $MnCl_2$ immediately before the Sequenase.

[d] For sequencing in the presence of dITP (Section 3.1.3), replace the dGTP in the termination mixes with 160 μM dITP, and increase the ddGTP concentration in the G-termination mix to 1.6 μM. For 7-deaza-dGTP inclusion, replace the dGTP in the termination mixes with 80 μM 7-deaza-dGTP but do not change the ddGTP concentration.

4 Gel electrophoresis and autoradiography

4.1 Denaturing polyacrylamide gel electrophoresis

High-resolution polyacrylamide gels separate strands which differ in length by one nucleotide, the amount of readable sequence being dependent on the length of the gel and the running time. The gel must be electrophoresed under de-naturing conditions to ensure that hairpin loops and other intra-strand base pairs do not form, because these alter the mobility of a polynucleotide and hence change the banding pattern that is seen on the autoradiograph.

The gels used in DNA sequencing are usually 20 cm wide and between 40 and 50 cm long, the precise size depending on the gel tank. The gel is poured between two glass plates separated by a pair of spacers, one placed along each long edge. One of the plates has a notch cut out of the top to allow efficient contact between the electrophoresis buffer and the gel. To maximize resolution, the gel is as thin as possible, usually either 0.2 mm or 0.4 mm. Spacers and combs can be purchased commercially or prepared from polystyrene sheets, with care taken to machine cut the comb so that the teeth are flat-bottomed. As many as 50 wells, 2 mm wide and 3 mm deep, can easily be accommodated if the gel is 20 cm wide. Sequence reading is easier if the lanes have no spaces between them, which can be achieved if a 'sharkstooth' comb is used. Samples should only be loaded into the central 16 cm of a 20-cm gel, because 'smiling' (slower running at the edges of the gel) occurs as a result of unequal heat distribution across the width of the gel. 'Smiling' makes it difficult for the band order to be read accurately in lanes at the sides of the gel.

4.1.1 Preparation of the polyacrylamide gel

Protocol 9 describes how to prepare a thin denaturing polyacrylamide gel. The gel plates must be as clean and dust-free as possible, or air bubbles will form during pouring. Bubbles and particulate matter in the gel matrix cause streaking of the samples when the gel is run. If bubbles form, then sometimes they can be dislodged by placing the gel in a vertical position and tapping the plates. Alternatively, a bubble can be moved to the side of the gel by manoeuvring it with a strip of polystyrene sheet.

The gel must be left to polymerize for at least 30 min before the comb is removed; if the comb is removed before polymerization is complete, poor quality wells are produced that result in a sequence of low quality. If the gel is not going to be used immediately then it can be stored overnight at 4 °C, with the comb in place, provided that the top of the gel is covered with Saran wrap to prevent it drying out.

Protocol 9

Preparation of a denaturing polyacrylamide gel

Equipment and reagents

- Commercial gel mould comprising glass plates, spacers and comb[a]
- Siliconizing solution (2% dimethyl-dichlorosilane in 1,1,1-trichloroethane) (Merck). **Caution: siliconizing solutions are toxic and flammable**
- Acrylamide solution (38% acrylamide, 2% bisacrylamide, purchased as a ready-prepared solution from various suppliers). **CAUTION: Acrylamide is a neurotoxin; wear gloves and do not mouth pipette** (polymerized polyacrylamide is non-toxic and can be disposed of as non-hazardous waste)

- Urea (ultrapure, e.g. Merck)
- 10× TBE (89 mM Tris-borate, pH 8.3, 2 mM EDTA). Dissolve 108 g Tris-base + 55 g boric acid + 9.3 g EDTA in deionized water and make up to 1 l. Store at room temperature. After a few weeks a white precipitate may form: discard when this occurs
- 10% ammonium persulphate solution. This can be stored at 4 °C for a few weeks in the dark
- TEMED
- Absolute ethanol

Method

1 Clean the glass plates thoroughly in warm water. Do not use abrasives or detergents.

2 In a fume cupboard, wipe a few millilitres of siliconizing solution over the inner surface of the notched plate. Leave to dry.[b]

3 Wipe the inner surfaces of both plates with absolute ethanol and polish thoroughly, taking care to remove dust particles from the plate surfaces.

4 Assemble the gel mould following the manufacturer's instructions.

5 Place 25.2 g urea in clean glass beaker. Add in the following order: 9.0 ml acrylamide

solution, 6.0 ml 10× TBE buffer, 24.0 ml water, 600 μl 10% ammonium persulphate solution and 60 μl TEMED.[c]

6 Mix until the urea is dissolved and immediately pour into the gel mould using a 50-ml disposable syringe. Start pouring the gel with the mould at about 45° to the horizontal, and gradually decrease this angle as the mould fills up with gel mix. Try to keep the flow of gel mix between the plates continuous, as this will minimize the risk of air bubbles.

7 Insert the gel comb and leave to polymerize for at least 30 min. Check that the top of the gel level does not drop, and top up with the remaining gel mix if necessary.

[a] It is also possible to use a home-made gel mould. Use heat-resistant tape to seal the edges and bottom of the mould and place bulldog clips directly over the spacers after the gel is poured to prevent leakage. Insert the comb to a depth of about 0.5 cm.

[b] Siliconization is needed to prevent the gel from adhering to the notched plate when the mould is disassembled. After disassembly, the gel will be transferred to filter paper (*Protocol 13*) and adherence to the backing plate can be a problem at this stage. However, siliconization of the backing plate is not recommended as this makes more difficult the manipulations that must be carried out while the gel is being prepared for autoradiography.

[c] This recipe gives 60 ml of a 6% gel. This volume is sufficient for most commercial gel moulds and can be scaled up or down if necessary. Gels with lower acrylamide percentages enable more sequence to be read but are less easy to handle. The recipe for 60 ml of a 4% gel is 25.2 g urea, 6.0 ml acrylamide solution, 6.0 ml 10× TBE buffer, 27.0 ml water, 600 μl 10% ammonium persulphate solution and 60 μl TEMED. Ammonium persulphate is the catalyst for polymerization and TEMED is the activator.

4.1.2 The gel apparatus

Many types of vertical gel apparatus for running sequencing gels are commercially available. Some include temperature control systems designed to dissipate the temperature gradient that develops during running the gel, causing smiling. Passive temperature control systems use a heat sink of some description: either an aluminium sheet that is placed in contact with the gel mould, or a construction that brings a large volume of buffer into contact with one of the gel plates. Active control involves pumping the buffer from a reservoir which is temperature-controlled around a chamber that contacts one side of the gel mould. Active control has the advantage of allowing gels to be run at precisely-controlled temperatures so that, for example, compressions can be melted and resolved. However, the extra expense of actively-controlled systems generally outweighs their advantages.

4.1.3 Gel loading and running

Protocol 10 describes the final stages of preparation of the gel and procedures used to load samples and carry out the electrophoresis. As soon as the comb is removed, urea begins to leach out of the wells, and this must be flushed away immediately before loading because its presence results in diffuse bands. The samples are heat denatured and should then be loaded as quickly as possible to prevent renaturation. Loading can be performed with a 20-μl automatic pipette, ideally using a tip with a flattened end; these are marketed by several companies especially for loading thin polyacrylamide gels. To avoid cross-contamination between wells, the pipette used to load the samples should be thoroughly rinsed in the lower buffer reservoir between each application.

Protocol 10

Loading a sequencing gel

Equipment and reagents

- Polymerized denaturing gel within a gel mould (from *Protocol 9*)
- 80°C water bath
- Gel apparatus (available from various suppliers)
- 1× TBE buffer (see *Protocol 9*)

Method

1 Carry out any final preparation of the gel mould as described in the manufacturer's instructions. For a home-made gel mould (*Protocol 9*, footnote a), use a razor blade to slit the tape covering the lower edge of the gel mould.

2 Place the gel mould in the gel apparatus and fill the reservoirs with 1× TBE buffer.

3 Remove the gel comb, and immediately flush out the wells with 1× TBE.[a]

4 Denature the samples by placing them in a 80°C water bath for 15 min. A decrease in sample volume owing to evaporation is not a problem.

5 Flush out the wells again, and then load the samples as quickly as possible without spillage between wells. Rinse the pipette thoroughly in the lower buffer reservoir between each loading.

6 When all samples are loaded, complete assembly of the gel apparatus and turn on the power supply. Run the gel at constant power: for a 50-cm gel run at 35–40 W (current = 25–32 mA; voltage = 1.3–1.5 kV) for about 2.5 h; for a 40-cm gel run at the same power for 1.5 h. By this time, the bromophenol blue marker dye should reach the end of the gel.

[a] If a sharkstooth comb is used then it is removed from the gel and then reinserted upside down. Refer to the manufacturer's instructions.

4.1.4 Maximizing the length of the readable sequence

The amount of sequence that can be determined in a single experiment is limited by the resolving power of the gel. A 40-cm, 6% gel will allow about 200 nucleotides of sequence to be read, and a 50-cm gel will give another 50 nucleotides. Two strategies can be used to increase the length of sequence still further. The simplest method is to use multiple loadings, in other words loading portions of the same sample at time intervals onto the same or different gels. For example, samples can be run on gels for 2, 4, and 6 h by loading at staggered intervals. The three resulting sequences will overlap and, in combination, can generate 450–500 nucleotides of readable sequence. During prolonged runs it is advisable to change the running buffer halfway through.

The second strategy for increasing the sequence length is to use a voltage gradient gel. Running samples on a linear gel with a uniform field strength produces a wide separation of bands near the bottom of the gel. This wastes gel space, and the ideal setup would be to have an even spacing of bands all the way up the gel. This can be achieved by having a gradient of field strength, decreasing down the gel. Thus. when smaller DNA molecules move further down the gel, they experience less driving force. Such a gradient can be set up in one of two ways:

(a) A wedge gel is produced with wedge-shaped spacers. Since the electrical resistance of a TBE/polyacrylamide gel decreases with cross-sectional area, a linear increase in the gel thickness from top to bottom results in a decrease in voltage gradient towards the bottom of the gel. The limit on the maximum thickness of the wedge is 1.5 mm, the bands being too fuzzy to read if the thickness is any greater. Thick gels are also difficult to dry down during preparation for autoradiography. The use of a wedge gel is described in *Protocol 11*.

(b) In a buffer gradient gel (10), the ionic strength increases towards the bottom of the gel, slowing down the migration of bands in this region. The gradient is generated within the pipette used to pour the gel, by drawing a sample from two gel formulations, with different TBE concentrations, and causing these to mix slightly. The gradient is restricted to the lower part of the gel mould, and its length and steepness can be altered by changing the volumes and concentrations of the different TBE mixes used. The degree of mixing within the pipette can also be varied to alter the gradient properties. Buffer gradient gels should not be run beyond the time needed for the bromophenol blue dye to reach the bottom of the gel, because the gel can become so hot that the plates crack. The procedure for a buffer gradient gel is given in *Protocol 12*.

With both a wedge and buffer gradient gel, it is possible to obtain 250 nucleotides of sequence from a 40-cm gel, and 300 nucleotides from a 50-cm gel. Both types of gel will take longer to run than a standard gel.

Protocol 11

Preparation and running of a wedge gel

Equipment and reagents

- All the equipment and reagents for *Protocol 9*, plus two additional 3 × 1 cm strips of polystyrene sheet
- Polystyrene cement

Method

1 Use polystyrene cement to attach a 3×1 cm strip of polystyrene sheet to one end of each spacer, thereby increasing the spacer thickness.

2 Assemble the gel mould with the thick end of the spacer at the bottom, and pour the gel, as described in *Protocol 9*.

3 Load the gel as described in *Protocol 10*.

Protocol 12

Preparation and running of a buffer gradient gel

Equipment and reagents

• All the equipment and reagents for *Protocol 9*

Method

1 Assemble the gel mould as described in *Protocol 9*, steps 1–4.

2 Prepare two gel mixes:

(a) Place 25.2 g urea in clean glass beaker. Add in the following order: 9.0 ml acrylamide solution, 3.0 ml 10× TBE, 27.0 ml water, 600 μl 10% ammonium persulphate solution and 60 μl TEMED. This gives 60 ml of a 0.5× TBE, 6% gel mix.

(b) Place 8.4 g urea in clean glass beaker. Add in the following order: 3.0 ml acrylamide solution, 5.0 ml 10× TBE, 4.0 ml water, 200 μl 10% ammonium persulphate solution and 20 μl TEMED. This gives 20 ml of a 2.5× TBE, 6% gel mix.[a]

3 With a 25-ml glass pipette, take up 12.5 ml of the 0.5× TBE gel mix, followed by 12.5 ml of the 2.5× TBE gel mix. This should be done as slowly as possible.

4 Allow five air bubbles to enter the pipette and pass up through the gel solutions. This will cause some mixing and generate the gradient.

5 Pour this gel mix into the mould as described in *Protocol 9*, step 6.

6 Fill the empty space that is left in the mould with some of the remaining 0.5× TBE gel mix.

7 Insert the comb and allow to polymerize (*Protocol 9*, step 7) and load the gel (*Protocol 10*).

[a] The addition of 200 μl 1% bromophenol blue to this solution will allow the gradient to be visualized and hence enable the success of the pouring process to be monitored.

4.2 Autoradiography

The banding pattern in the gel is visualized by autoradiography (see Chapter 4, Section 2.5). Because of the low emission of energy of ^{35}S, the gel must be dried to reduce its thickness prior to exposure of film, otherwise the signal will be severely quenched. Close contact between the gel and X-ray film is necessary if strong, sharp bands are to be obtained. The procedure for preparing and drying gels for autoradiography is given in *Protocol 13*.

Protocol 13

Preparation of gels for autoradiography

Equipment and reagents

- 80 °C vacuum oven
- 10% acetic acid
- Plastic mesh

- Whatman 3MM paper
- Autoradiography cassette
- X-ray film (see Chapter 4, Section 2.5.3)

Method

1 When electrophoresis is complete, disconnect the power supply and discard the TBE buffer (the lower buffer, in particular, will contain high levels of radioactivity and should be disposed of appropriately).

2 Remove the gel mould from the chamber and disassemble.

3 Prise the gel plates apart, using either scissors or a spatula, with the siliconized, notched plate uppermost. The gel should not stick to this plate.

4 Place the backing plate with the gel uppermost in a shallow tank (we routinely use seed trays), and gently cover with 10% acetic acid. Leave for 15 min at room temperature.[a]

5 Place a plastic mesh over the gel and carefully lift the plate and gel from the acetic acid solution. Drain by holding at about 20°. Take care that the gel does not slide off the plate.

6 Place the plate plus gel on a horizontal surface. Lay a piece of 3MM paper, cut to a suitable size, over the gel, and slowly peel the paper back. The gel should stick to the paper.

7 Cover the gel and paper with Saran wrap and dry the gel under vacuum at 80 °C for 30 min. It is essential that the gel is dried properly to prevent quenching of the radiation and the gel sticking to the X-ray film.[b]

8 Remove the Saran wrap from the dried gel, and place in a cassette next to the X-ray film.

9 Expose for 24 h, or longer, at room temperature.

10 Develop the film using the method recommended by the supplier.

[a] This 'fixing' step removes the urea from the gel. If the urea is not removed then it precipitates out when the gel is dried and quenches the radioactive signal during autoradiography. If a wedge gel is being used then extend the fixing step to 30 min.

[b] A wedge gel must be dried for at least 45 min.

4.3 Troubleshooting

Complete failure of a chain-termination sequencing experiment is a rare outcome but suboptimal results are a relatively frequent occurrence. These can be due to problems at all stages in the procedure from template preparation through to autoradiography. *Table 1* gives a guide to the most common causes of failure and how to remedy them in subsequent experiments.

5 Building a long contiguous sequence

Only a limited amount of sequence can be obtained from a single chain-termination experiment. Using the most sophisticated techniques described in the above protocols, a maximum of 500 bp can be determined from one experiment. If the template is double-stranded, and both forward and reverse universal primers are used, then up to 1000 bp of sequence could be obtained from one cloned insert, but this sequence would be low quality because, ideally, all regions of the template should be sequenced in both directions, with the sequences completely overlapping, so that ambiguities can be checked and any sequencing errors corrected. Clearly, strategies are needed that enable the internal regions of long cloned fragments to be sequenced.

5.1 Subcloning fragments of a long insert

The simplest approach is to excise a long insert from its vector, cut it into smaller pieces, and then to subclone these and sequence them individually. The resulting sequences will overlap and, hopefully, there will be sufficient overlaps for the entire sequence of the original insert to be deduced. This is the shotgun approach and it has been applied not only to cloned DNA fragments but to entire genomes (11). There are three ways of producing the subfragments:

(a) The insert DNA can be cut with restriction enzymes. Two sets of fragments must be generated, with different enzymes, because the fragments from a single enzyme abut rather than overlap. The enzymes should recognize 4-nucleotide restriction sites so that the resulting fragments are relatively small. This is a rapid and easy way to generate subfragments but, unfortunately, restriction sites are often inconveniently placed and gaps will still remain in the master sequence (these gaps being areas that are too far away from the nearest restriction site to be reached when the fragments are sequenced).

(b) More random fragmentation of the initial template can be achieved by sonication (12). The high-frequency vibrations generated by the sonicator probe randomly shear DNA molecules and the sizes of the resulting fragments can be controlled by changing the frequency and duration of the treatment. One disadvantage with this method is that sonication leaves 5′ and 3′ overhangs at the ends of the DNA fragments, and these must be removed or filled in by enzymatic treatment before the fragments can be cloned as blunt-ended molecules. Despite this drawback, sonication has become the method of choice for generating DNA fragments in genome sequencing projects.

Table 1 Troubleshooting guide for manual chain-termination sequencing[a]

Problem	Possible cause	Suggested remedy
Problems when running the gel		
Gel does not set or polymerizes slowly	Old reagents	Check the ammonium persulphate solution is fresh; do not store for more than 4 weeks at 4 °C. Check that the TEMED has not degraded
Bubbles when the gel is poured	Dirty plates	You cannot clean gel plates too thoroughly
	Too much siliconization	Follow protocols for siliconization accurately
	Gel solution was violently stirred	Stir slowly to dissolve urea
Difficulties when loading samples	Poor gel polymerization around the wells	Check the gel reagents are fresh
	Gel detached from the plates	Too much siliconization—follow procedure accurately
	Leaking sharkstooth comb	Make sure the points of the comb are not broken
Gel 'smiles' and other uneven running	Uneven heat conductance, so centre of gel gets hotter than the edges	Use a gel apparatus with passive or active heat distribution (see Section 4.1.2)
	Gel is not of even thickness	If the plate is clamped before pouring, then make sure the clamps are not too tight
	Buffer reservoirs are not aligned parallel to one another	Reassemble apparatus correctly
Gel sticks to plates	Not enough siliconization	Follow protocols for siliconization accurately
	Poor gel polymerization can make the gel sticky	Check the gel reagents are fresh
	Running buffer has degraded leading to overheating during electrophoresis	Check the TBE buffer is fresh; discard stock that contain a white precipitate
Problems with the banding pattern		
No bands	Major failure	Check all steps of the Protocols, especially the labelling step (correct label? fresh label?); check the primer has not degraded
Faint bands	Not enough DNA	Check the amount of DNA in the template preparation

Problem	Cause	Remedy
	Label was not active enough	Use fresh label
	Inefficient priming	Do not allow the annealed primer-template mix to cool down too quickly
	Quenching of signal	Follow accurately the instructions for preparation of gels for autoradiography
Bands in all positions	Problem with the polymerase	Replace with new batch of enzyme
	Not enough dNTPs in the sequencing reactions	Check the concentrations of the stock solutions
	Non-specific DNA breakdown	Dirty microfuge tube or pipette tip
Change in band intensity from top to bottom	Incorrect dNTP:ddNTP ratio	Faint bands at the bottom: increase the amount of ddNTP in termination mixes; faint bands at the top: decrease the amount of ddNTP in the termination mix
	Labelling step was suboptimal	Follow procedure accurately
Fuzzy bands	Running buffer has degraded leading to non-optimal pH and/or ionic strength	Check the TBE buffer is fresh; discard stock solutions that contain a white precipitate
	Gel overheated	Running buffer is not fresh—see above
		Applied voltage was too high; do not allow the gel to heat up above 60 °C
	Gel not fully polymerized	Check the gel reagents are fresh
		Leave the gel to polymerize for a longer period before loading the samples
	Gel not warmed prior to loading	Carry out a short pre-electrophoresis (30 min) prior to loading
	Loaded too much sample	Decrease the loading
	Buffer disequilibrium	Ensure both the gel and reservoirs contain 1× TBE
	Samples not fully denatured	Use fresh formamide dye mix
	Fuzzy bands in a localized area	Check that the film was tightly pressed on to the gel during exposure
Ghost bands	Two or more sequences are superimposed	Primer has annealed at a secondary site; follow primer annealing protocol accurately

Table 1 Continued

Problem	Possible cause	Suggested remedy
Compressions	Hairpin loops in the molecules being electrophoresed	Contaminating bacterial DNA and/or RNA is acting as primers and/or templates; ensure correct growth conditions are used for generation of M13 phage particles; if sequencing double-stranded DNA then clean up DNA by density gradient centrifugation
		Substitute dITP or 7-deaza-dGTP for dGTP in sequencing reactions
		Use a higher applied voltage so the gel heats up more
		Add an additional denaturant (e.g. formamide) to the gel
		Sequence the complementary strand—the compression may be strand specific
Wall or pile up of bands	Hairpin loop in the template DNA	Increase the temperature of the strand synthesis reactions
		Add single-strand binding protein to the strand synthesis reactions
Distortion of bands in the 450–550 nucleotide region of the gel	More than 0.5% glycerol in the samples	Ensure that the DNA polymerase is sufficiently diluted to reduce the glycerol content of the samples that are loaded to < 0.5%
Band are clear for only a limited size range	Not enough template DNA	Check the concentration of the template DNA; use more
Wavy bands in part of one lane	Bubble in the gel	Avoid bubbles
Streak from the top of the gel	Damaged well	Check the comb is undamaged
		Ensure that the gel is fully set before removing the comb
		Take care when removing the comb
Streak from part way down the gel	Bubble in the gel	Avoid bubbles
	Dirt particles on the plate	Clean plates thoroughly before use
	Dirt particles in the gel	Ensure reagents are good quality and glassware, etc., is clean
	Undissolved urea particles	Make sure urea is completely dissolved before pouring the gel
Spots	Dirt or undissolved urea	See above
Complex patterns on the film, or film partly fogged	Autoradiography problem	Refer to Chapter 4, Table 6

[a] Modified from Brown, T. A., *Molecular Biology Labfax*, Vol. 2, 2nd edn, pp. 43–47, (1998), by permission of the publisher, Academic Press, London.

(c) Random fragments can also be produced by treating a DNA molecule with deoxyribonuclease I (DNase I) in a buffer that contains Mn^{2+}. Under these conditions, DNase I makes clean double-stranded cuts that leave blunt-ended molecules that can be subcloned without further treatment. This technique appears straightforward but has not become popular because DNase I is a highly active enzyme and it can be difficult to limit its action sufficiently to produce fragments long enough to sequence. This problem is exacerbated by the relatively rapid decline in activity that occurs when DNase I is stored at $-20\,°C$, which means that the appropriate conditions have to be determined anew every time a cleavage is carried out.

5.2 Methods for progressively building up a contiguous sequence

A second approach to sequencing a long molecule is to generate the sequence from one end in a series of steps, each step extending the existing sequence by a few hundred base pairs. One way of doing this is to produce an initial sequence with a universal primer and then to design a new primer, one that will anneal towards the end of the sequence that has been obtained. This new primer will extend the initial sequence, enabling a third primer to be designed, and so on. Theoretically, the process could carry on *ad infinitum*. In practice, the varying annealing characteristics of different primers means that not all the sequences will be clear. It takes time to identify the modifications to the annealing conditions that are needed to obtain a good sequence with each primer and, as a result, the overall process is slow and tedious.

A second possibility is to make a series of deletions, of different lengths, at the ends of the DNA molecule to be sequenced, and then to clone these deleted fragments (13). Some fragments will have short deletions, and so will generate sequences that overlap with the sequences obtained with the universal primers. Others will have slightly longer deletions and enable the master sequence to be extended a bit further. Still others will have large deletions and enable the sequence from the internal regions of the DNA molecule to be accessed. Methods for doing this based on progressive deletion of the insert DNA with exonuclease III or nuclease *Bal*31 are given in ref. (14).

6 Automated DNA sequencing

Automated DNA sequencing underlies the large genome sequencing projects and is also an increasingly important facility in molecular biology research laboratories. For most laboratories, the automation refers only to the electrophoresis and sequence-reading stages of the experiment, with manual techniques still required to prepare the template DNA and carry out the sequencing reactions. Robotic systems that prepare the DNA and set up the reactions are available, but at present are found only in the most well-equipped facilities.

The main difference between automated and manual sequencing lies with the choice of label. The detection systems in all automated sequencers work not with radiolabels but with fluorescent markers. The label can be attached to the primer, in which case four separate sequencing reactions are set up as described in *Protocol 8*, but more usually the ddNTPs are labelled. This offers a major advantage over manual sequencing because a different fluorochrome can be used for each ddNTP, enabling all four sequencing reactions to be carried out in a single tube. Chains terminated with ddA therefore have a particular fluorescent signature, chains terminated with ddC have a different signature, and so on. This enables the sequence to be read directly as the bands pass the detector. The most sophisticated automated sequencers have 96 channels, using capillary rather than gel electrophoresis, and so allow 96 sequences to be determined in parallel. Each run takes 2 h, so 1000 sequences can easily be obtained in a single day. The resolving power of the capillary system is greater than for gel electrophoresis and 600 bp can be obtained per experiment, so 600 000 bp of sequence can be generated per machine per day, providing the capacity demanded by the genome sequencing projects.

The methodology described in *Protocols 7* and *8*, with a few modifications to account for the fluorescent labelling, could be used to produce samples suitable for loading into an automated sequencer. In practice, a different approach, called thermal cycle sequencing, is employed. To understand how this method works it is first necessary to be familiar with the principles of PCR. The relevant protocols are therefore given in Chapter 7, Section 4.

7 DNA sequence storage and analysis

All DNA sequences should be deposited in one of the international sequence databases so other researchers can make use of the information. This is requirement for publication in all of the reputable molecular biology journals. There are three databases (*Table 2*) which exchange new data on a daily basis so each one is a comprehensive collection of all known sequences.

Table 2 The three international DNA databases

Database	Internet address	Reference
Genbank	http://www.ncbi.nlm.nih.gov	15
EMBL	http://www.ebi.ac.uk	16
DDBJ	http://www.nig.oc.jp/home.html	17

Various software packages for sequence analysis are available commercially or online through the internet. The capabilities of these packages are continually being improved, so information about them quickly becomes out of date. In general, the types of analysis that can be carried out are as follows (8):

- searches can be made for open reading frames that might be protein coding regions

- restriction sites can be identified, providing a guide to future manipulations such as subcloning; some packages automatically generate the banding pattern that will be obtained when the DNA molecule is cut with a particular enzyme or combination of enzymes

- sequence features such as direct and inverted repeats can be located.

- comparisons can be made with other sequences in the databases; this is often the first analysis that is carried out as it is the best way of obtaining information about the possible function of a new gene that has been sequenced. If the gene has sequence similarity with a gene from another species, then the two genes might be homologous, meaning that they are derived from a common ancestor and so probably have similar functions. Packages for carrying out similarity searches include BLAST (18) and FASTA (19).

- alignments can be made between groups of sequences so that similarities can be clearly visualized and measured; these alignments can be used to generate phylogenetic trees that indicate possible evolutionary relationships between different sequences. Examples of software packages are CLUSTAL (20), PAUP (21) and PHYLIP (22).

Acknowledgements

Some parts of this chapter, notably Section 4, are based on the equivalent chapter from *Essential molecular biology: a practical approach*, 1st edn, written by Christopher J. Howe and E. Sally Ward.

References

1. Sanger, F., Nicklen, S., and Coulson, A. R. (1977). *Proc. Natl Acad. Sci. USA*, **74**, 5463.

2. Maxam, A. M. and Gilbert, W. (1977). *Proc. Natl Acad. Sci. USA*, **74**, 560.

3. Eckert, R. L. (1992). In *Current protocols in molecular biology* (ed. F. M. Ausubel, R. Brent, R. E. Kingston, D. D. Moore, J. G. Seidman, J. A. Smith, and K. Struhl), p. 7.5.1. John Wiley, New York.

4. Slatko, B. E., Heinrich, P., Nixon, B. T., and Eckert, R. L. (1993). In *Current protocols in molecular biology* (ed. F. M. Ausubel, R. Brent, R. E. Kingston, D. D. Moore, J. G. Seidman, J. A. Smith, and K. Struhl), p. 7.3.6. John Wiley, New York.

5. Birnboim, H. C. and Doly, J. (1979). *Nucl. Acids Res.*, **7**, 1513.

6. Tabor, S. and Richardson, C. C. (1987). *J. Biol. Chem.*, **262**, 15330.

7. Tabor, S. and Richardson, C. C. (1989). *J. Biol. Chem.*, **264**, 6447.

8. Brown, T. A. (ed.) (1998). *Molecular biology labfax*, 2nd edn, Vol. 2. Academic Press, London.

9. Tabor, S. and Richardson, C. C. (1989). *Proc. Natl Acad. Sci. USA*, **86**, 4076.

10. Biggin, M. D., Gibson, T. J., and Hong, G. F. (1983). *Proc. Natl. Acad. Sci. USA*, **80**, 3963.

11. Brown, T. A. (1999). *Genomes*, p. 70. BIOS Scientific Publishers, Oxford.

12. Bankier, A. T., Weston, K. M., and Barrell, B. G. (1987). In *Methods in enzymology* (ed. R. Wu), Vol. 155, p. 51. Academic Press, London.

13. Henikoff, S. (1987). In *Methods in enzymology* (ed. R. Wu), Vol. 155, p. 156. Academic Press, London.

14. Slatko, B. E., Heinrich, P., Nixon, B. T., and Voytas, D. (1991). In *Current protocols in molecular biology* (ed. F. M. Ausubel, R. Brent, R. E. Kingston, D. D. Moore, J. G. Seidman, J. A. Smith, and K. Struhl), p. 7.2.1. John Wiley, New York.

15. Benson, D. A., Boguski, M. S., Lipman, D. J., and Ostell, J. (1997). *Nucl. Acids Res.*, **25**, 1.

16. Stoesser, G., Sterk, P., Tuli, M. A., Stoehr, P. J., and Cameron, G. N. (1997). *Nucl. Acids Res.*, **25**, 7.

17. Tateno, Y. and Gojobori, T. (1997). *Nucl. Acids Res.*, **25**, 14.

18. Altschul, S. F., Gish, W., Miller, W., Myers, E. W., and Lipman, D. J. (1990). *J. Mol. Biol.*, **215**, 403.

19. Pearson, W. R. and Lipman, D. J. (1988). *Proc. Natl. Acad. Sci. USA*, **85**, 2444.

20. Thompson, J. D., Higgins, D. J., and Gibson, T. J. (1994). *Nucl. Acids Res.*, **22**, 4673.

21. Swofford, D. L. (1993). *PAUP: Phylogenetic Analysis Using Parsimony*. Smithsonian Institution, Washington.

22. Felsenstein, J. (1989). *Cladistics*, **5**, 164.

Chapter 7

The polymerase chain reaction

T. A. Brown

Department of Biomolecular Sciences, University of Manchester Institute of
Science and Technology, Manchester M60 1QD, UK

1 Introduction

The polymerase chain reaction is a deceptively simple but remarkably versatile
procedure that has applications in virtually all areas of molecular biology research. All that a PCR does is to make multiple copies of a target DNA sequence,
the amplification process enabling microgram quantities of DNA to be obtained
from a single starting molecule. Any segment of any DNA or RNA molecule can
be amplified, provided that the sequences on either side of the targeted region
are known. If genomic DNA is used as the starting material the PCR can be used
in an analogous fashion to cloning, to provide a pure sample of a single gene or
other DNA sequence, with the advantage that the gene is obtained after a few
hours work, rather than the weeks needed to prepare and screen a genomic
library. If RNA is the starting material then PCR can achieve the same objective
as cDNA cloning, again much more rapidly. Only the limitation that some
sequence information is needed in order to design appropriate primers for a
PCR, plus the difficulty in obtaining PCR products longer than 5–10 kb long, has
prevented PCR from completely replacing cloning as a means of gene isolation.

   This chapter begins with a description of the basic methodology for carrying
out a PCR and a discussion of the factors that influence the success, or otherwise,
of an amplification. Procedures used for cloning and sequencing PCR products
are covered in Sections 3 and 4, respectively.

2 Methodology for PCR

2.1 A summary of PCR

The polymerase chain reaction is a test-tube reaction that is carried out by mixing
together the appropriate reagents and incubating them in a thermal cycler (a
piece of equipment that enables the incubation temperature to be varied over
time in a pre-programmed manner). The basic steps in a PCR experiment are as
follows (*Figure 1*):

(a) DNA is prepared from the organism being studied and denatured by heating
   to 94 °C.

(b) A pair of oligonucleotides is added to the DNA, the sequences of these oligo-nucleotides enabling them to anneal either side of the gene or other DNA segment that is to be amplified, and the mixture is cooled to 50–60 °C so these oligonucleotides attach to their target sites.

(c) A thermostable DNA polymerase is added, together with a supply of deoxy-ribonucleotides, and the mixture is heated to the optimal temperature for DNA synthesis −74 °C if *Taq* DNA polymerase (the DNA polymerase I enzyme from the thermotolerant bacterium *Thermus aquaticus*) is used. The annealed oligonucleotides now act as primers for synthesis of new polynucleotides complementary to the template strands.

(d) The cycle of denaturation-annealing-extension is repeated 25–30 times, with the number of newly-synthesized DNA molecules doubling during each cycle. This exponential amplification results in synthesis of a large number of copies of the DNA sequence flanked by the pair of oligonucleotides.

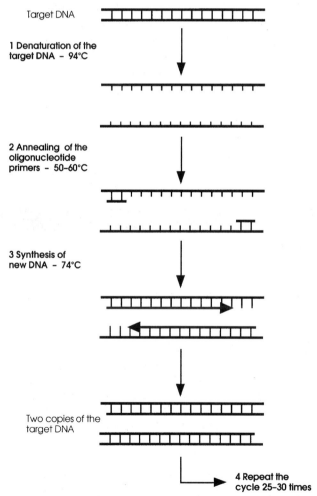

Figure 1 An outline of PCR.

2.2 Planning and carrying out a PCR

A successful PCR experiment is very much dependent on good planning. The design of the two oligonucleotide primers is absolutely critical to success or failure, as poor design results in no PCR products or, worse, synthesis of incorrect products. Each combination of template and primers affects the reaction in a subtle way that is manifested by variations in the optimal Mg^{2+} concentration for DNA synthesis. Further improvements in amplification efficiency and specificity can be achieved by modifying the parameters for the early cycles of the PCR and by adding one or more compounds that enhance primer annealing and reduce premature termination of strand synthesis.

With some PCR experiments the operator does not need to worry about any of these considerations. For some targets, especially clinically or forensically relevant human loci, standard primer pairs can be purchased commercially and used in accordance with instructions provided by the supplier. Other PCRs will require that a novel pair of primers are designed but might not necessitate extensive optimization of the reaction because the template is abundant and lacks sequence complexity, so mispriming at non-target sites is unlikely. An example is when PCR is used to check the construction of a recombinant DNA molecule or to screen a small library for a desired clone. However, most molecular biologists will, at some time or another, wish to use a novel pair of primers in an experiment where high yield and specificity are important and, under these circumstances, it is necessary not only to design the primers carefully but also to devote some time to optimizing the reaction.

Protocol 1 gives a basic procedure for PCR, suitable for amplification of fragments up to 3 kb. Longer fragments—up to 40 kb—are amplified by a modified procedure (see Section 2.2.9).

Protocol 1

Basic procedure for PCR

Equipment and reagents

- Thermal cycler with a hot lid (available for many suppliers)[a]
- 0.5 ml microfuge tubes
- 10× PCR buffer (100 mM Tris-HCl, pH 9.0, 500 mM KCl, 0.1% Triton X-100, prepared in sterile water)
- 20 mM $MgCl_2$ (prepared in sterile water)
- 40 mM dNTP solution (10 mM dATP, 10 mM dCTP, 10 mM dGTP and 10 mM dTTP, prepared in sterile water)

- Primers 1 and 2 (10 μM solutions of each primer in sterile water)[b]
- *Taq* DNA polymerase
- Template DNA[c]
- Gel loading buffer (95% deionized formamide, 20 mM EDTA, pH 8.0, 0.05% bromophenol blue, 0.05% xylene cyanol; prepare deionized formamide as described in Appendix 4, *Protocol 4*)

Method

1 Prepare a 'master mix' containing all of the components of the PCR except the template DNA. The amount of reagents to add depends on the number of reactions

that are being carried out.[d] The following recipe is sufficient for a single reaction and should be scaled up accordingly:[e] 5 μl 10× PCR buffer, 5 μl 20 mM MgCl$_2$, 1 μl 40 mM dNTP solution, 2.5 μl Primer 1, 2.5 μl Primer 2, sufficient water to bring the final volume to 49 μl, 2.5 units *Taq* DNA polymerase (total volume = 49 μl).

2   Transfer 49 μl of master mix to a 0.5 ml microfuge tube and add 1 μl template DNA.

3   Place each tube in a thermal cycler that has been programmed to provide the following heating regime:[f] 2 min at 94°C, 30 cycles of 1 min at the annealing temperature, 1 min at 74°C, 1 min at 94°C and 8 min at 74°C.

4   Terminate the reaction by adding 10 μl gel loading buffer to each tube.

5   Analyse the results by agarose gel electrophoresis (Volume I, Chapter 5). If the PCR has worked efficiently then a clear band should be visible after electrophoresis of a 10 μl aliquot of the reaction.

6   PCR products can be stored at 4°C for several months.

[a] Most commercial thermal cyclers have a 'hot lid' which ensures that the upper part of the microfuge tube is heated. This reduces condensation and ensures that all parts of the PCR are evenly heated. If a hot lid is not used, overlay the reaction with 50 μl of light mineral oil (Sigma M3516) to prevent evaporation. At the end of the reaction, after adding gel loading buffer, extract the oil by adding 300 μl chloroform, vortexing, centrifuging for 30 s, and collecting the aqueous layer.

[b] See Section 2.2.2 for primer design.

[c] The size of the target DNA dictates how much should be added to a PCR. For human genomic DNA aim to use 1 μg per reaction. For a recombinant plasmid, 10 pg is sufficient. However, the amount of template DNA that is added often depends on availability and successful PCRs are possible with substantially less than the amounts stated, assuming the reaction is fully optimized.

[d] For advice regarding controls, see Section 2.2.1.

[e] Final concentrations are 2 mM Mg^{2+} (see Section 2.2.5 for optimization procedures), 200 μM each dNTP (see Section 2.2.6 for relevant information), 0.5 μM of each primer.

[f] See Section 2.2.3 for an explanation of the temperature cycling and a description of how to determine the appropriate annealing temperature.

2.2.1 Controls and good working practices

Adequate controls are particularly important in a PCR experiment because the extreme sensitivity of the technique means that small amounts of contaminating template DNA can give rise to a PCR product that is visible when the results are examined by agarose gel electrophoresis. It is often assumed that the purpose of these controls is to monitor for cross-contamination of samples during DNA extractions and carryover of template DNA from one tube to another when PCRs are set up. According to this assumption a full range of controls is unnecessary because good technique will ensure that problems do not arise. However, the main source of contamination is aerosol droplets that contain PCR products from a previous experiment. It is very difficult to avoid the generation of such aerosols when microfuge tubes are opened or when PCR products are pipetted. As many as 10^9 product molecules can be present in a 0.1 μl aerosol droplet, compared with approximately 10^6 template molecules *per reaction* if the amounts of template DNA recommended in *Protocol 1* are used. The genuine template can therefore be completely swamped by old PCR products if the reaction tube is directly contaminated. Indirect contamination via one of the stock solutions used in template preparation or in setting up the PCR can also result in sufficient carryover for the genuine template molecules to be outnumbered by contaminating products.

Forensic scientists and biomolecular archaeologists who are forced to work with materials that yield very small quantities of template DNA have devised sophisticated operating procedures designed to minimize contamination of these templates with the products of previous reactions (1). These procedures are based on the physical separation of work areas used for preparing template DNA and setting up PCRs from those areas in which the amplifications are carried out and the products analysed. In extreme cases, the physical separation involves the two parts of the procedure being performed in different buildings, on the grounds that a biology department in which many groups are carrying out PCRs will have an unavoidable, continual background level of DNA aerosols that permeate all working areas. For most researchers, such extreme measures are unnecessary, but certain precautions should always be followed:

- *minimize handling of the solutions.* Reduce the number of pipetting steps to a minimum by using the 'master mix' procedure described in *Protocol 1* when many identical amplifications are to be performed.

- *use sterile technique.* Use a new pipette tip for each transfer. Store reagents in small bottles or tubes so that the pipette barrel does not enter the container. Keep tube caps closed.

- *autoclave reagents.* This breaks contaminating DNA into short pieces and also prevents the growth of organisms in the buffers.

- *prepare small aliquots of reagents.* Aliquots which are used for only a few days are less likely to be contaminated, and can be discarded at little expense if they are. Do not keep your entire stock of primer in a single tube.

- *separate workspaces.* Prepare template DNA and set up PCRs in a separate lab into which tubes containing PCR products are never taken. Use displacement pipettes (with disposable piston assemblies) that are used for no other purpose, as well as designated packs of microfuge tubes, etc. Avoid carrying PCR products into this clean room on clothing: change into a full-length laboratory overall on entering the room and change again when leaving. If these precautions are impossible because of space constraints then purchase a small glove-box cabinet for template DNA and PCR preparation.

Even when these precautions are followed there will still be occasions when contamination occurs. A set of control reactions should therefore be included in every experiment:

- a 'water blank'. This is a PCR set up with sterile water instead of template DNA, used to test for the absence of contamination in the PCR reagents.
- an 'extract blank', which checks for contamination of reagents used in preparation of the template DNA. An extract blank can be prepared by carrying out the procedure for template DNA preparation with no starting material, and then setting up the PCR with the resulting solution instead of genuine template DNA. A more satisfactory extract blank is prepared by starting with equivalent material that does not contain the primer annealing sites. For example, a clone containing a non-recombinant vector could be the starting material for preparation of an extract blank when PCR is being used to screen a clone library.
- a positive control, in which the template DNA has proven ability to yield the correct PCR product. This control is optional and, arguably, undesirable as omitting it will reduce the total amount of PCR product that is synthesized and hence reduce aerosol contamination. Its main purpose is to confirm that the reagents are active and the cycling parameters are appropriate, but this will be apparent if the test PCRs give the anticipated product. After electrophoresis, the sizes of the test products can be checked by comparison with the positive control, but this can equally well be done by comparison with DNA size markers. Note that correct size should not be taken as definitive proof that the correct product has been amplified and the product should always be sequenced to confirm its identity.

2.2.2 Primer design

The primers are the key component of a PCR and the success or failure of an amplification depends largely on the design and use of the primers. It is simple enough to synthesize a pair of oligonucleotides whose sequences enable them to anneal either side of the region to be amplified. It is much less straightforward to design a pair of primers that will give a specific and efficient amplification of the desired region of the template DNA. This assumes that there is some flexibility regarding the precise positions of the end-points of the amplified fragment, allowing a choice to be made between a series of alternative priming sites. With some PCRs, the nature of the experiment limits the choice of priming sites and then it will be necessary to pay more attention to the optimization and enhancement of the PCR (see Sections 2.2.5–2.2.7) in order to obtain a satisfactory result.

If there is some choice of primer sequences then the following points should be taken into consideration:

- length determines primer specificity. If a primer is too short then there is a greater chance of it annealing to more than one site in the template DNA. The approximate frequency of a particular sequence in a template DNA is represented by

4^n, where n is the length of the sequence. For example, a sequence 8 nucleotides will occur, on average, once every $4^8 = 65\,536$ bp. This calculation assumes that the GC content of the 8-nucleotide sequence is the same as the GC content of the template DNA, but even if this assumption is not met the result is still a good indication of the frequency of occurrence of a sequence. What this means is that an 8-nucleotide primer will have approximately 46 000 annealing sites (each one precisely complementary to the primer) in human DNA (3000 Mb). There is therefore a high chance that a pair of 8-nucleotide primers would amplify more than one fragment from human DNA. In contrast, a 17-nucleotide primer sequence occurs once every $4^{17} = 17\,179\,869\,184$ bp and would thus be expected to have a unique priming site in the human genome. In practice, primers of 15–20 nucleotides are suitable for most purposes. If they are > 30 nucleotides, then the hybridization rate, which increases with length (see Chapter 5, Section 3.3.10), becomes a factor and complete annealing may not occur during the time allowed within the thermal cycle.

- *primers should not be able to anneal to each other*. If the 3'-end of one primer can anneal to an internal site in the second primer, then the extension of that 3'-end by *Taq* polymerase will lead to a 'primer–dimer'. Dimerization is promoted by the high concentrations of primers during the early cycles of the PCR, and during this stage of the reaction can reduce synthesis of the amplification product by competing for the polymerase. The reduction in primer concentration caused by dimerization can cause a further decrease in product yield later in the reaction. Dimers appear on an agarose gel as one or more a diffuse, low molecular weight bands. Dimerization does not require extensive complementarity between the primers. As a rule of thumb, avoid using a primer pair if the four nucleotides at the 3'-end of one primer can form a perfect match with four nucleotides within the second primer.

- *primers should not be able to anneal to themselves*. If the four nucleotides at the 3' end of a primer can anneal to four nucleotides within the *same* primer, then once again a primer–dimer can be formed, although it is more likely that the 3'-end will loop back and form a hairpin structure (a primer *monomer* rather than dimer). The 3'-end is then extended by the polymerase, stabilizing the hairpin. Hairpin formation reduces the effective primer concentration and decreases yield of the desired product.

- *priming efficiency might be improved by a 3'-'anchor'*. Some researchers advocate the use of an 'anchor' sequence at the 3'-end of each primer. This is simply a run of two to four G or C nucleotides, which provides a relatively strong attachment between primer and template, G–C base pairs having three hydrogen bonds compared with the two for an A–T pair. The anchor ensures that the 3'-end of the primer is tightly annealed to the template so that strand synthesis, which begins at the 3'-end, can initiate efficiently. Intuitively, this appears to be a sound argument but there are conflicting data regarding the use of anchor sequences and some evidence that primers with GC-rich 3'-regions are less rather than more effective in PCRs.

A variety of computer programs are available for scanning primer sequences for potential dimer and hairpin formation. Some of these programs also take account of other factors such as melting temperature (Section 2.2.3) in attempts to design the best primer pair for a particular amplification. In practice, these programs provide few advantages over manual design of primers, the latter being quick, easy and usually successful if the factors described above are kept in mind.

2.2.3 Temperature cycles

During a PCR the reaction is cycled between the denaturation, annealing and extension temperatures, as shown in *Figure 1*.

Denaturation is usually carried out at 94 °C, at which temperature double-stranded DNA is rapidly converted to the single-stranded form. Extended incubations at this temperature can cause damage to the DNA, mainly through the water-induced cleavage of the β-glycosidic bond between a purine base and its sugar, leading to depurination and subsequent strand cleavage. In *Protocol 1*, the initial denaturation step is for 2 min, because at the start of the PCR the template DNA is completely double-stranded, but during the remainder of the reaction a denaturation time of 1 min is sufficient as the DNA is only partly double-stranded at this stage. This denaturation time can be reduced to 30 s without appreciable decrease in product yield if the amplified segment has a GC content < 50%.

The extension temperature is set at 72 °C, which is slightly below the optimum for *Taq* DNA polymerase (75 °C). A higher temperature is not used because there is a risk that the primers will detach from the template before they have been extended, especially if the ramp rate is high (see below). At 72 °C, the enzyme will polymerize approximately 6000 nucleotides of DNA/min, which means that an extension time of 1 min is adequate for most PCRs. If the amplified fragment is > 3 kb then increase the extension time to 2 min and if it is >6 kb then use 3 min. For amplification products > 10 kb use the special procedures described in Section 2.2.9.

The PCR is completed by an long extension step (8 min in the procedure given in *Protocol 1*). The intention is to ensure that all products are completely extended at the end of the reaction, by which time the enzyme might not be synthesizing DNA at the expected rate because one or other component of the reaction has become limiting.

Unlike the denaturation and extension temperatures, the annealing temperature is not the same for every PCR. Ideally, this temperature should be 2–4°C below the melting temperature—T_m (see Chapter 5, Section 3.2)—of the primer pair. The T_m of each primer can be estimated from its sequence, using the equation:

$$T_m = 4 \times (\text{number of GC base pairs}) + 2 \times (\text{number of AT base pairs})$$

This estimation is not entirely accurate and the actual T_m of a primer might be lower or higher, by as much as 5 °C, especially if the sequence includes a run of As and Ts (giving a lower T_m) or of Gs and Cs (higher T_m). The annealing temperature for a new pair of primers should therefore be optimized by setting up a series of PCRs with annealing temperatures ranging from 8 °C below to 4 °C above the estimated T_m, in 2°C increments. If the annealing temperature is too low, then mismatches between primer and template might be tolerated, increasing the number of potential priming sites and leading to spurious products.

The above discussion regarding annealing temperature assumes that the T_m is the same for both primers. Ideally, this will have been ensured when the primers were designed, the equation for estimation of T_m being used to select primer sequences that are 'balanced' with regards to annealing temperature. If the two primers cannot be balanced, then use the lower T_m to determine the annealing temperature.

As well as the temperatures themselves, the transfer periods between them are important. For most PCRs, the thermal cycler should be set at its maximum 'ramp rate' so that cycling between temperatures is as rapid as possible. Rapid ramp times are desirable because *Taq* DNA polymerase is not infinitely stable at high temperatures (its half-life is approximately 1.6 h at 95 °C; see ref. 2) and undergoes substantial decay during a PCR. Some thermal cyclers reduce ramp times to an absolute minimum by having three or more heat blocks between which the samples are shuttled by a robotic arm, rather than a single block that is heated and cooled between the appropriate temperatures. The single-block systems are satisfactory for normal usage, but note that with many cyclers the cooling rate drops off appreciably as the machine gets older. Most thermal cyclers can be programmed so that when the PCR is completed the reactions are taken immediately to 4 °C and held at that temperature for an unlimited period. This is useful if a PCR is set up late in the day and reaches completion half way through the night but, in general, it is better to transfer the samples to the fridge as soon as the reaction has finished. Frequent use of the cooling option can substantially decrease the working lifetime of a cycler.

2.2.4 Template DNA

Polymerase chain reactions can be carried out with relatively crude DNA preparations, though carry-over of acetate salts or SDS will reduce product yield. If the template DNA preparation is dilute, then *Protocol 1* can be modified so that up to 30 μl is used in the PCR (reduce the amount of water added to the master mix in step 1 to compensate for the increased volume of template DNA). If the template DNA is dissolved in TE buffer (Appendix 4, Section 2), then increasing the amount of template DNA in the PCR also increases the amount of EDTA, which will chelate Mg^{2+} and reduce the effective concentration of these ions in the reaction. It may therefore be necessary to adjust the amount of $MgCl_2$ that is added to the master mix (see Section 2.2.5).

Although relatively impure DNA can be used in PCRs, some DNA sources contain compounds that are very effective inhibitors of *Taq* polymerase and which completely prevent the PCR from working if they are co-purified with the template DNA. Environmental samples such as soils and some archaeological materials are notorious in this respect. Contamination of a template DNA preparation with a PCR inhibitor can be checked by adding 1 μl of the suspect DNA preparation to a PCR that also contains a template that is known to give a product. Failure of this 'spiked' PCR indicates that the suspect preparation does indeed include one or more compounds that are inhibiting the enzyme. Inhibitory preparations can be post-purified in a number of ways, for example by silica-binding (3).

The template DNA molecules do not need to have a high molecular weight, and indeed some fragmentation is desirable as smaller double-stranded fragments are more easily denatured. Clearly, the template DNA must be, on average, longer than the sequence that will be amplified. If the template is extensively fragmented then 'PCR jumping' might occur (4). In this process, an incomplete extension product from one cycle acts as a long primer during a later cycle, so that two template fragments contribute to the synthesis of a single product molecule. This phenomenon has been used as a means of obtaining relatively long PCR products from a highly fragmented DNA preparation (5); however, two problems can arise. First, Taq DNA polymerase sometimes attaches an adenosine nucleotide to the 3'-end of an extended strand, in a template-independent manner (see Section 3). This means that if a partially extended strand acts as a primer in a subsequent cycle then the resulting product might have an A inserted at the position corresponding to the 3'-end of the partially extended strand. The second problem arises if the locus being amplified is biallelic, in which case the sequences of the jumping products might be chimeric, comprising parts of the sequences of both alleles.

2.2.5 Optimization of the magnesium ion content

The interaction between Taq polymerase and the template DNA is magnesium-dependent, with the optimal concentration of Mg^{2+} ions determined by the sequence characteristics of the region to be amplified. The relationship between enzyme and DNA sequence is not sufficiently well understood for the ideal Mg^{2+} content of a PCR to be predicted, and the optimum should be determined experimentally whenever a new PCR is designed. This optimum can be at any point between 1.0 and 5.0 mM Mg^{2+}, so test PCRs should be set up covering this range in 0.5-mM increments. Table 1 shows the appropriate amounts of 20 mM $MgCl_2$ to add to the master mix in step 1 of Protocol 1, the water addition being reduced to compensate. The optimum Mg^{2+} concentration is assessed by comparing product yields on an agarose gel. Ideally, this is done after the appropriate annealing temperature has been determined (Section 2.2.3), but if a particularly sub-optimal Mg^{2+} concentration is used when the annealing temperature is tested then insufficient amplification will occur for the effect of temperature changes to be seen. Optimization of a PCR might therefore require judicious changes to Mg^{2+} concentration and annealing temperature in a single series of test PCRs.

When optimizing the Mg^{2+} concentration, do not be concerned if the product yields appear not to show a consistent variation across the range of concentrations that are tested. The graph of product yield versus Mg^{2+} concentration for a particular template sequence can be bimodal, with similar activities displayed by two different Mg^{2+} concentrations, separated by one or more intermediate

Table 1 Modifications to the master mix for magnesium ion optimization[a]

20 mM MgCl$_2$ addition (μl)[b]	Final concentration (mM)
2.50	1.0
3.75	1.5
5.00	2.0
6.25	2.5
7.50	3.0
8.75	3.5
10.00	4.0
11.25	4.5
12.50	5.0

[a] Refer to *Protocol 1*, step 1.

[b] Change the amount of water added to the master mix so that the final volume is 49 μl.

concentrations that give reduced activity. Also bear in mind the influence of EDTA concentration on the availability of Mg^{2+} ions (Section 2.2.4) and increase the magnesium addition if extra EDTA is introduced into the PCR with the template DNA.

2.2.6 dNTP concentration

Some guides to PCR recommend optimization of the dNTP concentration of the reaction, but this is usually unnecessary and can cause more problems than it solves. The Mg^{2+} optimum varies with different dNTP concentrations and attempts to optimize both parameters can be time-consuming and may never reach an satisfactory conclusion. The Mg^{2+} concentration is more critical and only a minor improvement will be obtained by changing the dNTP content of the reaction. It is a mistake a to add more dNTPs if product yield is low, because the dNTPs are rarely the limiting factor in a poor PCR (in *Protocol 1* each dNTP is at 200 μM, which is sufficient for synthesis of over 10 μg of DNA). Higher dNTP concentrations increase the number of errors that *Taq* polymerase makes when copying the template DNA, and if the concentrations are excessive then the polymerase is inhibited.

2.2.7 Addition of PCR enhancers

Various compounds can be added to a PCR to increase specificity and/or improve product yield. These enhancers are unnecessary for many PCRs but they are worth considering if the template is in short supply or specificity problems are encountered.

The most effective way of enhancing product yield is to add a compound that reduces decay of the polymerase during the thermal cycling. Possibilities are 2-mercaptoethanol or DTT at 10 mM final concentration, glycerol at 10% final concentration or tetramethyl ammonium chloride at 50 mM final concentration. Addition of tetramethyl ammonium chloride (a derivative of betaine) is also useful if the template DNA is GC-rich because it increases the denaturation rate

(6). The efficiency of primer annealing can be improved by adding $5\,\mu g/ml$ *Escherichia coli* single-stranded binding protein. Another factor to take into account is the accumulation of pyrophosphates (a product of dNTP polymerization) in the latter stages of a PCR because high concentrations of these ions promote de-polymerization of DNA. Roche market a thermostable pyrophosphatase which destroys pyrophosphates and counters this inhibitory effect.

Polymerase chain reaction enhancers of unknown composition are available from a number of companies. As well improving product yield, some of these are designed to increase the specificity of a PCR. If specificity is a problem then the first step should be to change to a 'hot-start' strategy (Section 2.2.8) and then, if the situation is still not ideal, to try one or other of the commercial preparations.

2.2.8 Alternative cycling regimes

Several modifications to the standard thermal cycling regime have been devised in attempts to improve the specificity of the PCR. Specificity problems result primarily from events that happen in the early stages of a PCR, including the period before the thermal cycling begins. Some template denaturation and primer annealing can occur when the reagents for the PCR are mixed together, even if this is done on ice. This annealing is non-specific because, at these low temperatures, primer–template hybrids that contain several mismatches are stable. *Taq* polymerase has some activity at room temperature and so can extend these mismatched primers, producing a longer hybrid. The extended part of this hybrid is, of course, fully complementary to the template. During the thermal cycling, these extended primers reanneal to their complementary sites, resulting in further copying of non-target parts of the template. Continuation of the process leads to synthesis of spurious amplification products.

The modified cycling regime called 'hot-start' (7,8) is designed to prevent DNA synthesis during the set-up stage, so that extension of misannealed primers does not occur and the resulting decrease in specificity is avoided. To carry out a hot-start, either the enzyme or the $MgCl_2$ is omitted from the initial mix, being added only after the tubes have reached the first denaturation temperature. In the absence of enzyme, or of its essential magnesium ion cofactors, no DNA synthesis is possible until after the primers have annealed to their authentic target sites during the first PCR cycle. *Protocol 2* describes this procedure.

Protocol 2

Hot-start PCR[a]

Equipment and reagents

- All the materials required for *Protocol 1*

Method

1 Prepare a 'master mix' containing the following components. The amount of reagents to add depends on the number of reactions that are being carried out. The

following recipe is sufficient for a single reaction and should be scaled up accordingly: 5 μl 10× PCR buffer, 1 μl 40 mM dNTP solution, 2.5 μl Primer 1, 2.5 μl Primer 2, sufficient water to bring the final volume to 44 μl, 2.5 units *Taq* DNA polymerase (total volume = 44 μl).

2   Transfer 44 μl of master mix to a 0.5 ml microfuge tube and add 1 μl template DNA.

3   Place each tube in a thermal cycler that has been programmed to provide the following heating regime: 2 min at 94 °C, 30 cycles of 1 min at the annealing temperature, 1 min at 74 °C, 1 min at 94 °C and 8 min at 74 °C

4   After 1 min at 94 °C, pause the thermal cycling and add 5 μl of 20 mM MgCl$_2$ to each tube.[b] Continue the thermal cycling.

5   Terminate the reaction and analyse the results as described in *Protocol 1*.

[a] See also the notes to *Protocol 1*.

[b] This addition assumes that 2 mM is the optimum Mg^{2+} concentration for the PCR. Adjust the water content of the master mix and the volume of the MgCl$_2$ addition if a different Mg^{2+} concentration is required (see Section 2.2.5 and *Table 1*).

Protocol 2 is a perfectly adequate way of performing a hot-start but it is sometimes considered undesirable to reopen the reaction tubes after the thermal cycling has begun, because of the possibility of contaminating the reactions with aerosol droplets containing old PCR products (Section 2.2.1). Reopening the tubes can be avoided in two ways. The first involves constructing a wax barrier over the main reaction mixture, with the enzyme pipetted on top. When the reaction heats up to the first denaturation temperature the wax melts and the enzyme enters the reaction. The second method is to use a complex consisting of *Taq* polymerase and a heat-sensitive antibody that prevents activity of the enzyme. When the first denaturation temperature is reached the antibody is degraded and the enzyme is activated (9). Commercial systems for carrying out both procedures are available from a variety of sources.

2.2.9 Long PCR

Protocol 1 is suitable for amplification of products up to 3 kb. The same procedure will give reasonable results with products up to 10 kb, provided that the Mg^{2+} concentration is optimized, the primer annealing temperatures are balanced, the extension period is increased to 2–3 min and product yield is enhanced by one of the methods described in Section 2.2.7. For products >10 kb a different amplification protocol must be used.

The main difference between normal PCR and long PCR is that the latter uses a combination of two thermostable polymerases (10). One of these polymerases has no proof-reading (3′→5′ exonuclease) activity and is present at 10- to 100-fold excess over the second polymerase, which does have the proof-reading function. It is not entirely clear why this combination improves the length of product that can be synthesized. Various enzyme combinations have been tried with differing degrees of success but it is important to choose a pair of enzymes with similar buffer requirements. Popular combinations are *Tth* and *Tli* polymerases, as used in *Protocol 3*, and *Taq* and *Pfu* polymerases.

Protocol 3 is based on a widely used method that has been described on the internet (11). Other than the enzymes, the main differences compared with the standard hot-start procedure is the use of acetate salts (*Tth* and *Tli* polymerases function more efficiently with acetate rather than chloride ions) and the inclusion of glycerol to enhance product yield (see Section 2.2.7). The extension temperature is set to the optimum for this enzyme pair, but it is not necessary to increase the extension time beyond 2 min per cycle. Products up to 40 kb can be synthesized.

Protocol 3

Long PCR with hot-start[a]

Equipment and reagents

- Thermal cycler with a hot lid (available for many suppliers)
- 0.5 ml microfuge tubes
- 10× PCR buffer (250 mM Tricine-KOH, pH 8.7, 850 mM potassium acetate, 10% dimethyl sulphoxide, prepared in sterile water)
- Glycerol
- 20 mM magnesium acetate (prepared in sterile water)
- 40 mM dNTP solution (10 mM dATP, 10 mM dCTP, 10 mM dGTP and 10 mM dTTP, prepared in sterile water)

- Primers 1 and 2 (10 μM solutions of each primer in sterile water)
- *Tth* DNA polymerase (various suppliers)
- *Tli* DNA polymerase (Promega or New England Biolabs, the latter under the tradename 'Vent')
- Template DNA
- Gel loading buffer (95% deionized formamide, 20 mM EDTA, pH 8.0, 0.05% bromophenol blue, 0.05% xylene cyanol; prepare deionized formamide as described in Appendix 4, *Protocol 4*)

Method

1 Prepare a 'master mix' containing the following components. The amount of reagents to add depends on the number of reactions that are being carried out. The following recipe is sufficient for a single reaction and should be scaled up accordingly: 5 μl 10× PCR buffer, 1 μl 40 mM dNTP solution, 2.5 μl Primer 1, 2.5 μl Primer 2, 5 μl glycerol, sufficient water to bring the final volume to 44 μl, 2.5 units *Tth* DNA polymerase, 0.1 units *Tli* DNA polymerase (total volume = 44 μl).

2 Transfer 44 μl of master mix to a 0.5 ml microfuge tube and add 1 μl template DNA.

3 Place each tube in a thermal cycler that has been programmed to provide the following heating regime: 2 min at 94°C, 30 cycles of 1 min at the annealing temperature, 2 min at 68°C, 1 min at 94°C and 8 min at 74°C.

4 After 1 min at 94 °C, pause the thermal cycling and add 5 µl of 20 mM magnesium acetate to each tube. Continue the thermal cycling.

5 Terminate the reaction and analyse the results as described in *Protocol 1*.

ᵃ See also the notes to *Protocols 1* and *2*.

2.2.10 PCR of RNA

RNA can also be used as a substrate for PCR provided that that the amplification protocol is preceded by a step in which the RNA is converted to cDNA. This involves treatment of the RNA with an enzyme that possesses reverse transcriptase activity, the overall procedure being called reverse transcriptase-PCR or RT-PCR. The technique is frequently used in gene expression studies as a means of determining if the transcript of a particular gene is present in the RNA extracted from a tissue or cell type. It is also used in a specialized method for mapping the ends of RNA transcripts, called 'random amplification of RNA ends' (RACE; ref. 12).

Reverse transcriptase polymerase chain reaction can be carried out in two stages, with the RNA first converted to double-stranded cDNA, as described in Chapter 3, Section 4, and then amplified by a standard PCR. Alternatively, the two stages can be combined through use of a thermostable DNA polymerase that can use either RNA or DNA as a substrate, and which therefore converts an RNA substrate into DNA in the first cycle of the PCR, and then amplifies the DNA in subsequent stages. *Tth* DNA polymerase is an example of an enzyme with the required activity (13) and several companies market kits for carrying out RT-PCR with this or a similar dual-activity polymerase. The main difficulty with RT-PCR is the need to distinguish between PCR products that derive from RNA templates and any that arise from amplification of trace amounts of genomic DNA in the RNA preparation. This is a significant problem in view of the extreme sensitivity of PCR and the possibility that a single contaminating molecule of genomic DNA can give rise to a PCR product. The safest option is to design the PCR so that the region of the target mRNA that is amplified spans an intron in the DNA sequence, so that RNA and DNA products have different lengths and hence can be clearly distinguished.

3 Cloning PCR products

Before cloning a PCR product it is necessary to purify the product from the reaction mixture, in particular to remove short primer–dimers (Section 2.2.2) which would be inserted into the vector at high efficiency and give rise to many spurious clones. Before purifying the product, the result of the PCR should be assessed by agarose gel electrophoresis. If the only visible band is the one corresponding to the desired product, then the product can be purified directly from the PCR mixture using any one of the various PCR purification kits that are available on the market (e.g. Qiagen QIAquick PCR Purification Kit). These systems remove low molecular weight molecules such as primers and short primer–dimers from the reaction mix, leaving behind the PCR product, provided that this is >100 bp. Note that the template DNA will still be present in the reaction and that this DNA will be co-purified with the PCR product. If the products are restricted before cloning (see below) then the identities of the inserts in the resulting clones should be checked because restriction fragments of the template DNA might also be cloned.

If examination of the agarose gel shows that as well as the desired PCR product there are one or more additional visible bands of 100 bp or longer, then direct purification from the reaction mix is unsuitable because these additional molecules will be co-purified along with the desired product. Instead, the desired product should be cut out of the gel and recovered from the resulting agarose slice. Again, many kits are available for doing this (e.g. Qiagen QIAquick Gel Extraction Kit).

It might be imagined that PCR products are blunt ended and so can be cloned by one of the methods described in Volume I, Chapter 7. Unfortunately, this is not the case. *Taq* DNA polymerase, and a few other thermostable polymerases, occasionally adds an additional nucleotide on to the 3′ of a strand that it has just synthesized, this addition being made in a template-independent manner. Often, but not always, the additional nucleotide is an adenosine and the addition is not made to every 3′-end. This means that PCR products have a mixture of ends: some have two blunt ends, some have one blunt end and a single-nucleotide 3′ overhang at the other end and others have single-nucleotide overhangs at both ends. Ligation into a cloning vector is therefore less straightforward than it might be. The two popular approaches are as follows:

(a) The single-nucleotide overhangs can be used as short sticky ends for attachment of the PCR product into a special vector that provides the appropriate cloning site. Examples are described in Volume I, Chapter 9, Section 3.7. These vectors are obtained in linear form with a single-thymidine overhang at each 3′-end and so can be ligated to those PCR products which have overhangs at both ends. Many PCR products can be cloned efficiently in this way, but occasionally a low yield of recombinants is obtained, presumably because a particular PCR product contains relatively few molecules with overhangs at both ends.

(b) The PCR products can be converted into genuine sticky-ended molecules by digestion with a restriction enzyme that cuts at restriction sites within the primer sequences. This approach is not limited to those occasions when the primer can be designed so that its annealing position spans a restriction site. It is also possible to include a 10–12 nucleotide extension at the 5′-end of the primer, non-complementary to the template DNA, containing the restriction site. An example would be 5′-GCGCGGATCC– followed by the primer sequence. In this example the underlined region is a *Bam*HI site. The addition of this extension to the primer has little effect on its annealing properties, and the extension should be excluded from the calculation when the annealing temperature of the primer estimated. When designing an extension it should be noted that not all restriction enzymes are able to cut at sites that are close to the end of a molecule. Digestion with *Bam*HI or *Eco*RI is efficient if the site is preceded by four nucleotides as shown in the example, but *Pst*I and *Sal*I are less effective (for more details see the New Englands Biolabs catalogue). Because of these problems, and because it is desirable to cut as many restriction sites as possible in order to generate many sticky ended molecules, this restriction should be carried out with an excess of enzyme. If the products have been purified from a 20-μl aliquot of the PCRs described in *Protocols 1–3*, then use 5 units of enzyme in a 2-h digestion. When designing primers with restriction site extensions, remember to chose an enzyme that does not have restriction sites within the amplified sequence as this would lead to partial products being cloned.

4 Sequencing PCR products

Cloned PCR products can be sequenced by the standard procedures described in Chapter 6, but this approach should be avoided if at all possible because the relatively high error rate of *Taq* polymerase means that the sequence of a cloned product might not be identical to the sequence of the original template. At least 10 clones should be examined in order to obtain a consensus. Even then, the sequence might be wrong because a polymerase error occurring in an early cycle of the PCR will lead to a preponderance of molecules with that error in the final product. All 10 clones might therefore give the same sequence, but all might be incorrect.

Direct sequencing of PCR products usually avoids this problem because each position within the sequence that is obtained represents the most frequent nucleotide at that position in the population of product molecules. Errors that lead to a small minority of molecules with an incorrect sequence will therefore be ignored. The direct sequence will therefore be correct, except on those unfortunate occasions when an early polymerase error becomes predominant in the final product.

4.1 Direct sequencing of PCR products

During direct sequencing, one strand of the PCR product is copied by *Taq* DNA polymerase in a thermal cycled reaction in the presence of dideoxynucleotides. The reaction is not an amplification because the products accumulate in a linear rather than exponential manner. The products are chain-terminated and so can be resolved in a polyacrylamide gel in order to reveal the sequence (see Chapter 6).

It is rarely possible to remove all the primers from a completed PCR, so during direct sequencing there can be a certain amount of copying of the strand that is not being sequenced, this copying being driven by primers that were carried over from the PCR. The chain-terminated molecules that result from this second-strand copying cause spurious bands to appear on the autoradiograph, sometimes preventing the authentic sequences from being discerned. Avoiding this problem is the key to direct sequencing. The possibilities are as follows:

- one of the primers used in the PCR can be tagged in such a way that it will be possible to purify the strands synthesized from this primer. This can be done by labelling with magnetic beads or with biotin. At the end of the PCR, the products are denatured and the labelled strands recovered with a magnetic device or by binding to avidin. These strands are then directly sequenced in the absence of any contaminating signal from the other strand

- one strand can be selectively digested with the enzyme λ endonuclease (14)

- a second PCR can be carried out with a large excess of one primer (typically 100:1) so that one strand is preferentially amplified (asymmetric PCR); the second strand is still present after the asymmetric PCR but in such low abundance that it does not interfere with the sequence readout.

- the sequence reaction can be carried out with a labelled primer, rather than labelling the chain terminated molecules uniformly through inclusion of a radioactive nucleotide. As a result, the only labelled chain-terminated molecules generated are those that are copies of the strand being sequenced: copies of the second strand are not labelled and do not appear on the autoradiograph.

The fourth of these possibilities is by far the most successful approach and is the only one that will be described here. The method is called thermal cycle sequencing. The version given in *Protocol 4* is based on ref. (15) and is designed for the thermostable polymerase called Vent (exo⁻) (New England Biolabs). Any equivalent thermostable polymerase (i.e. one that lacks both $5' \rightarrow 3'$ and $3' \rightarrow 5'$ exonuclease activities—see Chapter 6, Section 3.1.2)—can be used if the necessary changes are made to the buffer conditions.

Protocol 4

Direct sequencing of PCR products

Equipment and reagents

- Thermal cycler
- Oligonucleotide primer
- 10× kinase buffer (0.5 M Tris-HCl, pH 7.6, 0.1 M MgCl₂, 50 mM DTT, 1 mM spermidine)
- [γ-³²P]ATP at 3000 Ci/mmol and 10 mCi/ml
- 10× Vent buffer (200 mM Tris-HCl, pH 8.8, 100 mM KCl, 50 mM MgSO₄, 100 mM (NH₄)₂SO₄)

- T4 polynucleotide kinase
- 0.5 M EDTA, pH 8.0
- A sequencing mix (30 μM dATP, 100 μM dCTP, 100 μM dGTP, 100 μM dTTP and 900 μM ddATP, prepared in 1× Vent buffer). Prepare termination mixes from 20 mM stock solutions. Both the stock solutions and the termination mixes can be stored at −20°C for several months

- C-sequencing mix (30 μM dATP, 40 μM dCTP, 100 μM dGTP, 100 μM dTTP and 500 μM ddCTP, prepared in 1× Vent buffer)

- G-sequencing mix (30 μM dATP, 100 μM dCTP, 40 μM dGTP, 100 μM dTTP and 400 μM ddGTP, prepared in 1× Vent buffer)

- T-sequencing mix (30 μM dATP, 100 μM dCTP, 100 μM dGTP, 30 μM dTTP and 700 μM ddTTP, prepared in 1× Vent buffer)

- PCR product that is to be sequenced[a]

- Vent (exo⁻) DNA polymerase (New England Biolabs)

- Formamide dye mix (0.1% bromophenol blue, 0.1% xylene cyanol, 10 mM EDTA, pH 8.0, prepared in deionized formamide—see *Appendix 4, Protocol 4*)

Method

1   End-label the primer by the forward reaction with T4 polynucleotide kinase.[b] To do this, set up the following mix in a microfuge tube on ice: 10 pmol ends[c] oligonucleotide primer, 5 μl 10× kinase buffer, sufficient water to bring the final volume to 50 μl, 200 μCi (65 pmol) [γ-^{32}P]ATP (3000 Ci/mmol), 10 units T4 polynucleotide kinase (total volume = 50 μl). Incubate at 37°C for 30–45 min, then terminate the reaction by addition of 2 μl 0.5 M EDTA, pH 8.0.

2   Add 3 μl of the appropriate sequencing mix to each of four 0.5-ml microfuge tubes labelled A, C, G and T.

3   Set up the following mixture in a microfuge tube on ice: 5 μl PCR product[d], 5 μl end-labelled primer (from step 1), 1.5 μl 10× Vent buffer, sufficient water to bring the final volume to 15 μl, 2 units Vent (exo⁻) DNA polymerase (total volume = 15 μl).

4   Transfer 3 μl of the mix from step 3 into each of the tubes prepared in step 2.

5   Place each tube in a thermal cycler that has been programmed for 20 cycles of 20 s at 95°C, 20 s at 55°C and 20 s at 72°C.

6   Terminate the reaction by adding 4 μl formamide dye mix to each tube.

7   Prepare a denaturing polyacrylamide gel (Chapter 6, *Protocol 9*) and load and run the samples as described in Chapter 6, *Protocol 10*. Set up an autoradiograph as described in Chapter 6, *Protocol 13*.

[a] Before sequencing a PCR product it must be purified. The issues are the same as those described for cloning PCR products (see Section 3). If the PCR product to be sequenced is the only visible band on an agarose gel, then it can be purified directly from the PCR mixture (e.g. by using the Qiagen QIAquick PCR Purification Kit). If additional bands are seen in the gel then cut out the desired band and purify the product from the gel slice (e.g. by using the Qiagen QIAquick Gel Extraction Kit).

[b] For more details, see Chapter 4, Section 2.4.1.

[c] 10 pmol ends = 0.065 μg of a 20-mer oligonucleotide.

[d] The volume of PCR product may need to be adjusted to provide a clear sequence.

Table 2 Troubleshooting guide for PCR[a]

Problem	Possible cause	Suggested remedy
No amplification	Annealing temperature is too high	Reduce the annealing temperature in 2°C steps in a series of PCRs until amplification occurs
	Template DNA contains inhibitors of *Taq* DNA polymerase	Set up a PCR with control DNA and spike with an aliquot of DNA that will not amplify; if the control PCR is inhibited then clean up the template DNA
	Insufficient template DNA	Increase the amount of template DNA
	Mg^{2+} concentration is not optimal	Set up PCRs with a range of Mg^{2+} concentrations between 1.0 and 5.0 mM
	Incorrect primer sequences	Check the sequences. Be wary of errors in the sequence if it has been obtained from a DNA database
	Temperature display is inaccurate	Do not assume that the thermal cycler's temperature display gives the exact temperature within the reaction tube: check that the desired annealing temperature is actually being met
Products of the incorrect size	Annealing temperature is too low	Increase the annealing temperature in 2°C steps until the non-specific products are no longer synthesized
	Non-specific primer annealing is occurring during the initial stage of the PCR	Use hot-start PCR
	Non-specific primer annealing is occurring during the later stages of the PCR	Reduce the number of cycles so that the PCR does not continue after one or more of the reactants has been exhausted
	Primer artefacts	Reduce the concentration of primers, or redesign them so there is no self-complementarity
	Temperature display is inaccurate	Do not assume that the thermal cycler's temperature display gives the exact temperature within the reaction tube: check that the desired annealing temperature is actually being met
	Thermal cycling is too slow	Old or faulty thermal cyclers lose the ability to cool down and heat up sufficiently quickly; extended periods at lower temperatures increase formation of non-specific products
Smear of DNA in the gel	Too much template DNA	Reduce the amount of template DNA
	Too many cycles	Reduce the number of cycles
Smear shorter than the expected product	Incomplete extension	Increase the extension time, or check for contamination of the template DNA with inhibitors of *Taq* DNA polymerase
Unusual sequences when cloned products are examined	Point mutations	*Taq* polymerase lacks a proof-reading activity and so generates a relatively high number of point errors; optimizing the Mg^{2+} concentration may help
	Insertions of As	Indicates that jumping PCR has occurred owing to the template DNA being fragmented or the *Taq* polymerase being partially inhibited
	Sequences are chimeric, made up of parts of two or more different targets	Jumping PCR—see above

[a] Modified from Brown, T. A., *Molecular Biology Labfax*, Vol. 2, 2nd edn, pp. 69–70, (1998), by permission of the publisher Academic Press, London.

4.2 Thermal cycle sequencing of other DNA templates

Thermal cycle sequencing is not limited to sequence analysis of PCR products—the same approach can be taken for an any DNA template, including both single-stranded M13 clones or double-stranded plasmid DNA. The amount of template DNA that is used is much less than in normal chain-termination sequencing. For an M13 clone 0.05 pmol should be used, corresponding to 125 ng of an M13 clone with a 500 bp insert, and for a plasmid clone 0.1 pmol should be used, which is 100 ng of a pUC18 clone with a 500 bp insert. In practice, satisfactory results are usually obtained with 1 μl of single- or double-stranded template DNA prepared as described in Chapter 6, *Protocols 1, 2* or *5*. This template DNA is substituted for the PCR product in the mixture prepared in step 3 of *Protocol 4* above. If cloned DNA is used then the primer can be a universal one.

The thermal cycle method is also the basis of automated DNA sequencing. As described in Chapter 6, Section 6, fluorescent labels are used, either a single label attached to the primer, in which case the sequencing reactions are carried out as described in *Protocol 4*, or four different labels, one for each ddNTP. The latter strategy requires a few modifications to *Protocol 4*, but the principle of the technique is the same. If an automated sequencer is used, then the components of the sequencing reactions and the cycle times used for generation of the chain-terminated molecules must be strictly compatible with the specifications of the sequencer. The manufacturer's instructions should therefore be followed.

5 Problems with PCR

Table 2 provides a troubleshooting guide for PCR.

References

1. Cooper, A. and Poinar, H. N. (2000). *Science*, **289**, 1139.
2. Brown, T. A. (ed.) (1998). *Molecular biology labfax*, 2nd edn, Vol. 2. Academic Press, London.
3. Höss, M. and Pääbo, S. (1993). *Nucl. Acids Res.*, **21**, 3913.
4. Pääbo, S., Irwin, D. M., and Wilson, A. C. (1990). *J. Biol. Chem.*, **265**, 4718.
5. Allaby, R. G., Banerjee, M., and Brown, T. A. (1999). *Genome*, **42**, 296.
6. Rees, W. A., Yager, T. D., Korte, J., and von Hippel, P. H. (1993). *Biochemistry*, **32**, 137.
7. Erlich, H. A., Gelfand, D., and Sninsky, J. J. (1991). *Science*, **252**, 1643.
8. Chou, Q., Russell, M., Birch, D. E., Raymond, J., and Bloch, W. (1992). *Nucl. Acids Res.*, **20**, 1717.
9. Kellogg, D. E., Rybalkin, I., Chen, S., Mukhamedova, N., Vlasik, T., Siebert, P. D., and Chencik, A. (1994). *Biotechniques*, **16,** 1134.
10. Cheng, S., Fockler, C., Barnes, W., and Higuchi, R. (1994). *Proc. Natl Acad. Sci. USA*, **91**, 5695.
11. http://twod.med.harvard.edu/
12. Frohman, M. A., Dush, M. K., and Martin, G. R. (1988). *Proc. Natl. Acad. Sci. USA*, **85**, 8998.
13. Brown, T. A. (ed.) (1998). *Molecular biology labfax*, 2nd edn, Vol. 1. Academic Press, London.

14. Higuchi, R. G. and Ochman, H. (1989). *Nucl. Acids Res.*, **17,** 5865.
15. Slatko, B. E., Albright, L. M., Tabor, S., and Ju, J. (1999). In *Current protocols in molecular biology* (ed. F. M. Ausubel, R. Brent, R. E. Kingston, D. D. Moore, J. G. Seidman, J. A. Smith, and K. Struhl), p. 7.4A.11. John Wiley, New York.

Chapter 8

Protein expression in *Escherichia coli*

John M. Ward

Department of Biochemistry and Molecular Biology, University College London, Gower Street, London WC1E 6BT, UK

1 Introduction

There now exist a variety of host-vector systems for the synthesis of protein from recombinant DNA. Systems using bacteria such as *Bacillus subtilis*, lower eukaryotes such as *Saccharomyces cerevisiae* and *Pichia pastoris* (1), and higher eukaryotes such as *Spodoptera frugiperda* (2) insect cells and Chinese hamster ovary cells all have their advantages and disadvantages for protein expression. However, by far the most common host is still *Escherichia coli*. These expression systems are used for a variety of purposes. Very often, pure recombinant protein is needed for functional analysis, such as the determination of kinetic parameters, substrate specificity and inhibitor sensitivity. The structural analysis of proteins by X-ray diffraction and NMR techniques demands multi-milligram amounts of very pure proteins. High-throughput screening by pharmaceutical companies for novel drug discovery can use significant amounts of a protein target. At the small scale of growth of a few tens of millilitres to a litre or two, there are several highly efficient systems which can be used for the rapid production of recombinant protein, and these will be discussed in this chapter. Larger-scale growth and expression systems in *E. coli* involve a somewhat different set of problems and an overview of these will be given at the end of the chapter.

2 Factors affecting protein expression

The following factors affect the degree to which a recombinant gene will direct synthesis of protein in *E. coli*:

- strength of the promoter
- efficiency of the translational start process
- codon usage within the cloned gene
- rate of degradation of the recombinant mRNA and protein
- folding of the recombinant protein

The final amount of active protein that is obtained will be a combination of these factors and the concentration of cells in the culture.

2.1 Promoters

The early vectors for *E. coli* expression used natural promoters such as *lac*, *trp*, and λP_L. *Figure 1* shows the DNA sequences of the important parts of these promoters. These promoters allow good levels of expression and all three are controlled by a repressor. Expression of the *lac* and *trp* promoters is switched on by addition of the inducers IPTG or IAA, respectively. Expression from the λP_L promoter relies on destruction of the temperature-sensitive repressor cIts857 by a brief period of growth at 42 °C.

Amann et *al.* (3) showed that by taking a DNA fragment containing the –35 sequence of the *trp* promoter and fusing it to a DNA fragment containing the –10 sequence and operator sequence of the *lac* promoter, a promoter stronger than either of the parent promoters could be formed. This is the *tac* promoter and it retains the ability to be repressed by the *lac* repressor and induced by IPTG. The popular vector pKK223-3 (supplied by Amersham Pharmacia; see Volume I, Chapter 9, Section 3.5.2) was developed to take advantage of the *tac* promoter. It is a low to medium copy number vector but allows good expression of genes. Open reading frames inserted into the *Eco*RI site of pKK223-3 can make use of the 5′-AGGA-3′ sequence in the vector as a ribosome-binding site. The *Eco*RI site is positioned 5 bp downstream from the 5′-AGGA-3′ sequence and an ATG codon must be included in the cloned fragment at the *Eco*RI site to make this a translational start site. Open reading frames cloned in most of the other sites must have their own ribosome-binding sites and ATG codons as the ribosome-binding site provided by the vector is too far away. Another vector which has a similar hybrid promoter is the pTrc 99A vector (also from Amersham Pharmacia). This vector uses the hybrid *trc* promoter, which is another variant of the *lac* and *trp* promoters, again inducible with IPTG. pTrc 99 A has an *Nco*I site, containing an ATG sequence, located 6 bp downstream of a 5′-AGGA-3′ ribosome-binding site. An open reading frame that has been engineered to have an *Nco*I site at its

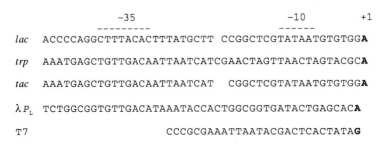

	–35		–10	+1

lac	ACCCCAGGCTTTACACTTTATGCTT	CCGGCTCGTATAATGTGTGG**A**
trp	AAATGAGCTGTTGACAATTAATCATCGAACTAGTTAACTAGTACGC**A**
tac	AAATGAGCTGTTGACAATTAATCAT	CGGCTCGTATAATGTGTGG**A**
λP_L	TCTGGCGGTGTTGACATAAATACCACTGGCGGTGATACTGAGCAC**A**
T7	CCCGCGAAATTAATACGACTCACTATA**G**

Figure 1 DNA sequences of the *lac*, *trp*, *tac*, λP_L and T7 promoters. Gaps have been placed within the sequences of the *lac* and *tac* promoters so that the –35 and –10 regions align with the equivalent parts of the other promoters. The T7 promoter does not have any homology to these –35 and –10 regions as it is recognized by the RNA polymerase encoded by T7 phage, whereas the other four promoters are recognised by the *Escherichia coli* RNA polymerase. The position at which mRNA synthesis begins is designated +1 and is shown in bold.

N-terminus can therefore be cut with *Nco*I and ligated into pTrc 99A in the correct position relative to the ribosome-binding site. One way to engineer the open reading frame would be to amplify it by PCR with a primer that contains an *Nco*I site as a 5′ overhang.

Recently, one of the late promoters from T7 phage has become very widely used for the expression of foreign genes in *E. coli*. (4; see *Figure 1*), for example in the pET series of vectors (see Volume I, Chapter 9, Section 3.5.2). The T7 promoter is only recognized by the RNA polymerase encoded by T7 phage, and not by the *E. coli* RNA polymerase. The system uses a host strain that carries the gene for the T7 RNA polymerase inserted into its chromosome within a defective λ phage genome. The T7 RNA polymerase gene is under control of the *tac* promoter so the system is again inducible by IPTG. Very efficient repression is possible, compared with the use the IPTG-inducible promoters on plasmids, since the copy number of the *tac* promoter in the chromosome is 1 and there are many *lac* repressor molecules available to bind to it. It is, however, possible for a few molecules of T7 RNA polymerase to be synthesized even when the system is repressed, which is a problem if the gene product is highly toxic. To reduce this problem, a host strain that expresses the T7 gene *lysS* can be used. This gene encodes the T7 lysozyme enzyme which binds to the T7 RNA polymerase and inactivates it. Upon induction of the *tac* promoter with IPTG there is sufficient synthesis of the T7 RNA polymerase to titrate out all the T7 lysozyme. Once some of the T7 RNA polymerase is made it will synthesize large numbers of RNA transcripts from the cloned gene and thus large amounts of protein are made.

2.2 Translation initiation

Efficient initiation of translation requires that the ribosome-binding site is correctly positioned relative to the ATG codon (5). The ribosome-binding site is the mRNA sequence which is thought to base-pair with a region near to the end of the 16S rRNA molecule. If the region of complementarity between the mRNA and rRNA is too long, then the ribosome binds strongly but there may not be sufficient release of the ribosome for translation to proceed. A sequence such as 5′-AGGA-3′ is sufficient, provided that it is correctly spaced 7–9 nucleotides from the ATG codon. The pET series of vectors use the sequence 5′-AAGGAG–8 bases–ATG-3′

Most of the recent vectors provide an efficient spacing between ribosome-binding site and initiation codon. With many vectors, the ATG codon is contained within the recognition sequence of a *Nco*I or *Nde*I restriction site. As described above, any open reading frame whether from a eukaryote or a prokaryote can be engineered by PCR to include one or other of these restriction sites at the N-terminus and then cloned in frame with this efficient translational start site.

3 Protein fusion systems

Some of the most widely-used expression systems for *E. coli* are those which employ fusion of the desired open reading frame to a well expressed protein. *Table 1* lists some of the more popular fusion systems.

Table 1 Protein fusion vectors

Plasmid	Fusion partner	Size of fusion partner	Properties of fusion partner	Source
pRIT, pEZZ	Protein A and Protein A domains	7 kDa for each Protein A domain, 30 kDa for entire Protein A	Stable folding domain, can promote folding and solubility of proteins fused to domains. Binds to IgG column; elution at low pH or with triethylamine	Amersham Pharmacia
pMAL	Maltose-binding protein	40 kDa	Binds to amylose column; elution with maltose	New England Biolabs
pGEX	Glutathione S-transferase	26 kDa	Binds to glutathione-Sepharose; elution with glutathione	Amersham Pharmacia
PinPoint Xa	Biotin carboxyl carrier protein of acetyl CoA carboxylase	42 kDa	Fusion tag becomes biotinylated, binds to modified avidin column; elution with biotin	Promega
pTrxFus, pThioHis	Thioredoxin	14 kDa	Thioredoxin promotes solubility of fused proteins. Binds to Thiobond	Invitrogen
pQE60, pQE70, pET14–16, 19–33	6× His	0.84 kDa	Smallest fusion tag. Binds to a chelated nickel ion on a metal chelate resin; mild elution with a change of pH 7 to 3.5, or with imidazole	pQE: Qiagen pET: Novagen
pTYB	Chitin-binding domain and intein	55 kDa	Binds to chitin beads; elution with chitobiose. Fusion partner is cleaved by the intein at 4°C with DTT	New England Biolabs

These protein fusion vectors offer many advantages for the small-scale expression and purification of a recombinant protein. The fusion partner or 'Tag' enables recombinant proteins to be isolated by an affinity interaction between the Tag and a molecule/component immobilized within a column, on a resin or even on magnetic beads. This allows the simple and rapid washing away of most or all of the non-recombinant *E. coli* proteins, followed by elution of the protein fusion, usually under mild conditions. For each of the fusion systems, the binding and elution conditions remain the same regardless of the nature of the recombinant protein that is fused to the Tag. This means that once the conditions for a particular system have been worked up in a laboratory, workers can become skilled at purifying many different fused proteins rapidly for further analysis. In each case, the vector provides translational initiation signals that have been engineered to give a very efficient translational start. These factors, combined with strong, controllable promoters, allows expression levels of up to 40% of the cell protein.

3.1 Histidine-tailed protein fusions

Protocol 1 describes the synthesis of a histidine-tailed fusion protein. A polyhistidine (polyHis) tail is a very versatile fusion because of its small size, which means it often does not need to be removed after purification. PolyHis tails have a high affinity for divalent metal ions such as nickel, copper and zinc (6) and this property is used in purification of the fusion protein. A resin containing NTA surface groups coordinates an ion such as Ni^{2+}, filling four of the six coordination sites. The histidine residues on the protein fusion wrap around a chelated metal ion in such a way that two of the imidazole side-chains fill the two spare coordination sites. This forms a very specific and stable interaction which withstands washing conditions that remove unbound proteins and allows mild elution conditions to be used. A change in pH or a wash buffer containing imidazole is used to selectively elute the polyHis-tailed protein. *Protocol 1* can be used for genes cloned in the pQE (Qiagen) and pET (Novagen) vectors which have *N*-terminal or *C*-terminal 6-histidine fusions.

Protocol 1

Expression and purification of a histidine-tailed protein

Equipment and reagents

- Sonicator (e.g. MSE Soniprep 150)[a]
- 5 ml HiTrap columns containing Chelating Sepharose (Amersham Pharmacia)
- Nutrient broth containing 100 μg/ml ampicillin (see Appendix 4)
- 100 mM IPTG

- Cell resuspension buffer (20 mM sodium phosphate, pH 7.2, 500 mM NaCl)
- Elution buffer (20 mM sodium phosphate, pH 3.5, 500 mM NaCl)
- Neutralization buffer (1 M Tris-HCl, pH 7.5)

Protocol 1 continued

Method

1 Inoculate a 200–1000 ml culture with recombinant *E. coli*. Use a 5 ml overnight culture grown in nutrient broth containing 100 µg/ml ampicillin to inoculate 200–1000 ml of the same medium.

2 Incubate the culture at 37 °C with vigorous shaking.

3 Measure the absorbance at 600 nm every 30 min.

4 When the optical density reaches 0.8 (usually after 3 h) induce by adding IPTG to a final concentration of 0.5 mM.

5 Incubate at 37 °C for a further 3–4 h.

6 Harvest the bacteria by centrifugation at 12 000 **g** at 4 °C for 5 min.

7 Resuspend the bacterial pellet in 10 ml of cell resuspension buffer.

8 Chill the cell suspension in an ice/water slurry and sonicate the cells at 8 µ amplitude. Use 5 cycles of 10 s sonication followed by 10 s cooling.

9 Centrifuge at 12 000 **g** for 15 min at 4 °C.

10 Equilibrate a 5 ml HiTrap column (previously charged with nickel ions following the manufacturer's instructions) with 10 column volumes of the cell resuspension buffer at a flow rate of 2 ml/min.

11 Load the clarified extract on to the equilibrated HiTrap column at a flow rate of 2 ml/min.

12 Wash the unbound proteins from the column with 20 ml of the resuspension buffer. Use a flow rate of 2 ml/min.

13 Elute the polyHis-tailed protein with 15 ml elution buffer at a flow rate of 1.3 ml/min.

14 Add a one-tenth volume of neutralization buffer to the eluted protein solution to adjust the pH to neutrality.

[a] The MSE Process Timer is a useful add on to the MSE Soniprep 150 as it allows the user to set the number of cycles and the on/off/cooling times.

In a typical experiment (7), 1 l of a culture containing a T4 lysozyme gene with a 4× His tail and expressed from a *tac* promoter produced 50 mg of recombinant polyHis fusion protein after induction. Chromatography of the extract in a 5 ml HiTrap column charged with copper ions gave 40–45 mg enzyme at a purity of >95%. The HiTrap columns (Amersham Pharmacia) can be charged with either Cu^{2+}, Zn^{2+} or Ni^{2+}, each of which gives good binding and elution. Qiagen also supply Ni^{2+} NTA resins in several different formats, such as spin columns, agarose beads and magnetic beads, allowing great flexibility in the chosen method for purification of histidine-tailed proteins.

The polyHis tail can be placed at the *N*-terminus or *C*-terminus of a protein. It is a small enough peptide to be introduced by site-directed mutagenesis of a gene already cloned in an expression vector. The histidine codon is CAC or CAT and six codons require the addition of only 18 bp of DNA. With 8–10 bp of sequence complementary to the target gene placed on either side of the six histidine codons to give stability during the mutagenesis reaction, a polyHis tail can be attached to almost any protein (7).

3.2 GST protein fusions

The GST fusion system is also widely used and works very well for the expression and purification of many proteins. There are ten pGEX vectors currently available which differ in their cloning sites. Six have a thrombin cleavage site to separate the fusion partner from the cloned protein and four have a Factor Xa cleavage site. The preparation of a GST fusion is described in *Protocol 2*.

Protocol 2

Expression and purification of a GST fusion protein

Equipment and reagents

- Sonicator (see *Protocol 1*), with a microprobe to fit into the standard 1.5 ml microfuge tubes
- Disposable 5 ml polypropylene columns (Pierce no. 29922)
- Nutrient broth containing 100 μg/ml ampicillin (see Appendix 4)
- 100 mM IPTG
- Extraction buffer (10 mM Tris-HCl, pH 7.5, 100 mM NaCl)
- Glutathione-agarose (Sigma G4510)
- GST elution buffer (50 mM glutathione [Sigma G 4251], 10 mM Tris-HCl, pH7.5)

Method

1 Inoculate a 5 ml culture of nutrient broth containing 100 μg/ml ampicillin with *E. coli* JM107 or BL21 containing recombinant pGEX-3X. Grow overnight at 37 °C with shaking.

2 Dilute 1 ml of the overnight culture into 10 ml of fresh nutrient broth containing 100 μg/ml ampicillin and grow for 2 h with shaking.

3 Add IPTG to a final of 0.5 mM. Grow for a further 4 h.

4 Centrifuge the cells at 12 000 **g** for 5 min at 4 °C.

5 Resuspend the cells in 1 ml extraction buffer, transfer to a 1.5 ml microfuge tube, place in an ice/water slurry and sonicate at 8 μ amplitude. Use 5 cycles of 10 s sonication followed by 10 s cooling.

6 Centrifuge at full speed for 5 min in a microfuge.

7 Remove the supernatant and add it to 1 ml of glutathione-agarose, in a clean 1.5 ml microfuge tube, to absorb the fusion protein. Mix by inverting the tube several times and pipette into a disposable 5 ml polypropylene column.

8 Wash off unbound protein with 10 ml extraction buffer. Apply the buffer 1–2 ml at a time using a 5 ml pipette.

9 Elute the GST fusion protein with 2 ml GST elution buffer.

10 The glutathione-agarose column can be washed with extraction buffer and stored at 4 °C for further use.

It is often possible to obtain 50 µg of GST fusion protein for every milliltre of the original growth volume so the concentration of protein in the elution buffer can be as high as 250 µg/ml.

3.3 Cleavage of a GST fusion protein

The protein of interest can be cleaved from GST by treatment with Factor Xa if pGEX-3X, pGEX-5X-1, pGEX-5X-2 or pGEX-5X-3 was used. The other pGEX vectors have a thrombin cleavage site. Thrombin recognizes the sequence Leu–Val–Pro–Arg↓Gly–Ser, which means that the recombinant protein has the amino acids glycine–serine at its N-terminus, which may not always be desirable. Factor Xa recognizes the sequence Ile–Glu–Gly–Arg↓X, where X is any amino acid, and so the recombinant protein is released from the fusion without an extra residue at the N-terminus.

Protocol 3 describes the cleavage of a GST fusion protein with Factor Xa. The dialysis step serves to remove any glutathione from the sample before applying it to the glutathione-agarose column. Glutathione will be present if the fusion protein has been eluted following Protocol 2 and would prevent the fusion tag from binding to the second glutathione-agarose column.

Protocol 3

Factor Xa cleavage of a GST fusion protein

Equipment and reagents

- Disposable 5 ml polypropylene columns (Pierce no. 29922)
- Factor Xa (New England Biolabs)
- Glutathione-agarose (Sigma G4510)

- 5× Factor Xa reaction buffer (100 mM Tris-HCl, pH 8.0, 500 mM NaCl, 10 mM CaCl$_2$)
- Wash buffer (10 mM Tris-HCl, pH 7.5, 100 mM NaCl)

Method

1 Add 1 µg of Factor Xa for every 50 µg of fusion protein.

2 Add 0.2 volumes of 5× Factor Xa reaction buffer and incubate at 25 °C for 1–3 h.

3 Place the fusion protein digest in dialysis tubing and dialyse against 500 ml of wash buffer for 1 h at 4 °C.

Protocol 3 continued

4   Load onto 1 ml glutathione-agarose in a 5 ml disposable column at room temperature.

5   Wash with 2 ml of wash buffer and keep the material which washes off.

6   Run an aliquot (e.g. 20 μl) in a denaturing SDS polyacrylamide gel to determine the extent of Factor Xa cleavage.[a]

[a] The GST should remain on the glutathione-agarose column when cleaved from the fusion protein. Factor Xa will run as two polypeptide chains of 30 kDa and 20 kDa when reduced and denatured. GST is 26 kDa.

4 Expression of proteins in the periplasm

The periplasmic space of *E. coli* is the cellular compartment which lies beyond the cytoplasmic membrane but is retained by the outer membrane. It is external to the reducing environment within the cell and thus allows the formation of disulphide bonds. Given that many mammalian proteins are secreted in their natural habitat and contain disulphide bonds, it makes sense to direct recombinant proteins to the periplasm. The periplasm is 20–40% of the cell volume (8,9) and contains only 4–8% of the total cell protein (10). These proteins include those that are involved in disulphide bond formation (11,12) and which aid in the folding of periplasmic proteins. Periplasmic secretion has been used extensively for the expression of antibody fragments (13) and other proteins such as ribonuclease A (14), HIV-1 receptors (15) and interleukin-2 (16).

Protocol 4

Analytical periplasmic protein release procedure

Equipment and reagents

- Sonicator (see *Protocol 2*)
- 50 mM Tris-HCl, pH 7.5
- Extraction buffer (200 mM Tris-HCl, pH 7.5, 1 mM EDTA, 20% sucrose, 500 μg/ml lysozyme)

Method

1   Remove 1 ml of culture[a] and measure the optical density at 600 nm.

2   Transfer the cells to a 1.5 ml microfuge tube and centrifuge at full speed in a microfuge for 5 min at room temperature.

3   Transfer the supernatant to a clean microfuge tube. Keep this and label it 'supernatant'.

4   Resuspend the pellet in 200 μl of extraction buffer. Vortex well to resuspend the pellet of cells. Leave at room temperature for 15 min.

5   Add 200 μl of distilled water, mix by inverting the tube several times and then leave at room temperature for 15 min.

6 Centrifuge at full speed for 10 min. Carefully remove all of the supernatant into a clean tube. Label this as 'periplasm'.

7 Resuspend the cell pellet in 500 µl of 50 mM Tris-HCl, pH 7.5. Vortex well and place on ice.

8 Sonicate at 8 µ amplitude, using 5 cycles of 10 s sonication followed by 10 s cooling. Centrifuge at full speed for 10 min. Keep the supernatant and label as 'cell associated'.

9 Assay an aliquot (20–50 µl) of each of the three fractions for enzyme activity or some property of the protein. Alternatively, run an aliquot of each fraction on an SDS polyacrylamide gel.

[a] From a growing culture, before and during induction.

This small-scale protocol can be used to analyse expression of a periplasmically located protein during induction of a culture in a shake flask or fermenter. The measurement of optical density at 600 nm allows a growth curve to be plotted. The time of induction and the extent of enzyme activity can also be plotted, giving a good guide to the best time for harvesting the culture. The secreted protein is contained in three fractions: cell associated, periplasmic and supernatant. The sum of the secreted protein in these three fractions is 100% and there is usually very little left that is cell associated. The periplasmic fraction is usually 70%. The proportion of enzyme or protein that leaks into the growth medium is the material labelled as 'supernatant'. This fraction is usually a small percentage of the total expressed protein, but it can sometimes contain a significant proportion of the total activity after prolonged growth and expression. If this happens then it is useful to know how much has leaked into the growth medium (and when) during growth and induction. This gives a guide for the length of growth and induction when carrying out a larger-scale periplasmic expression. The proportion that leaks from the periplasm can vary owing to several factors such as the nature of growth medium, the type of protein, and the degree of expression.

The small-scale protocol is usually used to indicate the best time for carrying out a larger scale periplasmic protein extraction. *Protocol 5* describes extraction from a culture volume of 5 l.

Protocol 5

Large-scale periplasmic release procedure

Equipment and reagents
- Sharples 1P continuous centrifuge
- Silverson homogenizer
- Fractionation buffer (1 mM EDTA, pH 8.0, 20% sucrose, 500 µg/ml lysozyme)

Protocol 5 continued

Method

1   Grow a 5-l culture expressing the secreted protein. This can be a 5 l fermenter culture or several shake flasks that add up to this volume.

2   Pass the 5 l culture through the Sharples 1P continuous centrifuge at 30 000 **g**.

3   Scrape the cell paste from the plastic liner into 1 l of fractionation buffer.

4   Resuspend the cells using a Silverson homogenizer.

5   Incubate for 15 min at room temperature.

6   Add 1 l of distilled water.

7   Incubate for 15 min at room temperature.

8   Pass through the Sharples 1P centrifuge at 30 000 **g**. The eluate contains the periplasmic protein.

5 Cofactor requirements

It should be remembered that the activities of many enzymes are dependent on the presence of cofactors. There are cofactors which are loosely bound, such as NADH, which can be added to the reaction mixture when assaying or using the enzyme. Other cofactors may need to be added at the time of cell breakage. Often, these are tightly bound cofactors which are needed to allow the enzyme to achieve its final, stable, folded configuration, and at the levels of protein expression that can be achieved in E. *coli* such cofactors can be limiting inside the cell. If such a cofactor is limiting for the amount of enzyme expressed, a proportion of the enzyme will be present in the apo form and may not be as tightly folded as the holoenzyme. The apoenzyme may be more prone to proteolysis. For example, cells over-expressing a transketolase (a thiamine pyrophosphate dependent enzyme) to levels of over 35% of the cellular protein at the time of cell breakage contained 80% holo- and 20% apoenzyme (17). Other cofactors which might become limiting if over expressed in E. *coli* are FAD in certain oxygenases, pyridoxal phosphate, metal ions such as zinc in proteases (e.g. hepatitis C protease) and the haem group in, for example, cytochrome P450. For some of these cofactors the medium can be supplemented with the compound: for example, γ-aminolevulinic acid can be added to the growth medium at a concentration of 40 μg/ml to increase the amount of haem available in the cell (18).

6 Problems when scaling up expression

When scaling up expression systems from the laboratory or bench scale (1–2 l) to tens or hundreds of litres, problems are often encountered. Frequently, these arise because the vector systems used for small-scale work do not have selection, replication and induction systems which are optimal in larger scale cultures.

One problem arises from the ampicillin resistance gene, which is the commonest antibiotic selection system used in cloning vectors. The ampicillin resistance gene codes for a β-lactamase enzyme, which is secreted through the cytoplasmic membrane and is retained in the periplasm of *E. coli*. The enzyme's action on its substrate is to cleave the β-lactam ring of the antibiotic. Some β-lactamase leaks from the periplasm into the medium of a growing culture, and the supernatant of an overnight culture contains significant amounts of β-lactamase activity. When protein expression is scaled up, a series of growth stages are usually carried out, each progressively larger than the last. An example would be a 500-ml shake flask to a 5-l fermenter to a 50-l final fermentation. At each stage, β-lactamase is being transferred along with the cells. The enzyme that has leaked from the cells is free in the supernatant and can immediately set to work destroying the ampicillin in the next batch of medium. After a few hours, and before maximal growth is reached in the new fermenter, the selective agent has been destroyed. This allows the growth of plasmid-free cells should they happen to arise during incubation of the culture.

Plasmid-free cells can indeed arise during the growth of pUC based replicons (19). The pUC plasmids lack a segment, *cer*, of the original replicon. This segment of DNA is a recombination sequence which allows a chromosomally encoded recombination system to convert multimers back into monomers. Multimers arise during the growth of these very high copy number plasmids simply due to recombination between the large number of identical sequences present in the cell. A multimer is counted as a single replicon and in a population of cells harbouring these plasmids there is an asymmetric 'tail' in the copy number distribution of plasmids such that there are cells in the culture with few or no plasmids (20). When these cells encounter a medium with little or no antibiotic, such as when β-lactamase has destroyed the ampicillin during the scaling up from one culture size to the next, plasmid-free cells can outgrow the whole culture. An answer to this is to use an alternative antibiotic resistance for selection, such as kanamycin, which is only active inside the cell. This, however, does not solve the root of the problem, which is aberrant multimer formation. A better solution is to make use of the natural multimer resolution system which ColE1 has. By putting *cer* back into a multicopy pUC18 recombinant containing a cloned transketolase gene, 100% of cells still contained the recombinant plasmid after five cycles of dilution into non-selective broth (21).

References

1. Scorer, C. A., Clare, J. J., McCombie, W. R., Romanos, M. A., and Sreekrishna, K. (1994). *Biotechnology*, **12**, 181.
2. Luckow, V. and Summers, M. (1988). *Biotechnology*, **6**, 47.
3. Amann, E., Brosius, J., and Ptashne, M. (1983). *Gene*, **25**, 167.
4. Studier, F. W., Rosenberg, A. H., Dunn, J. J., and Dubendorf, J. W. (1990). In *Methods in enzymology* (ed. D. V. Goeddel), Vol. 185, p. 60. Academic Press, London.
5. Gold, L., Pribnow, D., Schneider, T., Shinedling, S., Singer, B. S., and Stormo, G. (1981). *Annu. Rev. Microbiol.,* **35**, 365.

6. Hochuli, E., Bannwarth, W., Döbeli, H., Gentz, R., and Stüber, D. (1988). *Biotechnology*, **6**, 1321.

7. Sloane, R. P., Ward, J. M., O'Brien, S. M., Thomas, O. R. T., and Dunnill, P. (1996). *J. Biotechnol.*, **49**, 231.

8. Stock, J.B., Rauch, B., and Roseman, S. (1977). *J. Biol. Chem.*, **252**, 7850.

9. van Wielink, J. E. and Duine, J. A. (1990). *Trends Biochem. Sci.*, **15**, 136.

10. Beacham. I. R. (1979). *Int. J. Biochem.*, **10**, 877.

11. Wulfing, C. (1994). *Mol. Microbiol.*, **12**, 685.

12. Andersen, C. L. (1997). *Mol. Microbiol.*, **26**, 121.

13. Pluckthun, A. (1990). *Nature*, **347**, 497.

14. Tarragona-Fiol, A., Taylorson, C. J., Ward, J. M. and Rabin, B.R. (1992). *Gene*, **118**, 239.

15. Rochenbach, S. K., Dupuis, M. J., Pitts, T. W., Marschke, C. K., and Tomich, C.-S. C. (1991). *Appl. Microbiol. Biotechnol.*, **35**, 32.

16. Halfman, G., Brailly, H., Bernadac, A., Montero-Julian, F. A., Lazdunski, C., and Baty, D. (1993). *J. Gen. Microbiol.*, **139**, 2465.

17. Mitra, R. K., Woodley, J. M., and Lilly, M. D. (1998). *Enzyme Microbial. Technol.*, **22**, 64.

18. Park, S-J., Cotter, P. A., and Gunsalus, R. P. (1995). *J. Bacteriol.*, **177**, 6652.

19. Summers, D. K. and Sherratt, D. J. (1984). *Cell*, **36**, 1097.

20. Summers, D. K. (1991). *Trends Biotechnol.*, **9**, 273.

21. French C. and Ward J. M. (1995). *Biotechnol. Lett.*, **17**, 247.

Chapter 9

Transcript mapping

Colin P. Smith

Department of Biomolecular Sciences, University of Manchester Institute of Science and Technology, Manchester M60 1QD, UK

1 Introduction

A detailed study of the structure and expression of any gene will normally include the determination of the end-points of the mature transcript derived from it. This type of analysis is generally conducted in order to identify the positions of DNA sequences potentially involved in the initiation or termination of transcription, or to identify the position of processing sites within the transcript. A knowledge of the 'transcript map' can also be very useful for designing subsequent genetic manipulations, including, for example, the construction of a transcriptional fusion of the test gene's promoter to a reporter gene, or the cloning of the gene of interest into an expression vector.

Several methods are available for directly quantifying the amount of a specific transcript. These include northern hybridization analysis, slot-blotting and reverse transcription coupled with PCR (RT-PCR). However, these methods do not allow the mapping of transcript end-points (the subject of this chapter). Transcript mapping procedures can be classified into two groups:

- mapping on the basis of 'nuclease protection' (see Section 2)
- primer-extension mapping (more correctly called oligonucleotide-primed reverse transcription) (see Section 3)

In addition to allowing the precise determination of transcript end-points these methods also allow quantification of the specific transcript.

In nuclease protection mapping methods, a specific probe derived from a DNA clone, or directly by PCR, is hybridized, in solution, with a test RNA sample (see *Figure 1*). Normally, total cellular RNA or material fractionated through an oligo(dT) column is used as the substrate. Any one of a variety of probes may be used:

- radiolabelled or unlabelled PCR products
- radiolabelled or unlabelled DNA restriction fragments
- unlabelled single-stranded DNA templates (e.g. M13 clones)
- continuously-labelled (body labelled) single-stranded DNA or RNA fragments generated *in vitro* with DNA or RNA polymerases
- end-labelled chemically-synthesized oligonucleotides

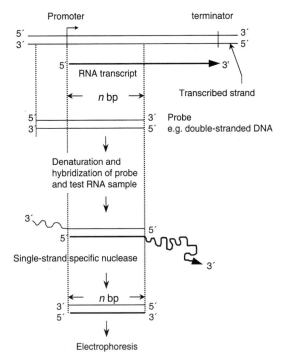

Figure 1 Mapping transcribed sequences by nuclease protection. Various types of probes may be used (as described in Section 2.2).

The hybridized probe and test RNA mixture is subsequently treated with a nuclease specific for single-stranded regions, which degrades any probe and test RNA sequences which have not formed a hybrid duplex. Thus, the resulting nuclease-resistant hybrid contains only those probe sequences which are transcribed from the genomic DNA sequence. The hybrid (or the single-stranded probe component) is then fractionated by gel electrophoresis along with suitable size markers in order to determine the size of the protected fragment and hence the physical location of the transcribed sequences.

1.1 Northern analysis versus nuclease protection mapping

It is important to appreciate that northern hybridization analysis (see Chapter 5, Section 2.7) and nuclease protection mapping yield different information. Consider, for example, the use of the two methods in analysing the genomic region illustrated in *Figure 2*. If either probe A or probe B (labelled by nick translation or random priming) were used to probe a northern blot, three transcripts, X, Y and Z, would be identified. Thus, northern hybridization provides information about contiguous transcribed sequences *outside* the probe fragment. The northern analysis does not, however, reveal the precise physical *locations* of the transcribed X, Y, and Z sequences *within* the probe fragments. In contrast, the use of probes A and B in nuclease protection mapping would, between them, allow the physical

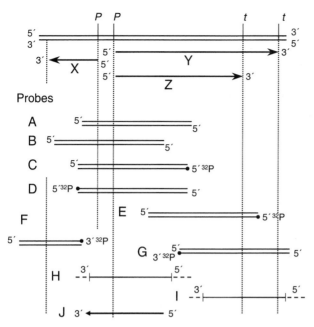

Figure 2 Probes commonly used in nuclease protection mapping. A hypothetical genomic region is illustrated. See Sections 1.1 and 2.2 for a discussion of the use of each type of probe fragment. Probes: A–E represent DNA restriction fragments, or DNA fragments generated directly by PCR; A and B, unlabelled fragments; C, D and E, fragments that are 5′-end labelled with ^{32}P; F and G, restriction fragments that are end-labelled with ^{32}P; H and I, single-stranded DNA templates (e.g. from M13 clones) that are unlabelled, 5′-end labelled with ^{32}P, or continuously labelled with ^{32}P or ^{35}S; J, RNA fragment continuously labelled with ^{32}P.

mapping of the transcribed sequences within the DNA fragments (assuming the distances between the restriction sites are known). Note, however, that no information would be revealed about transcription of contiguous sequences external to the probes (for example, transcripts Y and Z would be indistinguishable using probes A and B). In summary, northern analysis can provide a general picture of the number and size of transcripts originating, at least in part, from a cloned or PCR-amplified genomic region (information which would be useful for planning a transcript mapping strategy and which might also indicate that additional cloning was required), whereas nuclease protection mapping precisely locates the transcribed DNA sequences within the probe.

If northern analysis is being conducted with total cellular RNA preparations, a potential co-migration problem should be borne in mind. That is, if a transcript under investigation is of similar size to one of the ribosomal RNA species (even differing by several hundred base pairs) it will tend to be almost completely masked by the ribosomal RNA. The substantial amounts of rRNA bound to the northern blot appear to prevent hybridization of the probe to the co-migrating test transcript. Nuclease protection mapping, which employs liquid hybridization of the probe to RNA, clearly is not subject to such interference by rRNA, the latter being digested by the nuclease treatment prior to electrophoresis.

Finally, during the course of a northern hybridization experiment the RNA is more susceptible to degradation by contaminating ribonucleases than is the RNA in nuclease protection experiments, because in the latter the RNA enters a 'protective' hybrid in the initial stage of the procedure. In northern analysis, the RNA is particularly vulnerable during the filter hybridization stage. Thus, in practice, nuclease protection mapping is normally far more sensitive and reproducible than northern analysis.

2 Mapping by nuclease protection

The original procedure for nuclease protection mapping was developed by Berk and Sharp (1) who used DNA restriction fragments as probes and nuclease S1 as the single-strand-specific endonuclease. The conception of the method was based on an earlier observation that, in aqueous solutions containing high concentrations of formamide, RNA–DNA hybrids are significantly more stable than the corresponding DNA–DNA duplex (2). It is therefore possible to obtain selective hybridization of RNA to its DNA complement at a temperature above the melting temperature (T_m, see Chapter 5, Section 3.2) of the DNA duplex. Although the original procedure has been largely superseded by methods which employ end-labelled PCR products or various types of single-stranded probes (see Sections 2.2 and 2.5), the manipulations are essentially the same, although the more recent methods require less stringent hybridization conditions. The standard nuclease S1 mapping protocol is considered in detail below.

2.1 Standard nuclease S1 mapping

The procedure is described in *Protocol 1*. The notes that follow the procedure should be read in conjunction with the protocol.

It is best to store RNA preparations as isopropanol suspensions (0.15 M sodium acetate, pH 6.0, 50% isopropanol) at $-70\,^{\circ}$C. Unlike DNA, RNA does not aggregate in suspension. Therefore, following vortex mixing, a representative sample of RNA can be taken directly from the insoluble stock and added to the tube containing the probe DNA, previously adjusted to 0.15 M sodium acetate, pH 6.0, 50% isopropanol. If the nucleic acid pellet is dried thoroughly under vacuum it will be extremely difficult to dissolve completely in S1 hybridization solution. Thus, invert the tube over a tissue and leave for about 30 min to dry the pellet. Failure to dissolve the pellet completely results in substantial background signal from nuclease-resistant duplex DNA. After adding S1 hybridization solution place the tube in a 50$\,^{\circ}$C water bath for 30–60 min to assist dissolution of the pellet and then pipette the solution thoroughly for 1–2 min. The pellet is judged to be completely dissolved if the entire sample passes freely through a narrow orifice micropipette tip.

The hybridization temperature used for formation of hybrids is normally 1–3$\,^{\circ}$C above the T_m of the probe DNA fragment. If the probe is double-stranded then, in the initial mapping experiment, one needs to determine the T_m of a representative DNA fragment empirically, or calculate its theoretical T_m if the base composition is known (see Note (c) in *Protocol 1*). In order to obtain hybridization of all the probe-specific RNA to the probe fragment within a reasonable time (e.g. 4 h) a substantial molar excess of the probe is used. In initial experiments, use different quantities of probe and/or different times of incubation to establish conditions under which the hybridization goes to completion (i.e. when the intensity of the nuclease-resistant hybrid does not increase with increasing probe concentration or incubation time). Under these conditions the assay allows quantification of transcript levels. It is not recommended to leave the hybridization reaction overnight when one is determining RNA end-points at nucleotide resolution.

Towards the end of the hybridization period dilute the appropriate volume of S1 digestion buffer, add nuclease S1 to a final concentration of 200 units per 0.3 ml and place on ice close to the hybridization water bath. Collect together a microcentrifuge tube rack in a tray of wet ice, the vortex mixer, micropipette and tips. The success of the procedure relies on efficient nuclease digestion of the hybridization reaction. To achieve this, an excess of nuclease S1 is added in a large chilled volume to rapidly mix, cool and dilute the sample while initiating nuclease digestion. This step is particularly crucial if large quantities of RNA are used (e.g. 40 μg) and if the probe is a double-stranded DNA fragment. While the bulk of the sample tube is still submerged in the water bath, forcibly dispense 0.3 ml digestion buffer/nuclease S1 into the sample, mix immediately on a vortex mixer for 2 s, invert the tube and flick the base sharply with a finger, vortex mix again and place the tube in wet ice. When all samples have been processed, centrifuge for a few seconds and continue digestion at 37 °C.

For a more comprehensive discussion of the theory and practice of nuclease protection mapping, see refs. (3), (4) and (5).

Protocol 1

Standard nuclease S1 mapping protocol[a]

Equipment and reagents

- Water baths at 37 °C, 50 °C, and to cool from 85–40 °C

- Thermometers

- Vortex mixer

- Microcentrifuge

- Apparatus for agarose gel electrophoresis and Southern hybridization

- Probe DNA in TE buffer (see Appendix 4, Section 2), pH 8.0, or water

- Test RNA samples, as suspensions in 0.15 M sodium acetate, pH 6.0, 50% isopropanol, stored at −70 °C

- 3 M sodium acetate, pH 6.0, prepared by adding glacial acetic acid to 3 M sodium acetate until this pH is obtained

- Tray with wet ice

Protocol 1 continued

- S1 hybridization solution (40 mM PIPES, pH 6.4, 400 mM NaCl, 1 mM EDTA, 80% deionized formamide). For preparation of deionized formamide, see Appendix 4, *Protocol 4*. Store in small aliquots at −70 °C and discard once thawed

- S1 termination solution (2.5 M ammonium acetate, 50 mM EDTA). Store at 4 °C

- 5× S1 digestion buffer (1.4 M NaCl, 150 mM sodium acetate, pH 4.4, 22.5 mM zinc acetate, 100 µg/ml non-homologous DNA (sheared and denatured, see Appendix 4, *Protocol 3*). Store frozen in aliquots

- Nuclease S1 (Roche Diagnostics)

- Carrier tRNA (5 mg/ml)

Method

1 Co-precipitate the probe DNA (e.g. 25 fmol)[b] and test RNA (e.g. 40 µg total cellular RNA) in a 1.5 ml microcentrifuge tube by adding sodium acetate, pH 6.0, to 0.3 M and an equal volume of isopropanol.

2 Pellet the nucleic acid by centrifuging for 15 min at 4 °C.

3 Add 0.5 ml 80% ethanol, centrifuge for 2 min, remove all the supernatant and allow the pellet to dry by inverting the tube on a tissue (do **not** dry under vacuum).

4 Add 20 µl S1 hybridization solution and carefully dissolve the pellet by pipetting (this can take a long time).

5 Denature the nucleic acid by placing the tube, tightly capped, in an 85 °C water bath, submerged up to its neck. Incubate for 10 min.

6 Leaving the tube in place, turn down the temperature setting of the water bath allowing the temperature to drop gradually (e.g. over about 60 min) to the final hybridization temperature, which should be the T_m of the DNA–DNA duplex plus 1–3 °C.[c]

7 Incubate the sample for at least 4 h at the final temperature (keeping the lid on the water bath).

8 Add 0.3 ml of 1× S1 digestion buffer (cooled to 4 °C) containing 200 units nuclease S1 and mix immediately on a vortex mixer.

9 Incubate the sample at 37 °C for 30 min.

10 Add 75 µl S1 termination solution, vortex to mix, and add 2 µl carrier tRNA (5 mg/ml).

11 Add 0.4 ml isopropanol, mix, and centrifuge for 15 min at 4 °C to pellet the nucleic acid.

12 Rinse the pellet in 80% ethanol, dry and dissolve in electrophoresis loading buffer.

[a] The procedure is essentially as described in ref. 3.

[b] For example, 50 ng of a 3 kb double-stranded DNA fragment is 25 fmol.

[c] Determine the approximate T_m of the test-fragment empirically as described in note (c). Theoretically, the T_m of a DNA-DNA duplex is

Protocol 1 continued

$$81.5°C + 16.6\log M + 0.41(\%G + C) - \frac{500}{n} - 0.62(\%formamide)$$

where M is the molarity of monovalent cations, $\%G + C$ is the GC content, and n is the length in bp of the duplex DNA. For example, the calculated T_m for a 1-kb fragment of 50% $G + C$, in the given S1 hybridization solution, is 45.3 °C.

Notes

(a) Take care to treat and handle solutions and plasticware appropriately to minimise RNase contamination (see Volume I, Chapter 4, Section 2). It is recommended that the water and sodium acetate used for storing and co-precipitating the DNA and RNA is treated with DEPC, but DEPC treatment of the component solutions, tubes and tips used in *Protocol 1* is not necessary; indeed in the author's experience such treatment may reduce the quality of the results.

(b) The DNA probe fragment should be phenol extracted and alcohol-precipitated at least once prior to use. If restriction fragments or PCR products are purified from a gel following electrophoresis, visualize the bands with long-wavelength UV irradiation. The short-wavelength UV of many transilluminators induces the formation of pyrimidine dimers which form substrates for nuclease S1 in double-stranded nucleic acid.

(c) To determine experimentally the approximate T_m of a DNA fragment, subject replicate samples of the DNA (e.g. 0.1 μg), in S1 hybridization solution, to increasing temperature increments (e.g. 1 °C), treating a sample with nuclease S1 at each temperature. The T_m equation given above may be used as a guide for selecting the experimental temperature range. Fractionate the nuclease-treated samples on a standard agarose gel and stain with ethidium bromide. The approximate T_m of the fragment corresponds to the lowest temperature increment at which the DNA is degraded.

(d) It is essential to include negative controls in nuclease-mapping experiments in order to distinguish DNA–DNA reannealing artefacts from true hybrids. Relatively discrete bands of duplex DNA, smaller than the original probe fragment, are common artefacts in nuclease mapping gels. In the negative control samples include the same quantity of probe DNA and RNA (e.g. 25 fmol and 40 μg, respectively) and use non-homologous RNA or, ideally, RNA from equivalent cellular material in which the transcript being studied is not expressed, or in which the gene has been deleted. These negative controls are particularly important if the probe fragment is completely protected by RNA (e.g. see *Figure 3*). It is useful to electrophorese picogram quantities of the original probe fragments alongside the hybridization products for direct comparison.

(e) The optimal amount of nuclease S1 should be determined empirically. Sufficient nuclease should be used (200 units per reaction is common) to eradicate background hybridization while not affecting the hybrid profile or signal intensity.

2.2 Choice of probe in nuclease protection mapping

The types of probe chosen will often be determined by the amount of information that exists on the DNA sequence and transcriptional organization of the cloned region. It is essential to consider the polarity of transcription when selecting a probe. The nascent RNA is synthesized in a 5′ to 3′ direction. Therefore,

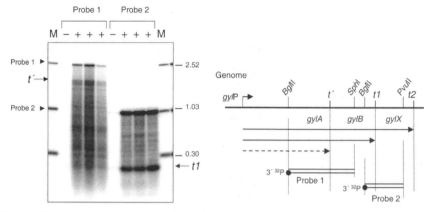

Figure 3 Nuclease S1 mapping of the 3′-termini of transcripts from the glycerol operon of *Streptomyces coelicolor*. The two 3′-end labelled restriction fragments shown were hybridized with RNA isolated from cultures either induced (+) or repressed (−) for transcription of the glycerol operon. Probe 2 identifies a discrete 3′-endpoint (*t1*) internal to the operon, whereas probe 1 identifies a series of degraded transcripts with 3′-ends extending up to *t′*. Note also the full-length protection of both probes. The negative controls (−), where no hybridization is observed, confirm that none of the nuclease resistant profile from either probe is attributable to DNA–DNA reannealing artefacts. Probes 1 and 2 were included in the size-markers (M) for direct comparison with the nuclease-treated samples. Samples were fractionated in 1.2% agarose (recirculated alkaline electrophoresis buffer, 3.2 V/cm, 4 h). Adapted from ref. (6).

owing to the antiparallel nature of the two strands in duplex nucleic acid, the transcribed DNA strand is read in a 3′ to 5′ direction. If nothing is known of the extent or the direction of transcription of the transcribed sequences in the cloned fragment it may be advisable first to use a selection of unlabelled restriction fragments as probes, since these will identify transcripts of both orientations (i.e. from both strands). This approach also has the advantage that no further subcloning or complicated and labour-intensive probe preparation is required. With reference to *Figure 2*, for example, probes A or B would identify segments of the three transcripts, X, Y and Z. Conversely, 5′-end-labelled probes C and D would, individually, identify transcripts Y/Z (in this case indistinguishable) and X, respectively. If an extensive restriction map of the cloned region is available, a series of overlapping restriction fragments may be used to map the extent and end-points of the transcribed sequences (see ref. (6) for an example of this approach).

Figure 4 Nucleotide-resolution mapping of transcription start-sites by nuclease protection and primer extension. (A) Nuclease S1 mapping of the start-site of the *Streptomyces coelicolor dnaK* operon. The probe was generated by PCR, and the internal primer was 5′-end labelled prior to amplification. The same labelled primer was used to generate the chain-termination sequencing ladder from a suitable template. In this method, the nuclease-resistant DNA fragments correspond directly to DNA fragments in the sequencing ladder. Arrows indicate the position of the transcription start site. Nuclease-resistant samples in lanes 1 and 2 were obtained from RNA isolated from control and heat-shocked cultures, respectively. (B and C)

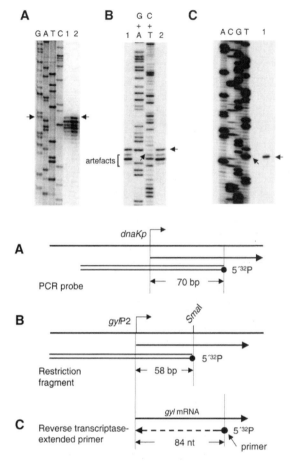

Mapping of the transcription start site of the *gyl*P2 promoter of the *Streptomyces coelicolor* glycerol operon by (B) nuclease S1 protection and (C) oligonucleotide-primed reverse transcription. In both experiments, the test RNA was isolated from cultures induced for transcription of the glycerol operon. Nuclease-resistant DNA fragments derived from a restriction fragment 5′-end labelled at the *Sma*I site are shown in lanes 1 and 2 (panel B) where they are fractionated alongside chemical cleavage ladders (produced by Maxam–Gilbert sequencing reactions) of the parent probe fragment. The oblique arrow indicates the fragment in the sequence ladder (T) which corresponds to the largest nuclease-resistant fragment (see Section 2.2.1 for the correction factor used). On the basis of the primer-extension mapping data (panel C) it is concluded that the smaller nuclease-resistant fragments in panel A constitute artefacts, presumably generated by 'nibbling' of the larger fragment by nuclease S1. Panel C (lane 1) shows the reverse-transcribed extension product derived from a 20-mer oligonucleotide, identifying the 5′-end of a single transcript (corresponding to that identified by the largest nuclease-resistant fragment in panel B). The primer-extended product is fractionated alongside chain-termination sequencing ladders obtained by priming a suitable clone with the same end-labelled oligonucleotide (as conducted in panel A). The oblique arrow indicates the fragment in the sequencing ladder (T) which corresponds directly with the extension product. Note that the DNA sequence ladders in panels A–C represent the complementary strands of the respective transcripts. All samples were electrophoresed in 6% or 8% polyacrylamide/7 M urea gels. (A) is adapted from ref. (7), (B) is adapted from ref. (9) and (C) is from unpublished data.

To determine the direction of transcription, uniquely end-labelled DNA fragments or single-stranded probes need to be used. Fragments are end-labelled at their 5′-termini using polynucleotide kinase for mapping the 5′-ends of transcripts (e.g. Probes C and D), or 3′-end labelled using Klenow polymerase to map 3′-termini (e.g. Probes F and G). An example of the latter is given in *Figure 3*. As a guideline use 1×10^5 d.p.m. of end-labelled probe (specific activity 2–4×10^6 d.p.m./pmol) against 40 μg total cellular RNA. If the DNA sequence of the region under investigation is known, then PCR provides the simplest method of generating uniquely 5′-end labelled probes because oligonucleotide primers can be end-labelled very efficiently using polynucleotide kinase and $[\gamma\text{-}^{32}P]$ATP. Thus, the 'downstream' primer, designed to be internal to the transcribed segment being analysed, is labelled and used in combination with an unlabelled 'upstream' primer to generate the probe by PCR (see *Figure 4A*). Restriction enzymes are used to generate suitable ends for 3′-end labelling; PCR products cannot be directly 3′-end labelled at one end.

One potential disadvantage of using only end-labelled probes in mapping is that internal RNA endpoints may not be detected. For example, if Probe E (*Figure 2*) is used, the 3′-terminus of transcript Z would not be seen. Clearly, in order to detect a transcript the labelled end of the probe must be protected by it; in the above case, transcript Z would hybridize to probe F but would not be visualized on the gel.

A low level of probe reannealing is a common artefact in nuclease protection mapping experiments when double-stranded probes are used. To distinguish this artefact from true full-length protection by RNA traversing the entire probe it is useful to incorporate additional non-homologous DNA at one end of the probe. This is straightforward if the region has been cloned. If the probe is produced by PCR then one of the primers can be designed to prime from adjacent vector DNA sequences (e.g. a 'universal' or 'reverse' primer could be used). Alternatively, a restriction fragment may be used which incorporates adjacent vector sequences, or even the entire vector. With such probes full-length protection of the probe can be immediately discounted as an artefact. The important problem of suppressing probe renaturation during nuclease protection experiments can be circumvented through the use of single-stranded probes. Their use enables one to adopt more defined and optimal conditions for hybrid formation than is possible with double-stranded probes (detailed in ref. 5). Unlabelled single-stranded probes such as M13 clones may be used, or alternatively the latter may be used as templates, *in vitro*, to generate continuously labelled single-stranded DNA fragments. Single-stranded probes are used to map both the 5′- and 3′-ends of transcripts (e.g. Probes H and I in *Figure 2*). Although continuously-labelled probes enhance the sensitivity of detection, the disadvantage of internal labelling, at least with ^{32}P, is that the integrity of the probe diminishes rapidly over a few days as a result of isotope decay, leading to strand scission.

If the 5′-end of the transcript under study is relatively well defined, and the DNA sequence of the genomic region is known, then a short 5′-labelled synthetic oligonucleotide may be used directly as a probe in the nuclease protection assay (8).

2.2.1 Fractionation and detection of nuclease-protected hybrids

When hybrids larger than about 300 bp are under investigation, the samples are normally fractionated by agarose gel electrophoresis, using either Tris-borate or Tris-acetate gels or denaturing alkaline gels (see Volume I, Chapter 5). If un-labelled probes are used in mapping, the fractionated hybrids are subjected to Southern transfer and are visualized following hybridization, on the filter, with a continuously labelled probe that includes the original test sequences (labelled by nick translation or random priming as described in Chapter 4). If end-labelled or internally labelled probe fragments are used for mapping, the fractionated hybrids are fixed in the agarose gel, which is then dried and subjected directly to autoradiography. To do this, soak the gel in at least three volumes of 7% TCA for 30 min (two changes), place the gel on pre-wetted filter paper (e.g. Whatman 3MM), cover with Saran wrap and dry the gel on a slab gel dryer for 30–60 min at about 60°C. Alternatively, directly after electrophoresis, carry out a standard Southern blot and expose the filter without further treatment.

Transcript mapping by agarose gel electrophoresis normally allows the determination of transcript end-points within a resolution of tens of base pairs. In order to extend the analysis to single-nucleotide resolution, nuclease-protected hybrids of less than 300 bp need to be generated using a radiolabelled probe and then analysed by fractionation on denaturing polyacrylamide–urea gels (see Chapter 6, Section 4.1). If the probe is generated by PCR with a 5′-end labelled 'downstream' primer then a corresponding sequencing ladder can be generated enzymatically by using the same labelled primer in a chain-termination sequencing reaction (see *Figure 4A*). If the sequencing is conducted with a 'cold' extension reaction, and no nucleotide analogues are used, then the nuclease-protected DNA fragments will correspond precisely to DNA fragments in the sequencing ladder. Note that chain-termination sequence ladders can only be used for nucleotide resolution mapping of the 5′-ends of transcripts. If the probe is a 3′- or 5′-end-labelled restriction fragment the hybrids are fractionated alongside a ladder of the chemical cleavage products from Maxam and Gilbert sequencing reactions (see Chapter 6, Section 1) carried out on the parent probe fragment, thus allowing determination of the 3′- or 5′-termini of transcripts at nucleotide resolution (see *Figure 4B*). There is a slight adjustment in aligning the nuclease-protected DNA fragments with fragments in the chemical cleavage sequencing ladder: the DNA component of the nuclease-resistant hybrid migrates 1 to 1.5 nucleotides *slower* than the band to which it corresponds in the sequencing ladder (see ref. 10 for a discussion).

2.3 Mapping of intron boundaries

In contrast to bacteria, eukaryotic genes generally contain introns. The mapping of transcripts from intron-containing genes is more complicated because the mature mRNA is not colinear with the genomic DNA from which it is derived (see *Figure 4A*). Thus, when mapping the 5′- and 3′-ends of such a mRNA by nuclease protection assays, discrete endpoints corresponding to intron bound-

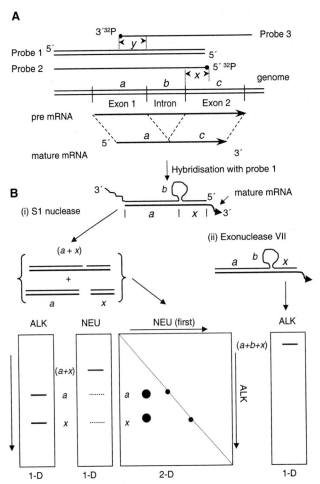

Figure 5 Methods for mapping intron boundaries. Analysis of an unlabelled DNA fragment, probe 1 (A) is represented in B. Information on transcript end-points can be derived from fractionating the nuclease-resistant hybrids on alkaline agarose gels (ALK) or neutral (NEU) gels, and by combining the two separations in a 2-dimensional (2-D) gel, the first dimension conducted under neutral conditions. See Section 2.3 for a discussion.

aries will also be observed. The different types of probes described in Section 2.2 can equally be applied to mapping intron boundaries (splice sites). As in any transcript mapping project, a series of overlapping (or specifically end-labelled) probe fragments should be used to unambiguously position the endpoints and splice sites of a transcript.

Three different probes are considered in *Figure 5* to illustrate approaches to mapping introns. In the standard nuclease S1 mapping procedure, Probe 1, an unlabelled restriction fragment or PCR product, would yield two 'nuclease-protected' DNA fragments of sizes a and x nucleotides (*Figure 5B*) when resolved by alkaline gel electrophoresis, blotted and probed with continuously labelled

Probe 1; the looped-out intron DNA in the DNA–RNA hybrid is degraded by the nuclease. On the other hand, if the same DNA fragment is specifically 5′-end labelled at one end (Probe 2) then a single protected DNA fragment of x nucleotides will be detected following alkaline gel electrophoresis and autoradiography of the gel. This would unambiguously map one intron boundary x bp from the labelled end. Similarly, Probe 3, specifically 3′-end labelled at the end within exon 1, would unambiguously map the other intron boundary, y bp from the labelled end. In principle, instead of arising by splicing, the endpoints detected in such an experiment could have been generated by initiation/termination of transcription, or by degradation of the mature mRNA. Their identification as true splice sites can be confirmed by comparing the sizes of the reaction products following neutral and alkaline gel electrophoresis. This is illustrated for Probe 1 in *Figure 5B*. The RNA strand opposite the DNA loop of the intron is not cleaved efficiently by nuclease S1 (particularly if the nuclease digestion is conducted at a lower temperature, such as 20 °C, rather than 37 °C; see ref. 3). Thus, the predominant DNA–RNA hybrid in the reaction comprises a continuous RNA strand annealed to DNA which is discontinuous at the splice site; this hybrid will be $(a + x)$ bp if separated by neutral gel electrophoresis (e.g. by using TBE buffer). However, the RNA strand opposite the 'nick' in the DNA strand will be cleaved in a smaller subpopulation of the hybrids, yielding hybrids of a bp and x bp. In the alkaline gel, the RNA component is degraded and therefore only the two DNA components of a and x nucleotides are observed. The observation of the larger hybrid in the neutral gel, corresponding in size to the sum of the individual DNA components, confirms that the endpoints have arisen through splicing. The two types of electrophoretic separation can be readily combined in a single two-dimensional gel analysis. Electrophoresis in the first dimension is conducted in neutral buffer. The gel is then equilibrated in an alkaline buffer, the gel is placed perpendicular to its original position, and electrophoresis is continued under denaturing conditions (see *Figure 5B* and ref. 3 for further details). Individual nuclease-protected DNA strands not held together by a 'linking' (spliced) RNA will migrate with the same relative mobility in *both* dimensions of the gel and hence will be positioned diagonally on the gel. In contrast, the DNA strands annealed to the intact spliced RNA $(a + x$ bp$)$ run together in the neutral dimension but then will be separated in the second (denaturing) dimension and will run below the diagonal line.

The existence of a splice site can also be independently confirmed by using *Escherichia coli* exonuclease VII (exo VII) for digestion of hybrids in place of nuclease S1, as illustrated in *Figure 5B*. Exo VII is specific for DNA and requires free 5′- or 3′-ends to initiate hydrolysis. Thus, the looped-out intron DNA in the hybrid duplex would escape digestion. The DNA component of the hybrid is then sized by alkaline gel electrophoresis. For example, Probe 1 would yield a DNA fragment of $(a + b + x)$ nucleotides (*Figure 5B*). If the intron is of a size that can be resolved by gel electrophoresis then its size can be independently determined by subtracting the sizes of the protected DNA fragments determined from nuclease S1 mapping of the same probe fragment. The use of exo VII is detailed in ref. (5).

2.4 Modifications to the standard nuclease S1 mapping protocol

Murray (11) introduced the use of the chaotropic salt, sodium trichloracetate (NaTCA), in place of formamide in the hybridization solution. In 3 M NaTCA, the difference between the thermal stability of RNA–DNA and DNA–DNA duplexes is far greater than in formamide and, moreover, NaTCA reduces the T_m of the DNA duplex further. Thus, unlike the standard protocol where the hybridization temperature is critical, the use of NaTCA enables one to use a wider range of hybridization temperatures where DNA–DNA renaturation is eliminated while RNA–DNA hybridization proceeds efficiently. The method is reported to be far more sensitive than the formamide-based protocol, and the use of single-stranded rather than double-stranded probes apparently offers no advantage. Murray (11) also reported that the use of mung bean nuclease in place of nuclease S1 results in fewer artefactual fragments in high-resolution nuclease protection mapping. Aldea *et al.* (12) have combined the use of NaTCA, M13 vectors and probes continuously labelled with ^{35}S in a rapid and simple method for mapping 5'-termini of transcripts at high resolution.

2.5 Ribonuclease protection mapping

Ribonuclease protection mapping exploits the use of highly processive bacterio-phage RNA polymerases to make continuously labelled RNA probes of very high specific activity *in vitro* (see Chapter 4, Section 2.3.3). Several vectors are commercially available for the production of such probes (e.g. the pGEM series from Promega; see Volume I, Chapter 9, Section 3.8); they typically have an MCS flanked on either side by different bacteriophage promoters. For example, pGEM-3Z contains a promoter for T7 polymerase on one side of the MCS, and a promoter for SP6 polymerase on the other side (*Figure 6A*); this has the great advantage that each strand of the cloned DNA fragment can be uniquely labelled by using the respective polymerases in different labelling reactions. In ribonuclease protection mapping the DNA region under investigation is cloned into such a vector, and prior to conducting the *in vitro* transcription reaction the plasmid is digested with a restriction enzyme to cleave the DNA downstream from the sequence to be transcribed (see *Figure 6A*); this template will yield 'run-off' transcripts of a fixed size. The *in vitro* transcription reaction is carried out in the presence of an $[\alpha\text{-}^{32}P]$-labelled nucleotide triphosphate and the resulting single-stranded RNA probe is then hybridized with the test RNA (as in the nuclease S1 mapping procedure). The sample is subsequently treated with a combination of RNase A and RNase T1 (single-strand-specific ribonucleases) to destroy RNA sequences that have not hybridized, and the RNA-protected RNA probe fragment is analysed on a denaturing polyacrylamide-urea gel. The procedure can be used to map 5'- and 3'- transcript termini. Ribonuclease protection mapping is described in *Protocol 2* and is considered in more detail in refs. (13) and (14). *Figure 6* illustrates the application of this technique for mapping transcripts in the chromosomal replication origin region (*oriC*) of *Streptomyces coelicolor*.

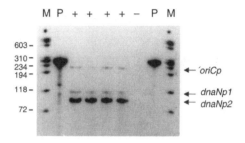

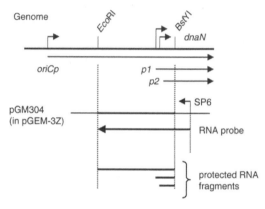

Figure 6 Ribonuclease protection mapping of transcripts in the replication origin (*oriC*) region of *Streptomyces coelicolor*. A high specific activity, single-stranded RNA probe of the region was generated by *in vitro* transcription of a cloned 218 bp *Eco*RI–*Bst*YI *oriC* fragment from the adjacent SP6 promoter in the plasmid vector pGEM-3Z. The plasmid, pGM304, was linearized with *Eco*RI prior to synthesizing the 261 nucleotide RNA probe. RNase-resistant fragments were fractionated in a 5% polyacrylamide/8 M urea gel. Lanes: M, end-labelled DNA size-marker, *Hae*III-digested φX174 DNA (sizes are given in bp); P, untreated probe RNA; '+', RNA samples from rapidly growing *S. coelicolor* cultures; '–', negative control, probe hybridized with tRNA. (G. B. Scott and C. P. Smith, unpublished data). See Section 2.5 for a discussion.

Protocol 2

Ribonuclease protection mapping

Equipment and reagents[a]

- Water baths at 85 °C, 45–50 °C, 37 °C and 30 °C

- Apparatus for polyacrylamide gel electrophoresis

- Labelled probe RNA in DEPC-treated water[b]

- Test RNA samples in DEPC-treated water, or as suspensions in 0.15 M sodium acetate, pH 6.0, 50% isopropanol, stored at −70 °C

- Hybridization solution (identical to S1 hybridization solution; see *Protocol 1*)

- RNase digestion buffer (10 mM Tris-HCl, pH 7.5, 5 mM EDTA, 300 mM sodium acetate)

- Ribonuclease A (7 mg/ml)

- Ribonuclease T1 (10 units/ml)

- 20% SDS

- Proteinase K (20 mg/ml)

Protocol 2 continued

- 25:24:1 Phenol–chloroform–isoamyl alcohol (see Appendix 4, *Protocol 6*)
- Labelled size markers (e.g. *Hae*III-digested φX174 DNA)
- Carrier tRNA (5 mg/ml)
- Loading dye (80% formamide, 2 mM EDTA, 0.1% xylene cyanol, 0.1% bromophenol blue)

Method

1 Co-precipitate approximately 1×10^5 c.p.m. of RNA probe and test RNA (e.g. 40 μg total cellular RNA) in a 1.5 ml microcentrifuge tube by adding sodium acetate (pH 6.0) to 0.3 M and an equal volume of isopropanol.[c]

2 Pellet the RNA in a microcentrifuge at 13000 r.p.m. for 15 min at 4°C. Discard the supernatant and allow the pellet to air-dry.

3 Dissolve the RNA in 30 μl of hybridization solution; ensure that the pellet is fully dissolved (see Section 2.1).

4 Denature the RNA by placing the tube in a 85°C water bath for 5 min.

5 Immediately transfer the tube to an adjacent water bath set at 45–50°C and incubate for at least 4 h.[d]

6 Add 350 μl RNase digestion buffer to the tube and mix on a vortex mixer.

7 Add 1 μl RNase A and 2.5 μl RNase T1, mix, and incubate at 30°C for 30 min.

8 Add 10 μl 20% SDS, mix, and then add 2.5 μl Proteinase K (20 mg/ml). Incubate at 37°C for 15 min.

9 Extract the sample with 400 μl phenol–chloroform–isoamyl alcohol and transfer the aqueous phase to a fresh microcentrifuge tube.

10 Add 2 μl carrier tRNA (10 μg), mix, and add 1 ml chilled absolute ethanol. Leave at −20°C for 30 min.

11 Pellet the protected RNA by centrifuging for 15 min at 4°C and discard the supernatant.

12 Dry the pellet by inverting the tube over a tissue. Dissolve the RNA in 5 μl loading dye and fractionate in a polyacrylamide–urea gel.

[a] A reliable 'RNase protection kit' containing all enzymes and reagents for this protocol is available from Roche (catalogue no. 1 427 580).

[b] The high specific activity RNA probe should be used as soon as possible, ideally on the same day it is synthesized, because it will degrade substantially within days. Comprehensive kits are commercially available for generating specific RNA probes (e.g. Riboprobe from Promega).

[c] A negative control should be included (e.g. tRNA) to detect any probe-specific artefacts, and to confirm efficient RNase treatment of the probe.

[d] Since the probe is not double-stranded the hybridization temperature is relatively unimportant compared with the standard nuclease S1 mapping protocol. However, the temperature should be high enough to avoid the formation of stable intramolecular secondary structures in the probe and test RNA samples. The negative control sample is important in this respect.

There are several advantages and disadvantages of this method over the standard nuclease S1 mapping procedure. The advantages include:

- it is very sensitive. The high specific activity of the continuously labelled RNA probe allows clear detection of poorly expressed transcripts that would not be detectable by nuclease S1 mapping
- the probe can identify small (regulatory) RNA molecules that would generally escape detection in nuclease S1 mapping experiments. This method is useful for 'scanning' both strands of a cloned region for short transcribed segments
- the probe is very simple to prepare, in large quantity, and at reproducible specific activity. The probe can generally be used directly, avoiding the need for gel-purification
- the probe is single-stranded, because only one strand of the cloned DNA is transcribed, and hence artefacts due to probe re-annealing do not occur. The annealing temperature is not so critical and the RNA–RNA duplexes are more stable than the corresponding RNA–DNA duplex

The disadvantages include:

- the short half-life of the probe. Since the RNA probes are internally labelled they break down rapidly and therefore, for optimal results, need to be used immediately
- each DNA segment to be analysed must first be subcloned into a specialized vector before the RNA probe can be generated
- the transcript endpoints cannot be mapped at nucleotide resolution because the protected RNA cannot be compared alongside a sequencing ladder

RNA fragments cannot be sized accurately against DNA size markers on a denaturing polyacrylamide–urea gel; as a guide, an RNA fragment co-migrating with a DNA fragment will be about 5% smaller than the DNA fragment. This method allows mapping of transcript ends to a resolution of 5–10 nucleotides.

3 Primer-extension mapping

Primer-extension mapping (or, more correctly, oligonucleotide-primed reverse transcription mapping) is a specialized technique that exploits retroviral reverse transcriptases for nucleotide-resolution mapping of the 5′-termini of transcripts. Before using this technique the nucleotide sequence proximal to the promoter of the transcribed region, and the approximate locations of the 5′-termini of transcripts, are normally already established.

In this method a chemically synthesized oligonucleotide (e.g. 20–40-mer) labelled at its 5′-end, or a small 5′-labelled restriction fragment complementary to sequences less than 200 nucleotides from the presumed 5′-end of the transcript, is hybridized with the test RNA preparation. The annealed DNA primer is then extended with reverse transcriptase, which transcribes the RNA template into DNA. Since the extension reaction proceeds ideally to the 5′-end of the transcript, the length of the primer-extended DNA product allows direct determination of the 5'-terminus. Typically, the reaction products are compared directly with chain-termination sequencing ladders that have been generated using the same end-labelled DNA primer on a DNA clone which contains the transcribed sequences being studied (see *Figure 4C*). The extended product should correspond precisely to one of the bands in the sequencing ladder.

Currently used protocols (e.g. refs. 5, 15) have been adapted from the method of Ghosh *et al.* (16). A general primer-extension mapping procedure is detailed in *Protocol 3*.

Protocol 3

Primer-extension mapping

Equipment and reagents

- Equipment for sequencing-resolution polyacrylamide–urea gel electrophoresis
- 5'-labelled oligonucleotide primer[a]
- Test RNA samples in DEPC-treated water, or as suspensions in 0.15 M sodium acetate, pH 6.0, 50% isopropanol, stored at −70°C
- S1 hybridization solution (see *Protocol 1*)[b]
- 3 M sodium acetate, pH 6.0, prepared by adding glacial acetic acid to 3 M sodium acetate until this pH is obtained
- 0.3 M sodium acetate, pH 6.0
- Loading dye (80% formamide, 2 mM EDTA, 0.1% xylene cyanol, 0.1% bromophenol blue)

- RT extension buffer, prepared just before needed and stored on ice. To make up 500 μl (enough for 10 reactions) mix the following: 100 μl 5× basal RT extension buffer (250 mM Tris–HCl, pH 8.3, 250 mM KCl, 30 mM MgCl$_2$), 25 μl 0.1 M DTT, 50 μl 20 mM dNTP mixture (i.e. 5 mM each of dATP, dCTP, dGTP and dTTP), 5 μl actinomycin D (4 mg/ml in 80% ethanol), 320 μl DEPC-treated water[c]
- AMV reverse transcriptase (20 units/μl)
- 25:24:1 phenol–chloroform–isoamyl alcohol (see Appendix 4, *Protocol 6*)

Method

1 Co-precipitate the 5'-labelled oligonucleotide primer (e.g. 50 000–100 000 d.p.m.) and RNA (e.g. 40 μg total cellular RNA), centrifuge, wash and dry the pellets as described in steps 1–3 of *Protocol 1*.

2 Add 20 μl S1 hybridization solution and dissolve the pellet by pipetting (as described in Section 2.1).

3 Denature the nucleic acid by incubating at 85°C for 10 min.

4 Incubate the sample at 30–60°C for 3–6 h.[d]

5 Add 200 μl 0.3 M sodium acetate (pH 6.0), mix, add 200 μl isopropanol, mix again and centrifuge for 15 min at 4°C.

6 Rinse the pellet with 80% ethanol and dry.

7 Dissolve the pellet in 50 μl RT extension buffer.

8 Add 20–40 units (1–2 μl) AMV reverse transcriptase and incubate the sample at 42°C for 1 h.

Protocol 3 continued

9 Extract the sample with phenol–chloroform–isoamyl alcohol and transfer the upper aqueous phase to fresh tube.

10 Precipitate the nucleic acid by adding sodium acetate, pH 6.0, to 0.3 M and 1 volume of isopropanol. Mix, centrifuge for 15 min, wash the pellet in 80% ethanol and dry.[e]

11 Dissolve the pellet in loading dye and fractionate the primer-extended fragments in a sequencing gel, alongside chain-termination sequencing ladders generated using the above end-labelled oligonucleotide as the primer (see Section 2.2.1 and Chapter 6)

[a] Typically, 25–50 fmol of oligonucleotide primer are used per reaction (e.g. 0.25–0.5 ng of a 30-mer), labelled to a specific activity of approximately 2×10^6 d.p.m./pmol.

[b] Some published protocols omit formamide from the hybridization solution.

[c] To minimize interference from contaminating ribonucleases either add vanadyl ribonucleoside complex (Life Technologies) to a final concentration of 1 mM or add RNAsin (Promega or Roche), as specified by the manufacturer.

[d] The optimal hybridization temperature needs to be determined empirically for the particular primer–RNA combination. In preliminary experiments use temperatures in the higher range if omitting formamide from the hybridization reaction.

[e] If the extension reaction contains a large quantity of RNA (e.g. 40 μg), it is advisable to remove the RNA before proceeding to step 11, otherwise the sample may be distorted on the sequencing gel. Use either RNase A or alkali treatment:

(i) Dissolve the pellet in TE, pH 8.0 (see Appendix 4, Section 2), and add heat-treated RNase A to 40 μg/ml. Incubate for 30 min at 37 °C. Phenol extract, add about 5 μg tRNA, and precipitate the nucleic acid as in steps 9 and 10.

(ii) Alternatively, dissolve the pellet in 150 mM NaOH, incubate for 1 h at 65 °C and neutralize with HCl. Add about 5 μg tRNA and precipitate the nucleic acid as in step 10.

There are several advantages and disadvantages of this method over nuclease protection mapping. The advantages include:

- there is no complicated probe preparation procedure
- ambiguities in 5′-end determination, which occur in nuclease protection mapping as a result of 'nibbling' at the ends of the hybrids (e.g. *Figure 4B*), are not encountered
- there is no uncertainty in reading the 5′-endpoints from the DNA sequence as in nuclease protection mapping when using chemical cleavage sequencing ladders (e.g. *Figure 4B*)
- primer extension allows mapping of 5′-termini through transcriptional discontinuities (e.g. splice sites); depending on the location of splice sites, this may prove difficult when genomic probes are used in nuclease protection mapping

The disadvantages include:

- the technique can map 5′-ends only
- 'promiscuous' (or heterologous) priming of other transcripts by the primer may occur

- 'fold-back' synthesis may occur, resulting in extension products that are larger than the 'true' extended products. It may be possible to distinguish promiscuous priming and fold-back artefacts from true products by their lack of direct correspondence with bands in the sequencing ladder, which is attributable to differences in their base composition.

- premature termination of the extension reaction (perhaps the greatest problem), which can be caused by secondary structures in the template RNA or by other sequence characteristics (e.g. GC-rich tracts). Thus several discrete, shorter reaction products may also be observed in the sequencing gel, confusing interpretation

Considering the potential artefacts in primer-extension mapping it is clearly an advantage to have a good idea beforehand of the expected size of the extension products. In the opinion of the author it is probably best, where possible, to first analyse the 5′-termini by nuclease protection mapping, and then adopt the primer-extension method for further refining the mapping to single-nucleotide resolution.

Additional points to consider in primer-extension mapping are:

(a) choose a primer which needs to be extended by only 50–100 nucleotides, to reduce the possibility of premature termination artefacts

(b) the inclusion of formamide in the hybridization solution is optional. The higher stringency of formamide-containing buffers may improve results by reducing heterologous priming and by minimizing RNA secondary structure formation, which can interfere with annealing of the primer

(c) the use of a relatively high temperature (42 °C) for the extension reaction and a high concentration of nucleotides (0.5 mM of each) helps to reduce pausing or premature termination

References

1. Berk, A. J. and Sharp, P. A. (1977). *Cell*, **12**, 721.
2. Casey, J. and Davidson, N. (1977). *Nucl. Acids Res.*, **4**, 1539.
3. Favaloro, J., Treisman, R., and Kamen, R. (1980). In *Methods in enzymology* (ed. L. Grossman and K. Moldave), Vol. 65, p. 718. Academic Press, London.
4. Sharp, P. A., Berk, A. J., and Berget, S. M. (1980). In *Methods in enzymology* (ed. L. Grossman and K. Moldave), Vol. 65, p. 750. Academic Press, London.
5. Calzone, F. J., Britten, R. J., and Davidson, E. H. (1987). In *Methods in enzymology* (ed. S. L. Berger and A.R. Kimmel), Vol. 152, p. 611. Academic Press, London.
6. Smith, C. P. and Chater, K. F. (1988). *Mol. Gen. Genet.*, **211**, 129.
7. Bucca, G., Ferina, G., Puglia, A. M., and Smith, C. P. (1995). *Mol. Microbiol.*, **17**, 663.
8. Reyes, A. A. and Wallace, R. B. (1987). In *Methods in enzymology* (ed. R. Wu and L. Grossman), Vol. 154, p. 87. Academic Press, London.
9. Smith, C. P. and Chater, K. F. (1988). *J. Mol. Biol.*, **204**, 569.
10. Hentschel, C., Irminger, J.-C., Bucher, P., and Birnstiel, M. L. (1980). *Nature*, **285**, 147.
11. Murray, M. G. (1986). *Anal. Biochem.*, **158**, 165.
12. Aldea, M., Claverie-Martin, F., Diaz-Torres, M. R., and Kushner, S. R. (1988). *Gene*, **65**, 101.

13. Krieg, P. A. and Melton, D. A. (1987). In *Methods in enzymology.* (ed. R. Wu). Vol. 155, p. 397. Academic Press, London.

14. Gilman, M. (1993). In *Current protocols in molecular biology.* (ed. F. M. Ausubel, R. Brent, R. E. Kingston, D. D. Moore, J. G. Seidman, J. A. Smith, and K. Struhl), p. 4.7.1. Wiley, New York.

15. Triezenberg, S. J. (1992). In *Current protocols in molecular biology.* (ed. F. M. Ausubel, R. Brent, R. E. Kingston, D. D. Moore, J. G. Seidman, J. A. Smith, and K. Struhl), p. 4.8.1. Wiley, New York.

16. Ghosh, P. K., Reddy, V. B., Swinscoe, J., Lebowitz, P., and Weissman, S. M. (1978). *J. Mol. Biol.*, **126**, 813.

Appendix 1
Health hazards and safety procedures

1 Introduction

Safety is an individual responsibility and you must ensure that you are aware of the health hazards and have identified the appropriate safety requirements before embarking on molecular biology experiments. Departmental and institutional officers should be consulted about the necessary precautions to take and you must be aware of any national health and safety regulations pertaining to the procedures that you will carry out. The following sections give an overview of molecular biology safety procedures and should be supplemented by reading the safety advice given in other chapters of this book and by referring to specialist publications.

2 Microbiological safety

Most molecular biology experiments are carried out with attenuated strains of *Escherichia coli* that are unable to survive outside the test-tube. Nevertheless, precautions to avoid contamination, as described in Volume I, Chapter 2, Section 1.3, should be observed whenever bacteria are being handled. For detailed information, consult refs. (1–8).

As well as general precautions to avoid infection with microorganisms, molecular biologists must also follow the additional precautions necessary when handling genetically manipulated bacteria. In most countries these procedures are laid down in law (9,10) and a license or other form of approval may be required before experiments can be carried out.

3 Radiochemicals

Radiochemicals labelled with ^{32}P or ^{35}S are often used in molecular biology experiments. These are not the most dangerous radionuclides in biological research, but they are used in large amounts in DNA analyses. Your national and local regulations (e.g. refs. 11–13) for handling and disposal of radiochemicals should be followed at all times. Various publications give general guidelines for the safe handling of radiochemicals (e.g. refs. 14–19) but the clearest and most succinct advice for molecular biologists is contained in the radiochemicals catalogue of Amersham Pharmacia (20) and can be summarized as follows:

- understand the nature of the hazard and get trained in the safety procedures before beginning work with radiochemicals
- plan ahead so that the minimum amount of time is spent handling radio-chemicals
- keep the maximum possible distance between yourself and sources of radiation
- use shielding that is appropriate for the type of radiation emitted by the source being handled
- keep radioactive materials in defined work areas
- wear appropriate clothing and dosimeters to monitor exposure
- monitor the work area frequently and follow appropriate procedures if a spill occurs
- keep waste accumulation as low as possible
- at the end of the experiment, monitor yourself, wash thoroughly, and monitor again

4 Chemical hazards

Many hazardous chemicals are used in molecular biology research. The container in which a hazardous chemical is supplied will carry labelling that describes the nature of the risk and the safety procedures that should be adopted. Most suppliers use a labelling system that combines an easily recognized symbol that indicates the general nature of the hazard (e.g. 'irritant') with a series of code numbers that provide specific risk and safety information (e.g. 'irritating to eyes and respiratory system'). The most comprehensive coding system, used by many companies, is the European Commission Risk and Safety Phrases. These phrases are defined in the suppliers' catalogues and safety literature (e.g. refs. 21–24) and there are also many books that give more general advice on the safe handling of hazardous chemicals (1, 25–37).

The *Molecular Biology Labfax* (1) contains a detailed list of chemicals used in molecular biology research, together with the appropriate health and safety advice for each one. The hazards fall into three main groups:

(a) *Organic solvents.* The most dangerous solvent used in molecular biology is phenol. It is no longer necessary to undertake the highly hazardous operation of phenol distillation (needed if solid phenol is purchased), because many companies now market phenol that does not need redistilling. Always wear gloves when carrying out phenol extractions and make sure that the gloves actually provide a barrier to phenol: some lightweight disposable gloves do not.

(b) *Mutagens and carcinogens.* Most chemicals that bind to DNA in the test-tube are mutagenic and/or carcinogenic. An example is ethidium bromide, which is used in large amounts to detect DNA in agarose gels. Some manuals recommend adding ethidium bromide to the electrophoresis buffer, but this should be avoided. It does not provide any major advantage over post-staining (unless you want to follow the progress of the electrophoresis) and it increases the chance of spillage. Check that your local regulations allow ethidium bromide solutions to be poured down the sink and, even if this is permitted, consider using a decontamination procedure (Appendix 4, *Protocol 5*; refs. 38–40) before disposal. Note that gels stained with ethidium bromide are dangerous and should not be placed in the general laboratory waste.

(c) *Toxic chemicals.* The most dangerous toxic chemical used in molecular biology is acrylamide which is lethal if swallowed and can also exert toxic effects by contact with the skin. The hazard is extreme with solid acrylamide monomer because the powder can form aerial contamination when being weighed out. For this reason, solid acrylamide should never be purchased: premixed acrylamide solutions should always be used.

5 Ultraviolet radiation

The most dangerous piece of equipment used by molecular biologists is the UV transilluminator for viewing agarose gels that have been stained with ethidium bromide. Standard laboratory models have recommended exposure levels of 4.5 s or less at 30.5 cm (12 inches). Unprotected exposure of skin can lead to severe burns and anyone foolish enough to look directly at the UV source without eye protection risks loss of sight. Always use a suitable screen when observing agarose gels and wear face and eye protection when the screen is removed for photography. Be particularly careful that your wrists and neck are protected when you are handling a gel on a transilluminator.

6 High-voltage electricity

When working in a laboratory it is very easy to concentrate on the special hazards of the research environment and to forget about the general hazards of daily life. The electricity used to run research equipment is equally as dangerous as that in your home, more so in some cases because higher voltages are involved. The power supplies used to run electrophoresis gels can kill. Commercial gel apparatuses are designed so that they cannot be disassembled before the power supply is disconnected. Home-made equipment should incorporate similar safeguards and broken gel tanks should never be used. Leaky gel apparatus is potentially lethal.

References

1. Brown, T. A. (ed.) (1998). *Molecular biology labfax*, 2nd edn, Vol. 2. Academic Press, London.
2. American Chemical Society, Committee on Chemical Safety (1985). *Safety in academic chemistry laboratories*. American Chemical Society, Washington.

3. Collins, C. H. (ed.) (1988). *Safety in clinical and biomedical laboratories*. Chapman and Hall, London.

4. Collins, C. H. (ed.) (1999). *Laboratory-acquired infections: history, incidence, causes and prevention*, 4th edn. Butterworth-Heinemann, Oxford.

5. Collins, C. H. (1995). *Collins and Lyne's microbiological methods.*, 7th edn. Butterworth-Heinemann, Oxford.

6. Health and Safety Commission (1993). *Control of substances hazardous to health; and control of carcinogenic substances*, 4th edn. HMSO, London.

7. Health Services Advisory Committee (1991). *Safe working and the prevention of infection in clinical laboratories*. HMSO, London.

8. Imperial College of Science and Technology (1974). *Precautions against biological hazards*. Imperial College, London.

9. Health and Safety Executive (1996). *A guide to the genetically modified organisms (contained use) regulations 1992, as amended in 1996*. HMSO, London.

10. Anonymous (1984). *Federal Register*, **49** (Part VI, Number 227), 46266.

11. Health and Safety Commission (1982). *The ionising radiations regulations*. HMSO, London.

12. Health and Safety Commission (1985). *The protection of persons against ionising radiation arising from any work activity*. HMSO, London.

13. Anonymous (1993). *The radioactive substances Act 1993*. HMSO, London.

14. Coggle, J. E. (1983). *The biological effects of radiation*, 2nd edn. Taylor and Francis, New York.

15. Martin, A. and Harbison, S. A. (1986). *An introduction to radiation protection*, 3rd edn. Chapman and Hall, London.

16. United Nations Scientific Committee on the Effects of Atomic Radiation (1996). *Sources and effects of ionising radiation*. United Nations, New York.

17. International Commission on Radioactive Protection (1977). *Recommendations of the International Commission on Radiation Protection*. ICRP Publications 26 and 27. Pergamon Press, Oxford.

18. Anonymous (1970). *Radiological health handbook*. US Department of Health, Education and Welfare, Washington, DC.

19. Meisenhelder, J. and Semba, K. (1996). In *Current protocols in molecular biology* (ed. F. M. Ausubel, R. Brent, R. E. Kingston, D. D. Moore, J. G. Seidman, J. A. Smith and K. Struhl), Unit A.1F. John Wiley, New York.

20. Anonymous (1999). *Life sciences catalogue*. Amersham International, Amersham.

21. Anonymous (1999). *Sigma-Aldrich material safety data sheets on CD-ROM*. Sigma-Aldrich Corp., Milwaukee.

22. Anonymous (1999). *BDH hazard data sheets, computer package*. BDH, Poole.

23. Lenga, R. E. (ed.) (1988). *The Sigma-Aldrich library of chemical safety data.*, 2nd edn. Sigma-Aldrich Corp., Milwaukee.

24. Lenga, R. E. and Votoupal, K. L. (ed.) (1993). *The Sigma-Aldrich library of regulatory and safety data*. Sigma-Aldrich Corp., Milwaukee.

25. Lewis, R. J. (1988). *Sax's dangerous properties of industrial materials*, 8th edn. Van Nostrand Reinhold, New York.

26. Bretherick, L. (ed.) (1986) *Hazards in the chemical laboratory.*, 4th edn. Royal Society of Chemistry, London.

27. National Research Council (1981). *Prudent practices for handling hazardous chemicals in laboratories*. National Academy Press, Washington, DC.

28. Royal Society of Chemistry (1992). *Hazards in the chemical laboratory*, 5th edn. Royal Society of Chemistry, Cambridge.

29. Furr, A. K. (ed.) (1990). *CRC handbook of laboratory safety*, 3rd edn. CRC Press, Boca Raton.

30. Lefevre, M. J. (1980). *First aid manual for chemical accidents*. Dowden, Hutchinson and Ross, Stroudsburg.

31. Pitt, M. J. and Pitt, E. (1985). *Handbook of laboratory waste disposal*. Wiley, New York.
32. Young, J. A. (ed.) (1987). *Improving safety in the chemical laboratory*. Wiley, New York.
33. Great Britain, Department of Education and Science (1976). *Safety in science laboratories*, 2nd edn. HMSO, London.
34. Pipitone, D. A. (1984). *Safe storage of laboratory chemicals*. Wiley, New York.
35. Collings, A. J. and Luxon, S. G. (ed.) (1982). *Safe use of solvents*. Academic Press, Orlando.
36. Anonymous (1975). *Toxic and hazardous industrial chemicals safety manual*. International Technical Information Institute, Tokyo.
37. Perbal, S. (1988). *A practical guide to molecular cloning*, 2nd edn. Wiley, New York.
38. Bensaude, O. (1988). *Trends Genet.*, **4**, 89.
39. Lunn, G. and Sansone, E. B. (1987). *Anal. Biochem.*, **162**, 453.
40. Quillardet, P. and Hofnung, M. (1988). *Trends Genet.*, **4**, 89.

Appendix 2
Equipment for molecular biology research

1 Introduction

Much of the equipment and other facilities needed for molecular biology research will already be present in the 'well-found' biology laboratory, but if you are moving into molecular biology from a different area then you will almost certainly need to obtain some specialist items. The lists that follow are intended for guidance and do not cover every possible experiment that you may wish to do. The detailed applications of these pieces of equipment are described in the relevant chapters.

2 General facilities

You will require access to the following general facilities:

- containment facilities that the meet the requirements of the genetic manipulation regulations that your laboratory is subject to (Appendix 1)
- microbiological facilities:
  - autoclaves for sterilizing media
  - clean areas for molecular biology work (laminar flow cabinets are useful but not essential)
  - 37°C room; incubators can be used as an alternative but it must be possible to incubate Petri dishes as well as cultures in test-tubes and flasks from 5 ml to 2 l
  - glassware cleaning facility
- Darkroom facilities:
  - red-light darkroom for autoradiography, including tanks and accessories for developing X-ray films
  - short-wavelength (302 or 366 nm) UV transilluminator
  - CCD detector or Polaroid camera for recording agarose gels
- cold room
- subzero storage facilities:
  - −80°C freezer
  - liquid nitrogen storage vessels

- Radiochemical handling and disposal facilities, primarily for ^{35}S and ^{32}P
- Wet-ice machine and dry-ice storage

3 Departmental equipment

These items can be shared by a number of groups:

- ultracentrifuge capable of 70 000 r.p.m., with at least one fixed-angle and one swing-out rotor plus accessories. Density gradient centrifugation can take 48 h so one machine is rarely sufficient
- refrigerated centrifuge capable of 25 000 r.p.m., plus rotors for 50- to 500-ml buckets
- UV spectrophotometer
- liquid scintillation counter
- sonicator with a variety of probes
- four-point balance
- platform incubators for 45, 55 and 65 °C
- vacuum drier (a freeze-drier is not essential)
- vacuum oven
- 100 °C oven
- automated DNA-sequencing machine (desirable but not essential)
- oligonucleotide synthesis machine. This is optional: commercial synthesis of oligonucleotides is cheap and rapid so an in-house facility is economical only if demand for oligonucleotides is high

4 Laboratory equipment

Each group carrying out molecular biology experiments will require these items, possibly in multiple numbers:

- bench-top centrifuge, capable of 5000 r.p.m., but not refrigerated
- 200 V power supply for agarose gels, electroelution, etc.
- horizontal gel apparatus (e.g. 10 × 10 cm, 15 × 10 cm, and 15 × 15 cm)
- microfuge
- automatic pipettes (e.g. Gilson P20, P200, P1000); each researcher will need a set
- water baths (non-shaking): ideally three, one set at 37 °C, one at 65 °C, and one varied for requirements
- two-point top pan balance
- vortex mixer
- magnetic stirrer

- pH meter
- Geiger counter or other type of radioactivity monitoring device
- microwave oven for melting agar and preparing agarose
- UV crosslinker
- thermal cycler for PCR
- hybridization oven or, as a cheap (and nasty) alternative, a bag sealer
- cassettes and intensifying screens for autoradiography
- light-box for viewing autoradiographs
- double-distilled water supply
- refrigerator
- $-20\,°C$ freezer
- fume cupboard
- computer (for virtually all molecular biology applications a networked PC or Macintosh is adequate)
- if DNA sequencing is being performed manually then you will also need:
  - vertical gel apparatus (e.g. 21×40 cm)
  - 3000 V power supply
  - gel drier
- for some types of hybridization analysis the following will be needed:
  - slot/dot blotter
  - electrotransfer or vacuum blotting unit

Appendix 3
Important *Escherichia coli* strains

1 Genotypes of important strains

A comprehensive listing of the genotypes of many *Escherichia coli* strains used in molecular biology research is given in ref. (1). This source also provides a detailed explanation of the nomenclature used when writing out a genotype as well information on the restriction and modification properties of important strains.

1.1 Strains for general purpose cloning

The following four strains are popular hosts for cloning vectors that do not employ Lac selection.

C600: *supE44 thi-1 thr-1 leuB6 lacY1 tonA21 mcrA*

DH1: *supE44 hsdR17 recA1 endA1 gyrA96 thi-1 relA1*

HB101: Δ(*gpt-proA*)62 *leuB6 supE44 ara-14 galK2 lacY1 rpsL20 xyl-5* Δ(*mcrCB-hsdSMR-mrr*) *mtl-1 recA13*

RR1: Δ(*gpt-proA*)62 *leuB6 supE44 ara-14 galK2 lacY1 rpsL20 xyl-5* Δ(*mcrCB-hsdSMR-mrr*) *mtl-1*

These four strains have been used in research for many years and versions with slightly different genotypes might be in circulation (1). For example, some strains of C600 are *mcrA*$^+$ and DH1 may also be *spoT1rfbD1*. HB101 and RR1 might be *leu*$^+$ and/or *thi-1*.

1.2 Strains for Lac selection

The following strains can be used for general-purpose cloning and also with plasmid and M13 vectors that utilize Lac selection. MV1184 and XL1-Blue are popular hosts for the generation of single-stranded DNA from phagemid vectors, using M13KO7 as the helper phage. The presence of the *recB recJ sbcC umuC uvrC* mutations in SURE helps to prevent rearrangement of cloned DNA sequences. SURE, XL1-Blue and XL1-Blue MRF′ are marketed by Stratagene; all three strains carry undescribed mutations that give colonies a more intensely blue coloration on agar plates containing X-gal and IPTG.

JM83: *ara* Δ(*lac-proAB*) *rpsL* (ϕ80 *lacZ*ΔM15)

JM109: *recA1 supE44 endA1 hsdR17 gyrA96 relA1 thi-1 mcrA* Δ(*lac-proAB*) F′[*traD36 proAB*$^+$ *lacI*^{q} *lacZ*ΔM15]

JM110: *dam dcm supE44 thi-1 leu rpsL lacY galK galT ara tonA thr tsx Δ(lac-proAB) F′[traD36 proAB⁺ lacIᑫ lacZΔM15]*

KK2186: *supE thi-1 endA1 hsdR4 sbcBC rpsL Δ(lac-proAB) F′[traD36 proAB⁺ lacIᑫ lacZΔM15]*

MV1184: *Δ(lac-proAB) rpsL thi (φ80 lacZΔM15) Δ(srl-recA)306::Tn10(tetʳ) F′[traD36 proAB⁺ lacIᑫ lacZΔM15]*

SURE: *mcrA Δ(mcrCB-hsdSMR-mrr)171 endA1 supE44 thi-1 gyrA96 relA1 lac recB recJ sbcC umuC::Tn5(kanʳ) uvrC F′[proAB⁺ lacIᑫ lacZΔM15 Tn10(tetʳ)]*

XL1-Blue: *recA1 endA1 gyrA96 thi-1 hsdR17 supE44 relA1 lac F′[proAB⁺ lacIᑫ lacZΔM15 Tn10(tetʳ)]*

XL1-Blue MRF′: *Δ(mcrA)183 Δ(mcrCB-hsdSMR-mrr)173 recA1 endA1 gyrA96 thi-1 supE44 relA1 lac F′[proAB⁺ lacIᑫ lacZΔM15 Tn5(kanʳ)]*

The genotype of JM110 is described by some sources as also being *hsdR17*, but this is probably incorrect. KK2186 is identical to the popular strain JM103, but KK2186 should be used instead of JM103 because some examples of the latter have accumulated additional mutations.

1.3 Strains for protein expression with pET vectors

The expression of protein from genes cloned in pET vectors requires that the host cells possess the gene for the T7 RNA polymerase (See Chapter 8, Section 2.1). This gene is carried by an engineered version of bacteriophage λ called λDE3 and the relevant strains contain this λ genome as a lysogen:

BL21(DE3): *hsdS gal (λcIts857 ind1 Sam7 nin5 lacUV5-T7 gene 1)*

JM109(DE3): *recA1 supE44 endA1 hsdR17 gyrA96 relA1 thi-1 mcrA Δ(lac-proAB) F′[traD36 proAB⁺ lacIᑫ lacZΔM15] (λcIts857 ind1 Sam7 nin5 lacUV5-T7 gene 1)*

TKB1: *dcm ompT hsdS gal (λcIts857 ind1 Sam7 nin5 lacUV5-T7 gene 1) (pTK tetʳ)*

The following strain carries the T7 *lysS* gene and allows a greater degree of control over synthesis of recombinant protein:

BL21(DE3)pLysS: *hsdS gal (λcIts857 ind1 Sam7 nin5 lacUV5-T7 gene 1) (pLysS camʳ)*

1.4 Strains for λ vectors

The following strains are frequently used with bacteriophage λ vectors.

DH5: *supE44 hsdR17 recA1 endA1 gyrA96 thi-1 relA1*

K802: *supE44 hsdR galK2 galT22 metB1 mcrA mcrB1*

NM646: *supE44 hsdR galK2 galT22 metB1 mcrA mcrB1 (P2cox3)*

Y1088: *Δ(lac)U169 supE supF hsdR metB trpR tonA21 mcrA proC::Tn5(kanʳ) (pMC9)*

Y1090: *mcrA araD139 Δ(lac)U169 lon-100 rpsL supF trpC22::Tn10(tetʳ) (pMC9)*

In the Y1088 and Y1090 genotypes '(pMC9)' denotes possession of pBR322 carrying *lacIᑫ*.

1.5 Strains for λ packaging extracts

The strains used as sources of the λ packaging extracts (Volume I, Chapter 8, *Protocol 6*) have the following genotypes.

BHB2688: N205 *recA* (λ *imm*434 *c*Its *b*2 *red*3 Eam4 Sam7)/λ

BHB2690: N205 *recA* (λ *imm*434 *c*Its *b*2 *red*3 Dam15 Sam7)/λ

2 Key features of *E. coli* genotypes

2.1 Recombination deficiency

The major recombination pathways in *E. coli* are coded by the *rec* genes. Strains that are Rec$^+$ are less suitable for cloning as they can multimerize plasmids and rearrange inserted DNA (2,3). There are at least nine different *rec* genes with host strains for molecular biology experiments often carrying one or more of *recA*, *recBC*, *recD*, *recF* and *recJ*. The mutation *sbcBC* may also be present to further reduce rearrangement of cloned DNA (4,5). One problem with *rec* strains is that they are relatively slow growing and may be difficult to transform. Strains used as hosts for bacteriophage λ vectors that are *red gam* (Volume I, Chapter 9, Section 4.1) must be RecA$^+$.

2.2 Host-controlled restriction-modification

There are four different *E. coli* restriction and/or modification systems that are relevant to the use of this bacterium in gene cloning (1):

(a) Most strains used for cloning are derived from *E. coli* K, which possesses the *Eco*K restriction-modification system (6). This is coded by the *hsd* genes, with an *hsdM* strain being methylation deficient, *hsdR* being restriction deficient, and *hsdS* lacking both functions. Strains used in cloning are usually *hsdR* or *hsdS*, so that there is no possibility of *Eco*K restriction sites in the cloned DNA being cleaved.

(b) Many host strains for cloning are also *mcrA* and *mcrBC* and so lack functional McrA and McrB restriction systems, which cleave DNA at short target sequences and which, if not mutated, could cleaved cloned DNA (6,7–9).

(c) Similarly, many strains are *mrr* and so lack the Mrr restriction system (6,8,10). Mrr is similar to McrA and McrBC, although the identity of the restriction sequence is not known.

(d) DNA adenine methylase and DNA cytosine methylase add methyl groups to target sequences (6,11–13). Strains used for cloning are sometimes *dam* and/or *dcm* so that the cloned DNA does not become modified. This avoids possible problems that might arise after the cloned DNA is purified, because the methylated DNA may not be cut by a restriction enzyme if the recognition sequence for the enzyme overlaps a methylation site and the enzyme action is inhibited by the presence of the methyl group.

Comprehensive details of the restriction-modification properties of *E. coli* strains are given in ref. (1).

2.3 Suppressor mutations

The amber suppressor mutations *supE* and *supF* result in insertion of glutamine and tyrosine, respectively, at UAG codons. These mutations provide a form of biological containment in that phages carrying amber mutations can survive only in a suppressing strain and so cannot escape to natural populations of bacteria.

2.4 Genotypes relevant to Lac selection

Plasmids carrying the *lacZ'* gene require special hosts. Usually, these carry a chromosomal deletion called Δ(*lac-proAB*), which spans the lactose operon and surrounding region. The deletion is partially complemented by an engineered F' plasmid, which carries *proAB*$^+$ (rescuing proline auxotrophy), and *lacZ*Δ*M15*, which is the *lac* operon minus the amino-terminal α-peptide of *lacZ* (amino acids 11–41) which is the part present in the *lacZ'* segment carried by the vector. The F' plasmid also carries *lacI*^{q}, a mutation that results in over-expression of the *lac* repressor (14) and reduces the background expression of the *lac* operon that occurs in uninduced cells. The presence of this mutation therefore makes the X-gal identification system more discriminatory by ensuring that non-recombinant cells are cream and not pale blue.

The genes described above provide all of the sequence information needed for utilization of lactose except for the a peptide coded by *lacZ'*. The presence of the *proAB*$^+$ genes on the F' plasmid means that retention of the F' can be selected for by maintaining the bacteria on a proline-deficient minimal medium. This is important because F' bacteria are unstable and can revert to F$^-$ if stored for long periods.

References

1. Brown, T. A. (ed.) (1998). *Molecular biology labfax*, 2nd edn, Vol. 1. Academic Press, London.
2. Yanisch-Perron, C., Vieira, J., and Messing, J. (1985). *Gene*, **33**, 103.
3. Bedbrook, J. R. and Ausubel, F. M. (1976). *Cell*, **9**, 707.
4. Kushner, S. R., Nagaishi, H., Templin, A., and Clark, A. J. (1971). *Proc. Natl Acad. Sci. USA*, **75**, 2276.
5. Leach, D. R.F. and Stahl, F. W. (1983). *Nature*, **305**, 448.
6. Raleigh, E. A., Lech, K., and Brent, R. (1989). In *Current protocols in molecular biology*. (ed. F. M. Ausubel, R. Brent, R. E. Kingston, D. D. Moore, J. G. Seidman, J. A. Smith, and K. Struhl), p. 1.4.6. John Wiley, New York.
7. Raleigh, E. A. (1992). *Mol. Microbiol.*, **6**, 1079.
8. Kelleher, J. and Raleigh, E. A. (1991). *J. Bacteriol.*, **173**, 5220.
9. Sutherland, E., Coe, L., and Raleigh, E. A. (1992). *J. Mol. Biol.*, **225**, 327.
10. Waite-Rees, P. A., Keating C. J., Moran, L. S., Stalko, B. E., Hornstra, L. J., and Benner J. S. (1991). *J. Bacteriol.*, **173**, 5207.

11. Hattman, S., Brooks, J. E., and Masurekar, M. (1978). *J. Mol. Biol.*, **126**, 367.
12. Marinus, M. G. and Morris, N. R. (1973). *J. Bacteriol.*, **114**, 1143.
13. May, M. S. and Hattman, S. (1975). *J. Bacteriol.*, **123**, 768.
14. Muller-Hill, B., Crapo, L., and Gilbert, W. (1968). *Proc. Natl. Acad. Sci. USA*, **59**, 1259.

Appendix 4
Recipes and general procedures

1 Recipes for *Escherichia coli* culture media

1.1 Liquid media

Recipes for 1 l of media:

- BBL broth: 18 g BBL trypticase peptone, 5 g NaCl
- DYT: 16 g Bacto-tryptone, 10 g Bacto-yeast extract, 5 g NaCl
- LB: 10 g Bacto-tryptone, 5 g Bacto-yeast extract, 10 g NaCl
- Nutrient broth: 25 g Bacto-nutrient broth
- NZY: 10 g NZ amine, 5 g Bacto-yeast extract, 2 g $MgSO_4$
- SOB: 20 g Bacto-tryptone, 5 g Bacto-yeast extract, 0.5 g NaCl
- Superbroth (SB): 32 g Bacto-tryptone, 20 g Bacto-yeast extract, 5 g NaCl, 5 ml 1 M NaOH
- Terrific broth (TB): 12 g Bacto-tryptone, 24 g Bacto-yeast extract, 4 ml glycerol. Prepare in 850 ml water, autoclave, cool, add 100 ml of a sterile solution of 0.17 M KH_2PO_4 + 0.72 K_2HPO_4, and make the final volume up to 1 l with sterile water
- Tryptone broth: 10 g Bacto-tryptone, 5 g NaCl
- YT: 8 g Bacto-tryptone, 5 g Bacto-yeast extract, 5 g NaCl

With all media the pH should be checked and if necessary adjusted to 7.0–7.2 with NaOH. Media should be sterilized by autoclaving at 121°C, 103.5 kPa (15 lbf/in^2), for 20 min.

1.2 Solid media

To prepare solid media add the required amount of Bacto-agar before autoclaving. For agar plates add 15 g agar per litre.

1.2.1 Soft agar overlays

To prepare soft or top agar overlays use the recipes given above but add a reduced amount of agar or agarose:

- LTS is DYT plus 6 g agar per litre

- NZY top agar is NZY plus 7 g agarose per litre
- YTS is YT plus 6 g agar per litre

1.3 Supplements

1.3.1 Supplements for λ phage

- maltose: prepare a 20% (w/v) stock solution and filter-sterilize. Store at room temperature and add to autoclaved media (final concentration 0.2%) immediately before use.
- $MgSO_4$: prepare a 1 M stock solution, autoclave and add to autoclaved media (final concentration usually 10 mM) immediately before use.

1.3.2 Antibiotics

All antibiotics must be prepared as filter-sterilized stock solutions and added to autoclaved media immediately before use. Agar media must be cooled to 50 °C before antibiotic supplementation as many are heat-sensitive.

Ampicillin:

- stock: 50 mg/ml in water
- working concentration: 40–60 μg/ml (plates) or 25 μg/ml (broth).

Kanamycin:

- stock: 35 mg/ml in water
- working concentration: 25–50 μg/ml (plates) or 25–70 μg/ml (broth).

Streptomycin:

- stock: 10 mg/ml in water
- working concentration: 40 μg/ml (plates) or 15 μg/ml (broth).

Tetracycline:

- stock: 5 mg/ml in ethanol
- working concentration: 15 μg/ml (plates) or 7.5 μg/ml (broth).

Note that tetracycline is light-sensitive and all solutions and plates should be wrapped in foil or stored in the dark. Magnesium ions antagonize the activity of tetracycline.

2 Buffers

- Phosphate-buffered saline (PBS): 8.0 g NaCl, 0.34 g KH_2PO_4, 1.21 g K_2HPO_4 per 1 l. pH should be 7.3. Sterilize by autoclaving
- SM (λ storage and dilution): 5.8 g NaCl, 2 g $MgSO_4.7H_2O$, 50 ml 1 M Tris-HCl, pH 7.5, 5 ml 2% (w/v) gelatin solution, water to make 1 l. Sterilize by autoclaving.
- TE buffers: 10 mM Tris-HCl, 1 mM EDTA, pH 8.0. The pH of the Tris-HCl determines the pH of the TE buffer.

3 General procedures

Protocols 1 to *5* describe five general procedures required as support for many different molecular biology procedures.

Protocol 1

Preparation of DNase-free RNase A[a]

Reagents

- Ribonuclease A (RNase A)
- 10 mM Tris-HCl, pH 7.5, 15 mM NaCl

Method

1 Dissolve solid RNase A to 10 mg/ml in 10 mM Tris-HCl, pH 7.5, 15 mM NaCl.

2 Heat to 100 °C for 15 min.

3. Cool slowly to room temperature. Store in aliquots at −20 °C.

[a] Commercial supplies of RNase A usually contain DNase activity.

Protocol 2

Preparation of siliconized glassware

Reagents

- Siliconizing solution (2% dimethyldichlorosilane in 1,1,1-trichloroethane) (Merck)

Method

1 Rinse the glassware in siliconizing solution. The solution can be reused several times. Avoid contact with the skin. Use a fume hood.

2 Allow the glassware to dry thoroughly (up to 2 h).

3 Rinse each item twice in double-distilled water.

Notes:

(a) DNA will adsorb to non-siliconized glassware, resulting in substantial losses. Glass wool can be treated in exactly the same way. Plasticware does not usually need to be siliconized, though some applications require it.

(b) If a desiccator can be dedicated to siliconization then instead of step 1 evaporate 1 ml of siliconizing solution under vacuum inside the desiccator with the glassware to be treated.

(c) **Caution: siliconizing solutions are toxic and flammable.**

Protocol 3

Preparation of non-homologous DNA for hybridization analysis

Reagents

- Salmon sperm DNA (Sigma Type III, sodium salt)
- 17-G syringe needle
- 25:24:1 (v/v/v) phenol–chloroform–isoamyl alcohol (see *Protocol 6*)
- cold absolute ethanol

Method

1   Prepare a 10 mg/ml solution of salmon sperm DNA in water.

2   Adjust the sodium concentration to 0.1 M and extract once with phenol and once with phenol–chloroform–isoamyl alcohol. This step can be omitted with most batches of salmon sperm DNA but should be included if poor results (i.e. high background hybridization) are obtained.

3   Pass through a 17-G syringe needle approximately 20 times. This should be done in a vigorous manner as the aim is to shear the DNA.

4   Add two volumes of cold absolute ethanol, leave on ice for 10 min, then spin in the microfuge for 30 min. Remove the ethanol, dry the pellet, and resuspend in the original volume of water. Again, this step may be unnecessary.

5   Place in a boiling water bath for 10 min, chill and then either:

 (a) use immediately at a final concentration of 100 µg/ml;

 (b) store at −20 °C in small aliquots. Immediately before use, place in a boiling water bath for 5 min and chill on ice.

Note:
Non-homologous DNA is often added to prehybridization and hybridization solutions to block non-specific binding sites on the filter surface (Chapter 5, Section 3). Usually the non-homologous DNA is obtained from salmon sperm, herring sperm or calf thymus.

Protocol 4

Preparation of deionized formamide

Reagents

- Formamide
- Whatman No. 1 paper
- Mixed bed ion-exchange resin (e.g. Bio-Rad AG 501-X8 or X8(D) resins)

Method

1   Add 5 g of mixed bed ion-exchange resin per 100 ml formamide.

Protocol 4 continued

2   Stir at room temperature for 1 h.

3   Filter twice through Whatman No. 1 paper.

Notes:

(a) Commercial formamide is usually satisfactory for use in molecular biology experiments (e.g. in hybridization solutions and loading buffers for gel electrophoresis) but if the liquid has a yellow colour it should be deionized until it becomes colourless, as described in this *Protocol*.

(b) The resin can be reused a number of times. The X8(D) resin contains a dye that indicates when it is exhausted.

(c) **Caution: formamide is a teratogen and an irritant. Avoid exposure and take appropriate safety precautions.**

Protocol 5

Decontamination of ethidium bromide solutions

Reagents

- Amberlite XAD-16 resin (Bio-Rad)

Method

1   Dilute the ethidium bromide solution to a concentration no greater than 100 µg/ml.

2   Add 0.5 g of Amberlite XAD-16 per 10 ml ethidium bromide solution.

3   Stir for 2 h for every 10 ml of solution (e.g. stir a 20 ml solution for 4 h and a 100 ml solution for 20 h).

4   Filter the mixture to collect the resin. The filtrate can be treated as non-hazardous waste and disposed of in the normal way. The resin must be treated as hazardous waste.

Protocol 6

Preparation of phenol and chloroform solutions

Reagents

- Water-equilibrated phenol
- 1 M Tris-HCl, pH 8.0
- 50 mM Tris-HCl, pH 8.0
- Chloroform
- Isoamyl alcohol (or more correctly 3-methyl-1-butanol)
- TE buffer, pH 8.0 (see *Section 2*)

Protocol 6 continued

Methods

A. Preparation of phenol

Caution: phenol is one of the most hazardous chemicals used in molecular biology (see Appendix 1, Section 4) and should be obtained ready for use either as an aqueous solution or buffered at pH 8.0. Aqueous phenol must be pretreated by buffering to pH 8.0 in the following way:

1 Remove and discard the upper aqueous layer from a 1 l bottle of water-equilibrated phenol. Replace the aqueous layer with 1 M Tris-HCl, pH 8.0, and mix the liquid phases by inverting the bottle several times.

2 Allow the layers to separate. If the lower layer is at pH 8.0, then remove and discard the upper aqueous phase. Replace this with 50 mM Tris-HCl, pH 8.0. Agitate the mixture and aliquot 1–10 ml samples in microfuge or Falcon tubes and store at −20 °C. If the lower layer is not at pH 8.0 then repeat step 1 until this pH is reached.

3 Thaw aliquots at 20 °C prior to use. It is the dense lower organic layer which is used for treatment of DNA. The phenol remains stable at 4 °C for about a month, but storage at room temperature or exposure to light enhances oxidation and the liquid becomes pink. Such samples must be discarded because they can lead to the modification and cleavage of DNA in solution. Oxidation can be inhibited by addition of 8-hydroxyquinoline at 0.1 % (w/v). This has the advantage of giving a bright yellow colour which identifies the organic layer during phenol extractions.

B. 24:1 (v/v) chloroform–isoamyl alcohol

1 Mix 96 ml chloroform with 4 ml isoamyl alcohol; store in a capped bottle.

C. 'Phenol-chloroform'

1 Thaw a 10-ml aliquot of phenol and transfer 5 ml of the lower phase to a universal bottle.

2 Mix in 5 ml of 24:1 (v/v) chloroform–isoamyl alcohol and 10 ml TE buffer, pH 8.0, and store at 4 °C. Use the lower organic layer.

Appendix 5
Restriction endonucleases

1 Details of restriction endonucleases

Numerous restriction endonucleases recognizing a variety of target sequences have been identified. A complete and up to date listing is provided by ref. (1). Extensive details of reactions conditions for different enzymes are given in ref. (2).

1.1 Recognition sequences of commonly–used enzymes

Table 1 gives the recognition sequences for many of the restriction endo-nucleases commonly used in molecular biology research. It is based on ref. (2) which contains a more comprehensive listing. Consult ref. 1 for the recognition sequences of enzymes not listed here or in ref. (2).

Table 1 Recognition sequences for commonly-used restriction endonucleases[a]

Enzyme	Recognition sequence	Type of end	Sequence of overhang
*Aat*I	AGG↑CCT	Blunt	
*Aat*II	GACGT↑C	3′ overhang	–ACGT
*Acc*I	GT↑MKAC	5′ overhang	MK–
*Acc*II	CG↑CG	Blunt	
*Acc*III	T↑CCGGA	5′ overhang	CCGG–
*Aci*I	CCGC(–3/–1)	5′ overhang	NN–
*Acs*I	R↑AATTY	5′ overhang	AATT–
*Acy*I	GR↑CGYC	5′ overhang	CG–
*Afa*I	GT↑AC	Blunt	
*Afl*I	G↑GWCC	5′ overhang	GWC–
*Afl*II	C↑TTAAG	5′ overhang	TTAA–
*Afl*III	A↑CRYGT	5′ overhang	CRYG–
*Age*I	A↑CCGGT	5′ overhang	CCGG–
*Aha*I	CC↑SGG	5′ overhang	S–
*Aha*II	GR↑CGYC	5′ overhang	CG–
*Aha*III	TTT↑AAA	Blunt	
*Ahd*I	GACNNN↑NNGTC	3′ overhang	–N
*Alu*I	AG↑CT	Blunt	
*Alw*I	GGATC(4/5)	5′ overhang	N–
*Alw*NI	CAGNNN↑CTG	3′ overhang	–NNN

267

Table 1 continued

Enzyme	Recognition sequence	Type of end	Sequence of overhang
*Aoc*I	CC↑TNAGG	5′ overhang	TNA–
*Aoc*II	GDGCH↑C	3′ overhang	–DGCH
*Aos*I	TGC↑GCA	Blunt	
*Apa*I	GGGCC↑C	3′ overhang	–GGCC
*Apa*LI	G↑TGCAC	5′ overhang	TGCA–
*Apo*I	R↑AATTY	5′ overhang	AATT–
*Apy*I	CC↑WGG	5′ overhang	W–
*Aqu*I	C↑YCGRG	5′ overhang	YCGR–
*Asc*I	GG↑CGCGCC	5′ overhang	CGCG–
*Ase*I	AT↑TAAT	5′ overhang	TA–
*Asn*I	AT↑TAAT	5′ overhang	TA–
*Asp*I	GACN↑NNGTC	5′ overhang	N–
*Asp*700I	GAANN↑NNTTC	Blunt	
*Asp*EI	GACNNN↑NNGTC	3′ overhang	–N
*Asp*HI	GWGCW↑C	3′ overhang	–WGCW
*Asu*I	G↑GNCC	5′ overhang	GNC–
*Asu*II	TT↑CGAA	5′ overhang	CG–
*Ava*I	C↑YCGRG	5′ overhang	YCGR–
*Ava*II	G↑GWCC	5′ overhang	GWC–
*Avi*II	TGC↑GCA	Blunt	
*Avr*II	C↑CTAGG	5′ overhang	CTAG–
*Axy*I	CC↑TNAGG	5′ overhang	TNA–
*Ba*lI	TGG↑CCA	Blunt	
*Bam*HI	G↑GATCC	5′ overhang	GATC–
*Ban*I	G↑GYRCC	5′ overhang	GYRC–
*Ban*II	GRGCY↑C	3′ overhang	–RGCY
*Ban*III	AT↑CGAT	5′ overhang	CG–
*Bbe*I	GGCGC↑C	3′ overhang	–GCGC
*Bbi*II	GR↑CGYC	5′ overhang	CG–
*Bbr*PI	CAC↑GTG	Blunt	
*Bbs*I	GAAGAC(2/6)	5′ overhang	NNNN–
*Bbu*I	GCATG↑C	3′ overhang	–CATG
*Bbv*I	GCAGC(8/12)	5′ overhang	NNNN–
*Bcg*I	(10/12)GCANNNNNTCG(12/10)[b]		
*Bcl*I	T↑GATCA	5′ overhang	GATC–
*Bcn*I	CC↑SGG	5′ overhang	S–
*Bfa*I	C↑TAG	5′ overhang	TA–
*Bfr*I	C↑TTAAG	5′ overhang	TTAA–
*Bgl*I	GCCNNNN↑NGGC	3′ overhang	–NNN
*Bgl*II	A↑GATCT	5′ overhang	GATC–
*Bln*I	C↑CTAGG	5′ overhang	CTAG–

Table 1 continued

Enzyme	Recognition sequence	Type of end	Sequence of overhang
*Blp*I	GC↑TNAGC	5′ overhang	TNA–
*Bmy*I	GDGCH↑C	3′ overhang	–DGCH
*Bpm*I	CTGGAG(16/14)	3′ overhang	–NN
*Bpu*AI	GAAGAC(2/6)	5′ overhang	NNNN–
*Bsa*I	GGTCTC(1/5)	5′ overhang	NNNN–
*Bsa*AI	YAC↑GTR	Blunt	
*Bsa*BI	GATNN↑NNATC	Blunt	
*Bsa*HI	GR↑CGYC	5′ overhang	CG–
*Bsa*JI	C↑CNNGG	5′ overhang	CNNG–
*Bsa*MI	GAATGC(1/−1)	3′ overhang	–NGAATGCN
*Bsa*OI	CGRY↑CG	3′ overhang	–RY
*Bsa*WI	W↑CCGGW	5′ overhang	CCGG–
*Bse*AI	T↑CCGGA	5′ overhang	CCGG–
*Bse*RI	GAGGAG(10/8)	3′ overhang	–NN
*Bsg*I	GTGCAG(16/14)	3′ overhang	–NNNN
*Bsi*CI	TT↑CGAA	5′ overhang	CG–
*Bsi*EI	CGRY↑CG	3′ overhang	–RY
*Bsi*WI	C↑GTACG	5′ overhang	GTAC–
*Bsi*YI	CCNNNNN↑NNGG	3′ overhang	–NNN
*Bsl*I	CCNNNNN↑NNGG	3′ overhang	–NNN
*Bsm*I	GAATGC(1/−1)	3′ overhang	–NGAATGCN
*Bsm*AI	GTCTC(1/5)	5′ overhang	NNNN–
*Bsm*BI	CGTCTC(1/5)	5′ overhang	NNNN–
*Bsm*FI	GGGAC(10/14)	5′ overhang	NNNN–
*Bso*BI	C↑YCGRG	5′ overhang	YCGR–
*Bsp*1286I	GDGCH↑C	3′ overhang	–DGCH
*Bsp*CI	CGAT↑CG	3′ overhang	–T
*Bsp*DI	AT↑CGAT	5′ overhang	CG–
*Bsp*EI	T↑CCGGA	5′ overhang	CCGG–
*Bsp*HI	T↑CATGA	5′ overhang	CATG–
*Bsp*MI	ACCTGC(4/8)	5′ overhang	NNNN–
*Bsp*MII	T↑CCGGA	5′ overhang	CCGG–
*Bsr*I	ACTGG(1/−1)	3′ overhang	–NACTGGN
*Bsr*BI	CCGCTC(−3/−3)	Blunt	
*Bsr*BRI	GATNN↑NNATC	Blunt	
*Bsr*DI	GCAATG(2/0)	3′ overhang	–NN
*Bsr*FI	R↑CCGGY	5′ overhang	CCGG–
*Bsr*GI	T↑GTACA	5′ overhang	GTAC–
*Bsr*SI	ACTGG(1/−1)	3′ overhang	–NACTGGN
*Bss*HII	G↑CGCGC	5′ overhang	CGCG–
*Bss*KI	↑CCNGG	5′ overhang	CCNGG–

Table 1 continued

Enzyme	Recognition sequence	Type of end	Sequence of overhang
*Bss*SI	CACGAG(–5/–1)	5′ overhang	NNNN–
*Bst*I	G↑GATCC	5′ overhang	GATC–
*Bst*BI	TT↑CGAA	5′ overhang	CG–
*Bst*EII	G↑GTNACC	5′ overhang	GTNAC–
*Bst*NI	CC↑WGG	5′ overhang	W–
*Bst*OI	CC↑WGG	5′ overhang	W–
*Bst*PI	G↑GTNACC	5′ overhang	GTNAC–
*Bst*UI	CG↑CG	Blunt	
*Bst*XI	CCANNNNN↑NTGG	3′ overhang	–NNNN
*Bst*YI	R↑GATCY	5′ overhang	GATC–
*Bst*ZI	C↑GGCCG	5′ overhang	GGCC–
*Cac*8I	GCN↑NGC	Blunt	
*Cfo*I	GCG↑C	3′ overhang	–CG
*Cfr*9I	C↑CCGGG	5′ overhang	CCGG–
*Cfr*10I	R↑CCGGY	5′ overhang	CCGG–
*Cfr*13I	G↑GNCC	5′ overhang	GNC–
*Cla*I	AT↑CGAT	5′ overhang	CG–
*Cpo*I	CG↑GWCCG	5′ overhang	GWC–
*Csp*I	CG↑GWCCG	5′ overhang	GWC–
*Csp*45I	TT↑CGAA	5′ overhang	CG–
*Cvn*I	CC↑TNAGG	5′ overhang	TNA–
*Dde*I	C↑TNAG	5′ overhang	TNA–
*Dpn*I	GA↑TC	Blunt	
*Dpn*II	↑GATC	5′ overhang	GATC–
*Dra*I	TTT↑AAA	Blunt	
*Dra*II	RG↑GNCCY	5′ overhang	GNC–
*Dra*III	CACNNN↑GTG	3′ overhang	–NNN
*Drd*I	GACNNNN↑NNGTC	3′ overhang	–NN
*Eae*I	Y↑GGCCR	5′ overhang	GGCC–
*Eag*I	C↑GGCCG	5′ overhang	GGCC–
*Eam*1105I	GACNNN↑NNGTC	3′ overhang	–N
*Ear*I	CTCTTC(1/4)	5′ overhang	NNN–
*Eco*NI	CCTNN↑NNNAGG	5′ overhang	N–
*Eco*RI	G↑AATTC	5′ overhang	AATT–
*Eco*RII	↑CCWGG	5′ overhang	CCWGG–
*Eco*RV	GAT↑ATC	Blunt	
*Ehe*I	GGC↑GCC	Blunt	
*Esp*I	GC↑TNAGC	5′ overhang	TNA–
*Fba*I	T↑GATCA	5′ overhang	GATC–
*Fdi*II	TGC↑GCA	Blunt	
*Fnu*DII	CG↑CG	Blunt	

Table 1 continued

Enzyme	Recognition sequence	Type of end	Sequence of overhang
*Fok*I	GGATG(9/13)	5′ overhang	NNNN–
*Fse*I	GGCCGG↑CC	3′ overhang	–CCGG
*Fsp*I	TGC↑GCA	Blunt	
*Hae*II	RGCGC↑Y	3′ overhang	–GCGC
*Hae*III	GG↑CC	Blunt	
*Hap*II	C↑CGG	5′ overhang	CG–
*Hga*I	GACGC(5/10)	5′ overhang	NNNNN–
*Hgi*AI	GWGCW↑C	3′ overhang	–WGCW
*Hha*I	GCG↑C	3′ overhang	–CG
*Hin*1I	GR↑CGYC	5′ overhang	CG–
*Hin*cII	GTY↑RAC	Blunt	
*Hin*dII	GTY↑RAC	Blunt	
*Hin*dIII	A↑AGCTT	5′ overhang	AGCT–
*Hin*fI	G↑ANTC	5′ overhang	ANT–
*Hin*P1I	G↑CGC	5′ overhang	CG–
*Hpa*I	GTT↑AAC	Blunt	
*Hpa*II	C↑CGG	5′ overhang	CG–
*Hph*I	GGTGA(8/7)	3′ overhang	–N
*Ita*I	GC↑NGC	5′ overhang	N–
*Kas*I	G↑GCGCC	5′ overhang	GCGC–
*Kpn*I	GGTAC↑C	3′ overhang	–GTAC
*Ksp*I	CCGC↑GG	3′ overhang	–GC
*Mae*I	C↑TAG	5′ overhang	TA–
*Mae*II	A↑CGT	5′ overhang	CG–
*Mae*III	↑GTNAC	5′ overhang	GTNAC–
*Mam*I	GATNN↑NNATC	Blunt	
*Mbo*I	↑GATC	5′ overhang	GATC–
*Mbo*II	GAAGA(8/7)	3′ overhang	–N
*Mfe*I	C↑AATTG	5′ overhang	AATT–
*Mfl*I	R↑GATCY	5′ overhang	GATC–
*Mlu*I	A↑CGCGT	5′ overhang	CGCG–
*Mlu*NI	TGG↑CCA	Blunt	
*Mnl*I	CCTC(7/6)	3′ overhang	–N
*Mro*I	T↑CCGGA	5′ overhang	CCGG–
*Msc*I	TGG↑CCA	Blunt	
*Mse*I	T↑TAA	5′ overhang	TA–
*Msl*I	CAYNN↑NNRTG	Blunt	
*Msp*I	C↑CGG	5′ overhang	CG–
*Msp*A1I	CMG↑CKG	Blunt	
*Mst*I	TGC↑GCA	Blunt	
*Mst*II	CC↑TNAGG	5′ overhang	TNA–

Table 1 continued

Enzyme	Recognition sequence	Type of end	Sequence of overhang
MunI	C↑AATTG	5' overhang	AATT–
MvaI	CC↑WGG	5' overhang	W–
MvnI	CG↑CG	Blunt	
MwoI	GCNNNNN↑NNGC	3' overhang	–NNN
NaeI	GCC↑GGC	Blunt	
NarI	GG↑CGCC	5' overhang	CG–
NciI	CC↑SGG	5' overhang	S–
NcoI	C↑CATGG	5' overhang	CATG–
NdeI	CA↑TATG	5' overhang	TA–
NdeII	↑GATC	5' overhang	GATC–
NgoMI	G↑CCGGC	5' overhang	CCGG–
NheI	G↑CTAGC	5' overhang	CTAG–
NlaIII	CATG↑	3' overhang	–CATG
NlaIV	GGN↑NCC	Blunt	
NotI	GC↑GGCCGC	5' overhang	GGCC–
NruI	TCG↑CGA	Blunt	
NsiI	ATGCA↑T	3' overhang	–TGCA
NspI	RCATG↑Y	3' overhang	–CATG
NspII	GDGCH↑C	3' overhang	–DGCH
NspIII	C↑YCGRG	5' overhang	YCGR–
NspIV	G↑GNCC	5' overhang	GNC–
NspV	TT↑CGAA	5' overhang	CG–
NspBII	CMG↑CKG	Blunt	
NspHI	RCATG↑Y	3' overhang	–CATG
NunII	GG↑CGCC	5' overhang	CG–
PacI	TTAAT↑TAA	3' overhang	–AT
PaeR7I	C↑TCGAG	5' overhang	TCGA–
PalI	GG↑CC	Blunt	
PflFI	GACN↑NNGTC	5' overhang	N–
PflMI	CCANNNN↑NTGG	3' overhang	–NNN
PleI	GAGTC(4/5)	5' overhang	N–
PmaCI	CAC↑GTG	Blunt	
PmeI	GTTT↑AAAC	Blunt	
PmlI	CAC↑GTG	Blunt	
PpuMI	RG↑GWCCY	5' overhang	GWC–
PshAI	GACNN↑NNGTC	Blunt	
PspAI	C↑CCGGG	5' overhang	CCGG–
PssI	RGGNC↑CY	3' overhang	–GNC
PstI	CTGCA↑G	3' overhang	–TGCA
PvuI	CGAT↑CG	3' overhang	–AT
PvuII	CAG↑CTG	Blunt	

Table 1 continued

Enzyme	Recognition sequence	Type of end	Sequence of overhang
*Rsa*I	GT↑AC	Blunt	
*Rsp*XI	T↑CATGA	5′ overhang	CATG–
*Rsr*I	G↑AATTC	5′ overhang	AATT–
*Rsr*II	CG↑GWCCG	5′ overhang	GWC–
*Sac*I	GAGCT↑C	3′ overhang	–AGCT
*Sac*II	CCGC↑GG	3′ overhang	–GC
*Sal*I	G↑TCGAC	5′ overhang	TCGA–
*San*DI	GG↑GWCCC	5′ overhang	GWC–
*Sap*I	GCTCTTC(1/4)	5′ overhang	NNN–
*Sau*I	CC↑TNAGG	5′ overhang	TNA–
*Sau*3AI	↑GATC	5′ overhang	GATC–
*Sau*96I	G↑GNCC	5′ overhang	GNC–
*Sca*I	AGT↑ACT	Blunt	
*Scr*FI	CC↑NGG	5′ overhang	N–
*Sdu*I	GDGCH↑C	3′ overhang	–DGCH
*Sex*AI	A↑CCWGGT	5′ overhang	CCWGG–
*Sfa*NI	GCATC(5/9)	5′ overhang	NNNN–
*Sfc*I	C↑TRYAG	5′ overhang	TRYA–
*Sfi*I	GGCCNNNN↑NGGCC	3′ overhang	–NNN
*Sfu*I	TT↑CGAA	5′ overhang	CG–
*Sgf*I	GCGAT↑CGC	3′ overhang	–AT
*Sgr*AI	CR↑CCGGYG	5′ overhang	CCGG–
*Sin*I	G↑GWCC	5′ overhang	GWC–
*Sma*I	CCC↑GGG	Blunt	
*Sml*I	C↑TYRAG	5′ overhang	TY–
*Sna*BI	TAC↑GTA	Blunt	
*Spe*I	A↑CTAGT	5′ overhang	CTAG–
*Sph*I	GCATG↑C	3′ overhang	–CATG
*Spl*I	C↑GTACG	5′ overhang	GTAC–
*Spo*I	TCG↑CGA	Blunt	
*Srf*I	GCCC↑GGGC	Blunt	
*Ssp*I	AAT↑ATT	Blunt	
*Ssp*BI	T↑GTACA	5′ overhang	GTAC–
*Sst*I	GAGCT↑C	3′ overhang	–AGCT
*Sst*II	CCGC↑GG	3′ overhang	–GC
*Stu*I	AGG↑CCT	Blunt	
*Sty*I	C↑CWWGG	5′ overhang	CWWG–
*Swa*I	ATTT↑AAAT	Blunt	
*Tai*I	ACGT↑	3′ overhang	–ACGT
*Taq*I	T↑CGA	5′ overhang	CG–
*Tfi*I	G↑AWTC	5′ overhang	AWT–

Table 1 continued

Enzyme	Recognition sequence	Type of end	Sequence of overhang
ThaI	CG↑CG	Blunt	
TseI	G↑CWGC	5′ overhang	CWG–
TspRI	CAGTG(2/–7)	3′ overhang	–NNNNNNNCAGTGNN
Tth111I	GACN↑NNGTC	5′ overhang	N–
TthHB8I	T↑CGA	5′ overhang	CG–
Van91I	CCANNNN↑NTGG	3′ overhang	–NNN
VspI	AT ↑ TAAT	5′ overhang	TA–
XbaI	T↑CTAGA	5′ overhang	CTAG–
XcmI	CCANNNNN↑NNNNTGG	3′ overhang	–N
XcyI	C↑CCGGG	5′ overhang	CCGG–
XhoI	C↑TCGAG	5′ overhang	TCGA–
XhoII	R↑GATCY	5′ overhang	GATC–
XmaI	C↑CCGGG	5′ overhang	CCGG–
XmaIII	C↑GGCCG	5′ overhang	GGCC–
XmaCI	C↑CCGGG	5′ overhang	CCGG–
XmnI	GAANN↑NNTTC	Blunt	
XorII	CGAT↑CG	3′ overhang	–AT

[a] The arrow indicates the position of the cut site in the 5′→3′ strand in a palindromic recognition sequence. If the sequence is not palindromic then the position of the cut site is indicated by the numbers in brackets. The first number is the position of the cut site relative to the recognition sequence in the 5′→3′ strand and the second number is the position of the cut site relative to the recognition sequence in the 3′→5′ strand. A positive number indicates that the cut site lies downstream of the recognition sequence and a negative number indicates that it lies upstream. Conventional abbreviations are used for denoting non-unique nucleotides: R = G or A; Y = C or T; M = A or C; K = G or T; S = G or C; W = A or T; B = not A (C or G or T); D = not C (A or G or T); H = not G (A or C or T); V = not T (A or C or G); N = A or C or G or T.

[b] BcgI cuts both upstream and downstream of the recognition sequence, releasing the latter within a 32 bp fragment with 3′ overhangs of two nucleotides (–NN) at either end.

References

1. http://www.neb.com/rebase/rebase.html
2. Brown, T. A. (ed.) (1998). *Molecular biology labfax*, 2nd edn, Vol. 1. Academic Press, London.

Appendix 6
DNA and RNA modification enzymes

1 Introduction

Many different enzymes for DNA and RNA manipulations are now available from commercial suppliers. A comprehensive list is provided in ref. (1) and most are mentioned in various chapters of this book. The following is a summary.

2 DNA polymerases

DNA polymerases synthesize DNA from deoxyribonucleotide subunits. Synthesis occurs in the $5'\rightarrow3'$ direction and is template-dependent with some enzymes and template-independent with others. Most template-dependent DNA polymerases also possess $5'\rightarrow3'$ and/or $3'\rightarrow5'$ exonuclease activities. Important DNA polymerases are:

- **DNA polymerase I** of *Escherichia coli* possesses a $5'\rightarrow3'$ DNA-dependent DNA polymerase activity as well as $5'\rightarrow3'$ and $3'\rightarrow5'$ exonuclease activities. It is used for DNA labelling by nick translation and for second-strand cDNA synthesis

- **Klenow polymerase** is derived from DNA polymerase I by removal of the enzyme segment responsible for the $5'\rightarrow3'$ exonuclease. The enzyme is used for a number of applications, notably DNA labelling by random priming or end-filling, and conversion of 5'-overhangs to blunt ends

- **Sequenase** is a modified version of the DNA polymerase of bacteriophage T7. It has no exonuclease activity and is ideal for DNA sequencing by the chain-termination method

- *Taq* **DNA polymerase** from *Thermus aquaticus* is heat-stable and so can be used in PCR and certain specialized DNA sequencing applications. A number of other thermostable DNA polymerases are also available and are used in various applications: these include *Pfu* DNA polymerase from *Pyrococcus furiosus* and *Pwo* DNA polymerase from *Pyrococcus woesei*

- **T4 DNA polymerase** has a highly-active $3'\rightarrow5'$ exonuclease which enables it to carry out an end-replacement reaction that is useful in DNA labelling. It is also used to convert 3'-overhangs into blunt ends

- **Reverse transcriptases** are obtained from various sources, including avian myeloblastosis virus (AMV) and Moloney murine leukaemia virus (M-MuLV).

They are RNA-dependent DNA polymerases, so copy an RNA template into DNA (the basis of cDNA synthesis)

3 RNA polymerases

The most important RNA polymerases used in molecular biology research are three bacteriophage enzymes that recognize their own promoter sequences with high fidelity and which can be used to synthesize large amounts of RNA in *in vitro* reactions. These three enzymes are the SP6, T3 and T7 RNA polymerases.

4 Nucleases

A great variety of nucleases are used to manipulate DNA and RNA molecules. Restriction endonucleases are described in *Appendix 5*. Other important nucleases are:

- **Bal 31 nuclease** possesses a complex set of activities that enable it to progressively shorten double-stranded blunt-ended molecules
- **S1 nuclease** is single-strand specific and more active on DNA than RNA, so can be used to trim non-hybridized regions from DNA-RNA duplexes. Its use in S1 nuclease mapping allows the termini of transcripts to be located, along with intron boundaries
- **RNase H** degrades the RNA component of a DNA–RNA hybrid and plays an important role in cDNA synthesis
- **DNase I** is an endodeoxyribonuclease that has a number of important applications. It can be used to introduce nicks in double-stranded DNA molecules prior to labelling by nick translation, and can detect protein binding sites in nuclease protection experiments
- **RNase A** is an active ribonuclease that is used to remove RNA from DNA preparations

5 Ligases

The T4 DNA ligase is usually used in construction of recombinant DNA molecules, but the *E. coli* ligase (which requires NAD rather than ATP as the cofactor) is also used. There is also a T4 RNA ligase.

6 End-modification enzymes

The ends of DNA and RNA molecules can be altered by treatment with an end-modification enzyme. Examples are:

- **Alkaline phosphatase** removes 5′-phosphate groups from single- and double-stranded DNA molecules, which prevents their self-ligation. Restricted cloning vectors are often treated with alkaline phosphatase. The resulting phosphatased ends cannot ligate to one another, but can ligate to the non-phosphatased ends of another DNA molecule. This means that recircularization of the cloning vector can occur only when new DNA is inserted, so only recombinant molecules are synthesized. BAP, from *E. coli*, and CIP have been used extensively in the past but both are relatively resistant to heat treatment and so it can be difficult to inactivate them after the phosphatase reaction has been carried out. More recently, heat-labile alkaline phosphatases from arctic bacteria and arctic shrimps have been introduced

- **T4 polynucleotide kinase** adds phosphates to 5′-hydroxyl termini and is used in a DNA labelling procedure
- **Terminal deoxynucleotidyl transferase** adds a single-stranded tail onto a (usually) blunt-ended molecule. This reaction has been used to add sticky ends to blunt-ended DNAs prior to construction of recombinant molecules

Reference

1. Brown, T. A. (ed.) (1998). *Molecular biology labfax*, 2nd edn, Vol. 1. Academic Press, London.

Appendix 7
List of suppliers

Agfa-Gevaert Ltd., 27 Great West Road, Brentford, Middlesex TW8 9AX, UK
Tel: 0208 231 4900
Agfa Corp., 100 Challenger Road, Ridgefield Park, NJ 07660, USA
Tel: 001 201 440 2500
Fax: 001 201 440 5733

Amersham Pharmacia Biotech UK Ltd, Amersham Place, Little Chalfont, Buckinghamshire HP7 9NA, UK (see also Nycomed Amersham Imaging UK; Pharmacia)
Tel: 0800 515313
Fax: 0800 616927
URL: http//www.apbiotech.com/

Anderman and Co. Ltd, 145 London Road, Kingston-upon-Thames, Surrey KT2 6NH, UK
Tel: 0181 5410035
Fax: 0181 5410623

Beckman Coulter (UK) Ltd, Oakley Court, Kingsmead Business Park, London Road, High Wycombe, Buckinghamshire HP11 1JU, UK
Tel: 01494 441181
Fax: 01494 447558
URL: http://www.beckman.com/
Beckman Coulter Inc., 4300 N. Harbor Boulevard, PO Box 3100, Fullerton, CA 92834-3100, USA
Tel: 001 714 8714848
Fax: 001 714 7738283
URL: http://www.beckman.com/

Becton Dickinson and Co., 21 Between Towns Road, Cowley, Oxford OX4 3LY, UK
Tel: 01865 748844
Fax: 01865 781627
URL: http://www.bd.com/
Becton Dickinson and Co., 1 Becton Drive, Franklin Lakes, NJ 07417-1883, USA
Tel: 001 201 8476800
URL: http://www.bd.com/

Bio 101 Inc., c/o Anachem Ltd, Anachem House, 20 Charles Street, Luton, Bedfordshire LU2 0EB, UK
Tel: 01582 456666
Fax: 01582 391768
URL: http://www.anachem.co.uk/
Bio 101 Inc., PO Box 2284, La Jolla, CA 92038-2284, USA
Tel: 001 760 5987299
Fax: 001 760 5980116
URL: http://www.bio101.com/

Bio-Rad Laboratories Ltd, Bio-Rad House, Maylands Avenue, Hemel Hempstead, Hertfordshire HP2 7TD, UK
Tel: 0181 3282000
Fax: 0181 3282550
URL: http://www.bio-rad.com/
Bio-Rad Laboratories Ltd, Division Headquarters, 1000 Alfred Noble Drive, Hercules, CA 94547, USA
Tel: 001 510 7247000
Fax: 001 510 7415817
URL: http://www.bio-rad.com/

Boehringer-Mannheim (see Roche)

Calbiochem-Novabiochem Corp., PO Box 12087, La Jolla, CA 92039-2087, USA
Tel: 001 800 854 3417
Fax: 001 800 776 0999

CN Biosciences (UK) Ltd., Boulevard Industrial Park, Padge Road, Beeston, Nottingham NG9 2JR, UK
Tel: 0115 943 0840
Fax: 0115 943 0951

CP Instrument Co. Ltd, PO Box 22, Bishop Stortford, Hertfordshire CM23 3DX, UK
Tel: 01279 757711   Fax: 01279 755785
URL: http//:www.cpinstrument.co.uk/

Dupont (UK) Ltd, Industrial Products Division, Wedgwood Way, Stevenage, Hertfordshire SG1 4QN, UK
Tel: 01438 734000
Fax: 01438 734382
URL: http://www.dupont.com/
Dupont Co. (Biotechnology Systems Division), PO Box 80024, Wilmington, DE 19880-002, USA
Tel: 001 302 7741000
Fax: 001 302 7747321
URL: http://www.dupont.com/

Eastman Chemical Co., 100 North Eastman Road, PO Box 511, Kingsport, TN 37662-5075, USA
Tel: 001 423 2292000
URL: http//:www.eastman.com/

Fisher Scientific UK Ltd, Bishop Meadow Road, Loughborough, Leicestershire LE11 5RG, UK
Tel: 01509 231166   Fax: 01509 231893
URL: http://www.fisher.co.uk/
Fisher Scientific, Fisher Research, 2761 Walnut Avenue, Tustin, CA 92780, USA
Tel: 001 714 6694600
Fax: 001 714 6691613
URL: http://www.fishersci.com/

Fluka, PO Box 2060, Milwaukee, WI 53201, USA
Tel: 001 414 2735013
Fax: 001 414 2734979
URL: http://www.sigma-aldrich.com/
Fluka Chemical Co. Ltd, PO Box 260, CH-9471, Buchs, Switzerland
Tel: 0041 81 7452828
Fax: 0041 81 7565449
URL: http://www.sigma-aldrich.com/

Gibco-BRL (see Life Technologies)

Hybaid Ltd, Action Court, Ashford Road, Ashford, Middlesex TW15 1XB, UK
Tel: 01784 425000
Fax: 01784 248085
URL: http://www.hybaid.com/
Hybaid US, 8 East Forge Parkway, Franklin, MA 02038, USA
Tel: 001 508 5416918
Fax: 001 508 5413041
URL: http://www.hybaid.com/

HyClone Laboratories, 1725 South HyClone Road, Logan, UT 84321, USA
Tel: 001 435 7534584
Fax: 001 435 7534589
URL: http//:www.hyclone.com/

ICN Biochemicals Division, 1263 South Chillicothe Road, Aurora, Ohio 44202-8064, USA
Tel: 001 330 562 1500
Fax: 001 330 562 2642
ICN Biochemicals, Cedarwood, Chineham Business Park, Crockford Lane, Basingstoke, Hampshire RG24 8WD, UK
Tel: 0125 670 7744
Fax: 0125 670 7334

Ilford Imaging UK Ltd., Town Lane, Mobberley, Knutsford, Cheshire WA16 7JL, UK
Tel: 01565 684000
Ilford Imaging USA Inc., West 70 Century Road, Paramus, NJ 07653, USA
Tel: 001 201 265 6000

Invitrogen Corp., 1600 Faraday Avenue, Carlsbad, CA 92008, USA
Tel: 001 760 6037200
Fax: 001 760 6037201
URL: http://www.invitrogen.com/
Invitrogen BV, PO Box 2312, 9704 CH Groningen, The Netherlands
Tel: 00800 53455345
Fax: 00800 78907890
URL: http://www.invitrogen.com/

Kodak, Kodak House, Station Road, Hemel Hempstead, Herts HP1 1JU, UK
Tel: 01442 844648
Fax: 01442 844842

Life Technologies Ltd, PO Box 35, 3 Free Fountain Drive, Inchinnan Business Park, Paisley PA4 9RF, UK
Tel: 0800 269210
Fax: 0800 243485
URL: http://www.lifetech.com/
Life Technologies Inc., 9800 Medical Center Drive, Rockville, MD 20850, USA
Tel: 001 301 6108000
URL: http://www.lifetech.com/

Merck Sharp & Dohme Research Laboratories, Neuroscience Research Centre, Terlings Park, Harlow, Essex CM20 2QR, UK
URL: http://www.msd-nrc.co.uk/
MSD Sharp and Dohme GmbH, Lindenplatz 1, D-85540, Haar, Germany
URL: http://www.msd-deutschland.com/

Molecular Dynamics, 928 East Arques Avenue, Sunnyvale, CA 94086-4520, USA
Tel: 001 800 333 5703
Fax: 001 408 773 1493

Millipore (UK) Ltd, The Boulevard, Blackmoor Lane, Watford, Hertfordshire WD1 8YW, UK
Tel: 01923 816375
Fax: 01923 818297
URL: http://www.millipore.com/local/UKhtm/

Millipore Corp., 80 Ashby Road, Bedford, MA 01730, USA
Tel: 001 800 6455476
Fax: 001 800 6455439
URL: http://www.millipore.com/

New England Biolabs, 32 Tozer Road, Beverley, MA 01915-5510, USA
Tel: 001 978 9275054

Nikon Inc., 1300 Walt Whitman Road, Melville, NY 11747-3064, USA
Tel: 001 516 5474200
Fax: 001 516 5470299
URL: http://www.nikonusa.com/
Nikon Corp., Fuji Building, 2–3, 3-chome, Marunouchi, Chiyoda-ku, Tokyo 100, Japan
Tel: 00813 32145311
Fax: 00813 32015856
URL: http://www.nikon.co.jp/main/index_e.htm/

Novagen, Inc., 601 Science Drive, Madison, WI 53711, USA
Tel: 001 608 238 6110
Fax: 001 608 238 1388

Nycomed Amersham Imaging, Amersham Labs, White Lion Rd, Amersham, Buckinghamshire HP7 9LL, UK
Tel: 0800 558822 (or 01494 544000)
Fax: 0800 669933 (or 01494 542266)
URL: http//:www.amersham.co.uk/
Nycomed Amersham, 101 Carnegie Center, Princeton, NJ 08540, USA
Tel: 001 609 5146000
URL: http://www.amersham.co.uk/

Pall, Europa House, Havant Street, Portsmouth, Hampshire PO1 3PD, UK
Tel: 0239 230 3303
Fax: 0239 230 2506
Pall Technical Center, 25 Harbor Park Drive, Port Washington, NY 11050, USA
Tel: 001 516 484 3600
Fax: 001 516 484 3651

Perkin Elmer Ltd, Post Office Lane, Beaconsfield, Buckinghamshire HP9 1QA, UK
Tel: 01494 676161
URL: http//:www.perkin-elmer.com/

Pharmacia, Davy Avenue, Knowlhill, Milton Keynes, Buckinghamshire MK5 8PH, UK (also see Amersham Pharmacia Biotech)
Tel: 01908 661101
Fax: 01908 690091
URL: http//www.eu.pnu.com/

Pierce
Pierce, PO Box 117, Rockford, IL 61105, USA
Tel: 001 800 842 5007
Fax: 001 815 968 7316

Perbio Science UK Ltd., Century House, High Street, Tattenhall, Cheshire CH3 9RJ, UK
Tel: 01829 771 744
Fax: 01829 771 644

Promega UK Ltd, Delta House, Chilworth Research Centre, Southampton SO16 7NS, UK
Tel: 0800 378994
Fax: 0800 181037
URL: http://www.promega.com/
Promega Corp., 2800 Woods Hollow Road, Madison, WI 53711-5399, USA
Tel: 001 608 2744330
Fax: 001 608 2772516
URL: http://www.promega.com/

Qiagen UK Ltd, Boundary Court, Gatwick Road, Crawley, West Sussex RH10 2AX, UK
Tel: 01293 422911
Fax: 01293 422922
URL: http://www.qiagen.com/
Qiagen Inc., 28159 Avenue Stanford, Valencia, CA 91355, USA
Tel: 001 800 4268157
Fax: 001 800 7182056
URL: http://www.qiagen.com/

Roche Diagnostics Ltd, Bell Lane, Lewes, East Sussex BN7 1LG, UK
Tel: 0808 1009998 (or 01273 480044)
Fax: 0808 1001920 (01273 480266)
URL: http://www.roche.com/
Roche Diagnostics Corp., 9115 Hague Road, PO Box 50457, Indianapolis, IN 46256, USA
Tel: 001 317 8452358
Fax: 001 317 5762126
URL: http://www.roche.com/
Roche Diagnostics GmbH, Sandhoferstrasse 116, D–68305 Mannheim, Germany
Tel: 0049 621 7594747
Fax: 0049 621 7594002
URL: http://www.roche.com/

Sarstedt GmbH, Postfach 12 20, D–51582 Nümbrecht, Germany
Tel: 022 93 30 50   Fax: 022 93 30 51 22
Sarstedt Ltd., 68 Boston Road, Beaumont Leys, Leicester LE4 1AW, UK
Tel: 01162 359023   Fax: 01162 366099
Sarstedt Inc., PO Box 468, Newton, NC 28658-0468, USA
Tel: 001 828 465 4000
Fax: 001 828 465 0718

Schleicher and Schuell Inc., Keene, NH 03431A, USA
Tel: 001 603 3572398

Shandon Scientific Ltd, 93–96 Chadwick Road, Astmoor, Runcorn, Cheshire WA7 1PR, UK   Tel: 01928 566611
URL: http//www.shandon.com/

Sigma–Aldrich Co. Ltd, The Old Brickyard, New Road, Gillingham, Dorset SP8 4XT, UK
Tel: 0800 717181 (or 01747 822211)
Fax: 0800 378538 (or 01747 823779)
URL: http://www.sigma-aldrich.com/
Sigma Chemical Co., PO Box 14508, St Louis, MO 63178, USA
Tel: 001 314 7715765
Fax: 001 314 7715757
URL: http://www.sigma-aldrich.com/

Stratagene Inc., 11011 North Torrey Pines Road, La Jolla, CA 92037, USA
Tel: 001 858 5355400
URL: http://www.stratagene.com/
Stratagene Europe, Gebouw California, Hogehilweg 15, 1101 CB Amsterdam Zuidoost, The Netherlands
Tel: 00800 91009100
URL: http://www.stratagene.com/

United States Biochemical (USB), PO Box 22400, Cleveland, OH 44122, USA
Tel: 001 216 4649277

Index

acrylamide 247
activated paper 110, 111, 113
agar overlays 261-2
alkali blotting 129-31
alkaline phosphatase 17, 24-5, 26-7, 102-3, 276-7
  controls 27-8
γ-aminolevulinic acid 219
ampicillin 262
ampicillin resistance gene 220
amplification
  cDNA library 61-2
  genomic library 34-9
anchor sequence 193
annealing 168-9, 170-1, 194-5
antibiotics 262
apoenzyme 219
aqueous hybridization solution 115
aryl azide 102
aryl phosphate-substituted 1, 2-dioxetane 102-3
autoradiography 69, 71, 92-4, 97-9, 100, 149, 177-8
  gel preparation 178
avidin 102

bacterial artificial chromosomes 15
bacterial colony screening 123-4
bacteriophage RNA polymerases 81-5
Bal 31 nuclease 276
*Bam*HI 18

base composition
  hybridization rate 138
  thermal stability 134
beta (β) emissions 71, 74, 147
biotin, cDNA library construction 41
biotin labelling 68, 101-5
  detection 102-3,104-5
  efficiency check 103
  hybridization analysis 103-4
  photobiotin 102
BLAST 185
blue-light sensitive film 98
blunt end creation 49-50
bromophenol blue 127
buffer gradient gel 176, 177
buffers 21, 262

carcinogens 246-7
chain-termination DNA sequencing 11, 157, 166-72
  Klenow polymerase 167
  nucleotides 168
  primer 166-7
  Sequenase 167, 168-72
  troubleshooting 179, 180-2
chemical cleavage 11, 158
chemical hazards 246-7
chip technology 154
chloroform preparation 265-6
chromosomes
  in situ hybridization 151, 154
  mechanically stretched 154
clone identification 9

clone libraries 3-5; *see also* complementary DNA libraries; genomic libraries
cloning 2-5
  direct selection 3
  single restriction fragment 2-3
CLUSTAL 185
cofactors 219
colloidal gold 102
colony replica filters 120-4
  hybridization to 145
complementary DNA (cDNA)
  fractions
    processing 53-4
    quantification 55
  library, *see* complementary DNA libraries
  ligation into λ vector 57
  preparation for cloning 48-55
    blunt end creation 49-50
    ligation to *Eco*RI adapters 50
    phosphorylation of *Eco*RI adapters 50-1
    *Xho*I digestion 51
  size fractionation 51-3
  synthesis 46-8
complementary DNA (cDNA) libraries, 5, 41-62
  amplification 61-2
  characterization 59-61
  clone as probe 8
  establishment 55-9
  packaging 57-8
  plating and titre determination 58-9
  probing for abundant clone 8
  storage 61-2